Electrical Power Systems:
A Conceptual Approach

Electrical Power Systems: A Conceptual Approach

Editor: Louis Baker

New York

Published by NY Research Press
118-35 Queens Blvd., Suite 400,
Forest Hills, NY 11375, USA
www.nyresearchpress.com

Electrical Power Systems: A Conceptual Approach
Edited by Louis Baker

© 2022 NY Research Press

International Standard Book Number: 978-1-63238-847-6 (Hardback)

Cataloging-in-Publication Data

Electrical power systems : a conceptual approach / edited by Louis Baker.
p. cm.
Includes bibliographical references and index.
ISBN 978-1-63238-847-6
1. Electric power systems. 2. Electric power production. 3. Electric power transmission. I. Baker, Louis.
TK1001 .E44 2022
621.31--dc23

Contents

Preface

An electrical power system refers to a network that uses, supplies and transfers electric power using electrical components. An electrical grid is an electric power system. It constitutes of generators, transmission systems and a distribution system. A power system ideally consists of power sources, loads, conductors, power electronics, capacitors and reactors, protective devices, etc. Power systems can be of different types depending on their design and mode of operation. Some common types are residential and commercial power systems. This book is a compilation of chapters that discuss the most vital concepts and emerging trends in electrical power systems. It aims to shed light on some of the unexplored aspects of power systems and the recent researches in this technology. With state-of-the-art inputs by acclaimed experts of electrical, electronic and power engineering, this book targets students and professionals.

The researches compiled throughout the book are authentic and of high quality, combining several disciplines and from very diverse regions from around the world. Drawing on the contributions of many researchers from diverse countries, the book's objective is to provide the readers with the latest achievements in the area of research. This book will surely be a source of knowledge to all interested and researching the field.

In the end, I would like to express my deep sense of gratitude to all the authors for meeting the set deadlines in completing and submitting their research chapters. I would also like to thank the publisher for the support offered to us throughout the course of the book. Finally, I extend my sincere thanks to my family for being a constant source of inspiration and encouragement.

Editor

Feasibility Study of PV-Wind-Fuel Cell Hybrid Power System for Electrification of a Rural Village in Ethiopia

Mikias Hailu Kebede ⓘ[1] **and Getachew Bekele Beyene**[2]

[1]*Electrical and Computer Engineering Department, Debre Berhan University, Debre Berhan, Ethiopia*
[2]*Addis Ababa Institute of Technology (AAiT), Addis Ababa University, Addis Ababa, Ethiopia*

Correspondence should be addressed to Mikias Hailu Kebede; hailumikias@yahoo.com

Academic Editor: Gorazd Stumberger

As the energy consumption is increasing in an alarming rate and peoples and international communities are well aware of environmental protection, alternative (i.e., renewable and fuel cell based) distributed generation (DG) systems have attracted increased interest. Wind-based and photovoltaic- (PV-) based power generation are two of the most promising renewable energy technologies. Fuel cell (FC) systems also show great potential in DG applications due to their fast technological development and the merits they have, such as high efficiency, zero or low emissions (of pollutant gases), and flexible modular structure. In this work, the techno-economic feasibility study (using HOMER) of emission-free hybrid power system of solar, wind, and fuel cell power source unit for a given rural village in Ethiopia called Nifasso (latitude of $9°58'40''$N and longitude of $39°50'3''$E with an estimated population of 1059) that can meet the electricity demand in a sustainable manner has been studied. The main power for the hybrid system comes from the solar and wind energy while the fuel cell and rechargeable batteries are used as a secondary and primary energy back up units, respectively. We can say storage as primary and secondary based on the sequence of operation. Hence, when there is shortage, first the battery discharges to fulfill the load demand and if the battery reaches to its allowable minimum capacity, it will stop further discharging and the fuel cell will operate so as to convert the stored hydrogen into electricity. In the result, different feasible alternative solutions have been obtained with a narrow range of COE which are better than the previously studied PV-wind-Genset hybrid set ups.

1. Introduction

Most people in the rural villages use kerosene for lamp, diesel for water pumping and flour mills, fire wood for cooking, and dry cells for radio and tape recorders. Desertification of the land is getting worse and worse due to deforestation and backward agricultural practices. Hence, the communities that live in the rural villages are suffering from lack of electricity. Therefore, women are forced to do their day-to-day domestic activities such as cooking, using fuel wood which leads to a rapid growth of deforestation; they travel long distances to fetch water; and they also use kerosene lamp at night. Additionally, due to lack of access to electricity, the communities in the rural villages are not able to benefit from social services such as clinics and schools sufficiently.

Therefore, village electrification is a vital step for improving the socioeconomic conditions of rural areas and crucial for the country's overall development. The villages' welfare is one of the main aims of the rural electrification program. Enormous benefits can be achieved in irrigation, food preservation, crop processing, agriculture, and rural small-scale industries [1, 2]. It creates employment opportunities for the villages' youth and promotes a better standard of life. Hence, availability of electricity reduces poverty and helps economic development by enhancing the health, education, and water supply (for drinking and irrigation) needs of the rural population [3–5].

Keeping in mind the above facts, the authors of this work believes that designing and implementing cost-effective renewable energy-based hybrid systems to supply electric loads is a best candidate solution.

Most feasibility studies in Ethiopia which have been conducted till now on the hybrid systems are PV/wind/Genset and PV/wind/hydro types. Hence, this work tries to look the other possibilities by incorporating fuel cell systems as secondary back-up units. But, several authors have studied the wind/PV/fuel cell, PV/fuel cell, and wind/fuel cell hybrid systems for different case studies in the world. For instance, Debnath et al. in [6] describe a hybrid power generation system suitable for remote area for agricultural application. The hybrid system is studied using HOMER software. The load has a peak value of 13 kW with a 2.2% day-to-day and hour-to-hour variation. To satisfy this load demand, a hybrid PV/FC/battery/electrolyzer system with 80 kW PV and 10 kW FC has been suggested with a COE of 0.431$/kWh [6].

2. Brief Note on HOMER

Hybrid Optimization Model for Electric Renewables (HOMER) software helps to find the least cost combination of components that meet a required load, based on an hourly analysis of the input variables, such as wind and solar data. For systems that meet the yearly load, the life-cycle cost is also estimated by the software. HOMER can be applied to a number of system designs: grid-connected or off-grid, stand-alone or distributed generation, and conventional or renewable technologies. The renewable technologies can be classified into three groups as power sources, storage, and loads [7].

In the category of power sources, the most common types are the following: photovoltaic (PV); wind turbine; hydropower; generators with different prime movers such as diesel, biogas, or coal-fired; electric utility grid; and microturbine or fuel cell. In the storage class, bank of batteries and hydrogen can be mentioned. With regard to loads, there are two types: primary and deferrable loads. The primary load is the electrical load that must be met at a specified time (e.g., lighting), and the deferrable load is the load that need not be met within a specified time but should be met within a certain time period (e.g., water pumping) [7].

The software is sufficiently intelligent to identify the proper timings of energy supplied to the components which are connected to the system. For systems which include batteries and a generator, the software can decide the times at which the batteries should be charged and when the generator should be operated. It also gives the deferrable load a lower priority than the primary load but a higher priority than charging the batteries. For this, there are two dispatch strategies that HOMER follows. A dispatch strategy is a set of rules by which the operation of the generator and the batteries is controlled whenever there is a shortage of energy from the renewable resources. There are two types of dispatch strategy: load following and cycle charging. The load following strategy enables HOMER to serve the deferrable load under two conditions. These are (a) when the storage tank is empty and (b) when the system produces excess electricity. Under a cycle charging strategy, the generator serves the deferrable load when it is able to produce more electricity than that needed by the primary

load. If the storage tank is empty, then the deferrable load is considered a primary load and all the power sources serve the deferrable load as much as possible [7].

The software also provides a feature for carrying out sensitivity analysis, which enables the evaluation of the economic and technical feasibility of several technological options. This feature can also be used when there is doubt over the exact value of a certain input, such as the annual average wind speed, annual average solar radiation, diesel price, or the price of PV cells. Furthermore, when the data represent a range of applications, this feature can be used. The sensitivity analysis performs energy balance calculations on hourly basis for a whole year (8,760 hours). It compares the electric load for each hour to the energy that the system can supply during that hour. While carrying out the sensitivity analysis, the optimization process is repeated for each input value in the range so that the effect that changes in the value have on the results can be examined [7].

Nonetheless, this is not as simple as it may seem; as the number of sensitivity variables increases, the computational time of the software also increases, which could be considered a limitation or a challenge to using the software. When using the sensitivity analysis feature of the software, several sizes of each component must be considered in order to meet the load and the computation time is dependent on how many of them are used. To minimize the computation time, an iterative process needs to be followed. This is done by first considering just a small number of sizes and/or variables over a relatively large range to decrease the initial running time. After each successive run, a greater number of options and variables are added within the range in order to increase the resolution. Indeed, this takes quite a long time and can be considered as a limitation of the software [7].

Furthermore, similar to the several sensitivity variables that can be input into the sensitivity table, there is another search space table to which can be applied different sizes and quantities of the different system components, such as the size of PV array, generator, inverter, and the quantities of batteries and wind turbines. Again, this further lengthens the computation time of the software, as the software tries each and every component size and quantity. As in the previous case, the simulation can first be run coarsely, by minimizing the number of variables within the range. The results are then refined by adding a greater number of variables within the range. It should also be noted that the greater the number of variables within a certain range supplied to HOMER, the better the result. However, attention needs to be paid to the computation time [7].

After running the simulation, the results are given as a list of feasible system configurations, sorted by life-cycle cost. From the results, the least-cost systems, which are displayed in the first few rows of the list, can be chosen for implementation. The designer can also scan for other feasible systems in the list and decide to take any particular setup by evaluating the pros and cons of the setup against cost, renewable resource contribution, future price trend of the components, and so on. Following on from this, the technology options, component costs, and available resources were input to the software. HOMER used the inputs to provide different feasible system configurations, which were sorted according to their net present cost [7].

3. Load Estimation

Deciding on the load is one of the most important steps in the design of a hybrid system [8–10]. Primary load, which must be met immediately, and deferrable load, which must be met within a certain time, are both considered in this work. Electric load in the rural villages of Ethiopia can be assumed to be composed of lighting; radio and television, water pumps, health post, and primary schools load [7, 11]. The work indicated in [12] considered only lighting, radio, and television as community loads. Another study in Ethiopia [13] assumes additional electricity demand for cooking and for flour mills to the load together with the load considered in [7]. In this paper, barbers, shops, and church loads are incorporated additionally.

The daily power demand and energy need for the community of 289 households are estimated by considering basic domestic appliances such as television (70 W), CFLs of 11 W and 15 W for lighting, radio/tape (5 W), VCD/DVD player (15 W), refrigerator (70 W), "electric mitad" (2.5 kW), cell phone (2.5 W), and stove (1.5 kW).

The primary school consists of 8 classrooms, a director office, teacher's staff room, toilets, and a guard house. Each classroom is assumed to be installed with four 15 W CFLs (compact fluorescent lamps); one 15 W CFL is suggested for the teacher's staff room; one 15 W CFL is assumed for the director office; two 11 W CFLs are allowed for the toilet; one 11 W CFL for the guard house, and additional 4 CFLs (of 15 W) for external lighting are also considered. Evening classes are conducted from 18:00 to 21:00. Additionally, four computers (LCD desktop type of 200 W), one in the director office and three in the teacher's staff room; one printer (360 W); one photocopy machine (1000 W); tape/radio (5 W); one GSM wireless telephone (2 W); and two ceiling fan (80 W, each one in the director office and one in the teacher's staff room) are proposed.

For the health centers, having 3 rooms, 15 W CFL per room, one 15 W CFL for external lighting, one 11 W for the toilet, and one 11 W CFL for guard house are considered. When we come to the appliances used, it is based on the typical category I rural health clinic presented in [14]. Accordingly, basic small health post clinical equipment's such as vaccine refrigerator (134 W), a 15 W capacity microscope, a 135 W capacity centrifuge, a 2 kW water heater, a 400 W incubator, a 75 W television, a GSM wireless telephone (2 W), and three ceiling fans (each of 80 W in each room) are suggested.

For the community, two flour milling machines of 12.5 kW capacity each working from 9:00 to 12:00 and 14:00 to 18:00 with four 15 W CFLs for internal lighting (from 18:00 to 20:00) and two 15 W CFLs for external lighting (from 18:00 to 6:00) are assumed.

For the church, six 15 W CFLs in the internal lighting, one 15 W CFL in the "Bethlehem," one 15 W CFL in the office, four 11 W CFLs in the "meeting hall," four 15 W CFLs for the external lighting, and one 11 W for the guard house

are assumed. Besides these, a GSM wireless telephone of 2 W is proposed.

There are a total of three shops assumed, and each shop is expected to have a radio/tape of 5 W and 15 W CFL. The CFLs are supposed to operate from 18:00 to 21:00.

In the community, two barbers are considered each having two hair clipping machines of 12 W. Additionally, one 15 W CFL for lighting between 18:00 and 21:00 is suggested for each.

The total primary load in weekend is expected to be less than weekdays. Because, some loads such as milling machines, school loads, and clinics will not operate. Hence, by considering some load reduction in weekends, the 24-hour load trend curve indicated in Figure 1 was obtained.

Water pumping system is required for households, schools, and health care centers. A minimum of 100 liters of water per day per family and 2400 liters/day for each pair of one health center and one primary school is suggested [7, 13]. Therefore, to accomplish this, we need to have three 1.528 kW of 40.416 l/m capacity (a total need of 4.584 kW) water pumps so as to pump water for the community which operates for four hours with the energy demand of 18.336 kWh. Again, we need to have one 1.119 kW of 20 l/m capacity water pump for the school and clinic that operates for two hours with the energy demand of 2.238 kWh. Hence, the total peak deferrable load is 5.703 kW, which needs 20.574 kWh energy.

A water storage capacity of 4 days is suggested requiring a storage capacity of 73.344 kWh for the community and 8.952 kWh for primary school and health centers. Hence, the total storage capacity is equal to 82.296 kWh. The storage capacities can be obtained by multiplying number of days with energy demand of each water pump [15, 16].

Up to 30% deferrable load reduction can be expected in the rainy season [7, 12, 17]. 15% load decrease for June, January, and September while 30% for July and August are assumed [13, 15]. There are no classes in July and August (annual break) and in January (semester break). With these assumptions, the monthly power consumption and energy demand of the water pumps are indicated in Figure 2.

4. HOMER Model of the Hybrid System

In this work, a hybrid system based on hydrogen technology is considered which needs a hydrogen producing unit (electrolyzer), a hydrogen storing unit (storage tanks), and a hydrogen utilizing unit (PEM fuel cell). However, the system is based on intermittent energy sources and is likely to experience large minutely, hourly, and daily fluctuations in energy input. Thus, it should be emphasized that the main purpose of the hydrogen storage system is to store energy over a long period of time (hour to hour and season to season). Other small short-term energy storage such as a battery must also be included to supply power during transient load conditions. Figure 3 shows the model and the hybrid setup in HOMER.

4.1. Resource Inputs. The monthly average wind speed for the study area obtained from [18] and National

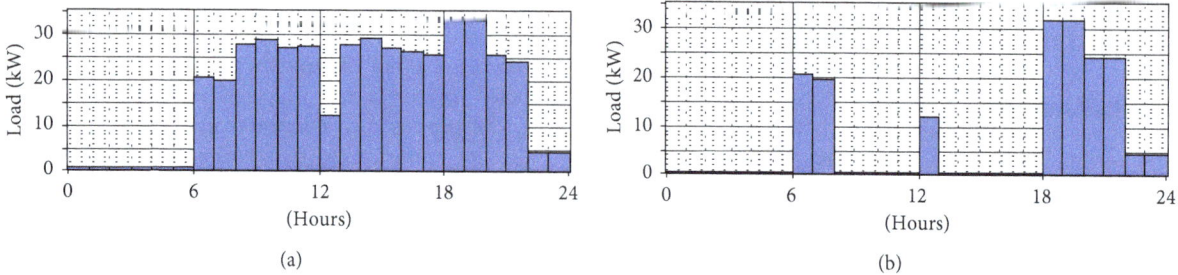

(a)   (b)

FIGURE 1: Primary load daily profiles: (a) weekday; (b) weekend.

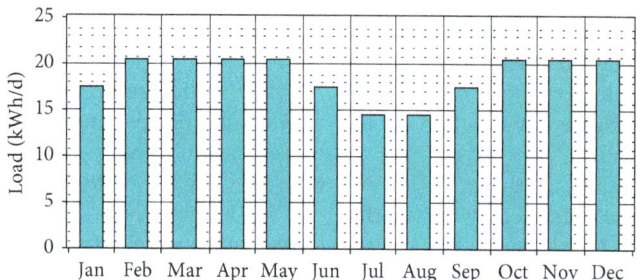

FIGURE 2: Monthly deferrable load profiles.

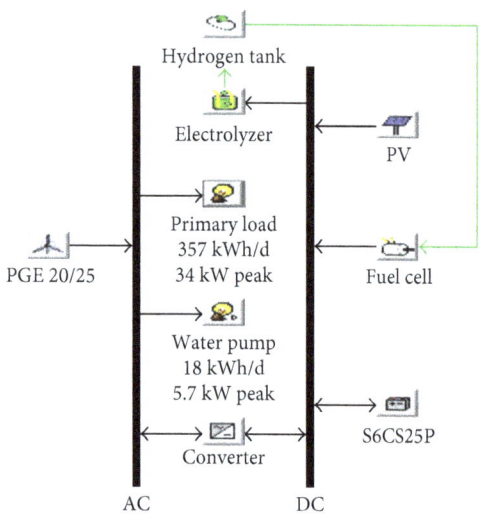

FIGURE 3: HOMER model of the hybrid power system.

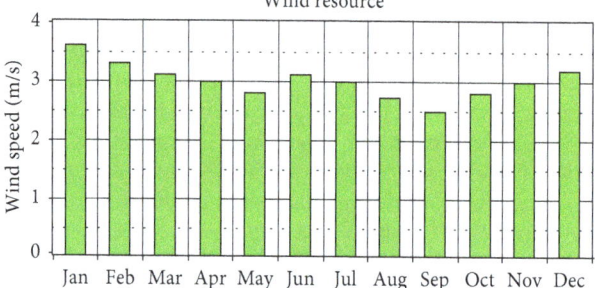

FIGURE 4: Monthly average wind speeds at 10 m.

Meteorological Service of Ethiopia were fed into HOMER. Figure 4 shows the average wind resource profile of each month at 10 m above the surface of the earth.

The probability density function of the wind speed synthesized by HOMER is also given in Figure 5. The software estimates a Weibull $k = 1.87$ and $c = 3.39$ m/s.

In a similar way, the solar energy potential for the study area was fed into HOMER, and this is shown in Figure 6. This figure also shows the clearness index which HOMER generates for the analysis. The clearness index is the fraction of solar radiation transmitted through the atmosphere which strikes the surface of the earth, and hence, it is a measure of the clearness of the atmosphere. It is a dimensionless

number between 0 and 1, defined as the surface radiation divided by the extraterrestrial radiation [7, 16, 19].

Many literatures reported that typical values for the monthly average clearness index range from 0.25 (a very cloudy month) to 0.75 (a very sunny month) [7, 20]. And this fact can be seen clearly in Figure 6. Note also that, in Figure 6, clearness index falls in between 0.4 and 0.8, which is the indication of the availability of a relatively strong solar potential in the study area.

The monthly temperature variation which is fed into the HOMER software is indicated in Figure 7. HOMER uses the ambient temperature to calculate the PV cell temperature and to consider temperature effect in the PV system.

4.2. Summary of Additional Inputs for Simulation. The capital, replacement, and operational and maintenance cost of each component in the hybrid system are given in Table 1. The table also contains the components' average operational life year and the size and/or quantities considered in the

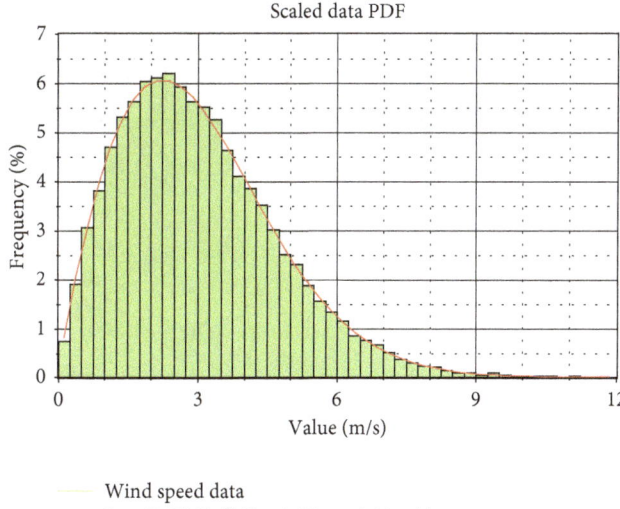

FIGURE 5: Wind speed probability density function at 10 m.

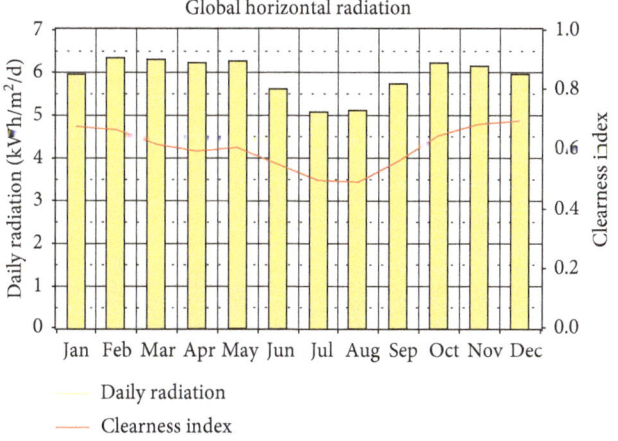

FIGURE 6: Monthly average radiations.

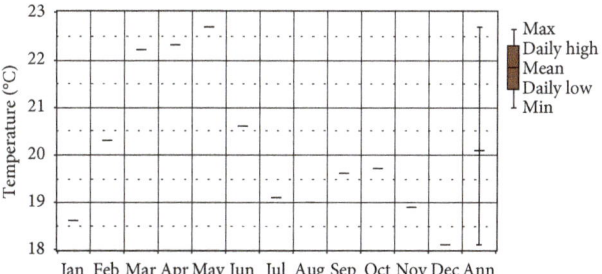

FIGURE 7: Monthly average ambient temperatures.

study. Therefore, as the software searches the optimum system combination, it uses the input values summarized in the table. Besides this input values, sensitivity variables are also defined for PV and FC costs.

5. Simulation Results and Discussions

5.1. Optimization Results. Figure 8 shows some optimization results (which are part of the numerous alternative solutions

from rank 1 up to 26) that are the possible configurations able to feed the system total load. Despite the numerous alternatives with equal renewable fraction, the choice of optimal system type is restricted by the varying nature of initial capital, net present cost, excess electricity, and COE of each set up. Accordingly, the one which is marked in blue color (the fifth rank) in which fuel cell operates for 2,731 hours/year is the best optimum PV-wind-fuel cell hybrid system configuration. Because, from rank 1 up to rank 4, the system configurations are PV only and PV-fuel cell types with small changes in COE and NPC from the selected one. Moreover, in the fifth rank, due to the availability of all power sources (PV, wind, and fuel cell), the reliability of the hybrid system will also be improved with only small addition of cost.

A hybrid power system containing PV, wind, and fuel cell power sources marked in blue color in Figure 8 can be summarized by Table 2.

As it can be seen clearly in the fifth rank that is marked in blue in Figure 8, the capacity shortage is reported as 0.01 which is nearly zero. Furthermore, the system report shows that there is an excess electricity of 43,460 kWh/yr (18.9%). This excess electricity is not discouraging as it can be used to serve the load growth that might come in near future. Besides this, this excess electricity can be utilized as a compensation for the losses in the distribution network.

Figure 9 shows the detailed net present cost share by each component of the optimal hybrid set-up supported by a bar graph. Hence, the net present cost share of each component in decreasing order can be recognized as PV, battery, wind turbine, converter, electrolyzer, hydrogen tank, and fuel cell.

Figure 10 displays the monthly average electricity production by each power unit. Hence, as it can be seen clearly, most of the electricity production is being shared by the PV followed by the wind turbine. But, the power production share by the fuel cell, even though it is significant, is small as compared with the PV and wind turbine power unit outputs. This is also presented in Table 3.

The monthly average hydrogen production by the electrolyzer is given in Figure 11. Recall that an electrolyzer converts the excess electricity into hydrogen by the electrolysis of water. As the simulation result reports that the total hydrogen production and total hydrogen consumption were found to be 684 kg/yr and 670 kg/yr, respectively. Besides this, the hydrogen tank autonomy (expressed in hours) which is defined as the ratio of the energy capacity of the hydrogen tank to the electric load was reported as 32.

5.2. Sensitivity Analysis Result. Figure 12 shows sensitivity of the optimal system type under the imposed price variation of sensitive components. The PV capital cost multiplier on the Y axis and FC capital cost multiplier on the X axis are considered as sensitivity variables. In the graph, the total net present costs (NPCs) of the most cost-effective systems are also displayed.

6. Conclusions

In short, this work addressed a techno-economic analysis of PV/wind/FC hybrid power system preceded by community load estimation and resource collection.

TABLE 1: Summary of additional inputs for HOMER simulation.

	PV array	Wind turbine (PGE20/25)	Fuel cell	Electrolyzer	Converter	Battery (Surrette 6CS25P)	Hydrogen tank
Unit	kW	kW	kW	kW	kW	Ah	kg
Size	1	25	1	1	1	1156	1
Capital cost ($)	2,000	55,000	3,000	1,000	700	855	1,300
Replacement cost ($)	2,000	36,667	2,000	667	700	555	867
O&M cost ($/yr)*	0 (0%)	1,100 (2.5%)	$0.02/hr	5 (0.5%)	0 (0%)	8.55 (1%)	13 (1%)
Size considered	0, 50, 80 100, 110, 150, 200	—	0, 5, 10, 15, 20, 25	0, 5,10, 15, 20, 25, 30, 50, 100	0, 30, 40, 50	—	0, 5, 10, 15, 20, 25, 30, 50
Quantities considered	—	0, 1, 2, 3	—	—	—	0,80, 100, 120, 140, 160, 180	—
Life time	25 yrs	25 yrs	40,000 hrs	25 yrs	15 yrs	9,645 kWh	25 yrs

*Considering O&M cost as 0–10% of the capital cost is a standard procedure.

		PV (kW)	PGE25	FCell (kW)	S6CS25P	Conv. (kW)	Elec. (kW)	H2 Tank (kg)	Disp. Strgy	Initial Capital	Operating Cost ($/yr)	Total NPC	COE ($/kWh)	Ren. Frac.	Capacity Shortage	FCell (hrs)
		150			160	40			CC	$464,800	5,796	$538,893	0.310	1.00	0.01	
		150			160	40			LF	$464,800	5,796	$538,893	0.310	1.00	0.01	
		150		10	100	40	30	10	LF	$486,500	4,436	$543,207	0.313	1.00	0.01	2,057
		150		5	120	40	20	10	LF	$478,600	5,087	$543,624	0.312	1.00	0.01	2,939
		110	1	5	120	40	25	15	LF	$465,100	6,215	$544,550	0.313	1.00	0.01	2,731
		150		10	100	40	25	15	LF	$488,000	4,469	$545,128	0.314	1.00	0.01	2,072
		150		5	120	40	15	15	LF	$480,100	5,091	$545,175	0.314	1.00	0.01	2,844
		150			160	40		5	CC	$471,300	5,846	$546,032	0.314	1.00	0.01	
		150			160	40		5	LF	$471,300	5,846	$546,032	0.314	1.00	0.01	
		150		5	120	40	30	5	LF	$482,100	5,064	$546,832	0.315	1.00	0.01	2,836
		150			160	50			CC	$471,800	5,982	$548,270	0.316	1.00	0.01	
		150			160	50			LF	$471,800	5,982	$548,270	0.316	1.00	0.01	
		150		5	140	40	10	5	LF	$479,200	5,462	$549,019	0.316	1.00	0.01	2,271
		150		5	120	40	25	10	LF	$483,600	5,124	$549,097	0.315	1.00	0.01	2,994
		110	1	5	120	40	30	15	LF	$470,100	6,246	$549,951	0.316	1.00	0.01	2,759
		150		10	100	40	30	15	LF	$493,000	4,516	$550,727	0.316	1.00	0.01	2,114
		150		5	120	40	20	15	LF	$485,100	5,148	$550,903	0.316	1.00	0.01	2,989
		110	1	5	120	40	25	20	LF	$471,600	6,275	$551,814	0.317	1.00	0.01	2,774
		150		5	120	40	15	20	LF	$486,600	5,146	$552,379	0.318	1.00	0.01	2,867
		150		10	100	40	25	20	LF	$494,500	4,528	$552,380	0.318	1.00	0.01	2,089
		150		10	100	50	30	10	LF	$493,500	4,622	$552,585	0.318	1.00	0.01	2,057
		150		5	120	50	20	10	LF	$485,600	5,273	$553,001	0.318	1.00	0.01	2,939
		110	1		180	40			CC	$456,900	7,528	$553,128	0.318	1.00	0.01	
		110	1		180	40			LF	$456,900	7,528	$553,128	0.318	1.00	0.01	
		150			160	40		10	CC	$477,800	5,896	$553,171	0.319	1.00	0.01	
		150			160	40		10	LF	$477,800	5,896	$553,171	0.319	1.00	0.01	

FIGURE 8: Some of the optimization results.

TABLE 2: System architecture and cost summary.

System architecture		Cost summary	
110 kW PV array	120 × Surrette 6CS25P	Total NPC	$544,550
1 × PGE 20/25	25 kW electrolyzer	Initial capital	$465,100
5 kW fuel cell	15 kg hydrogen tank	Operating cost	$6,215/yr.
40 kW converter	LF dispatch strategy	Levelized COE	$0.313/kWh

Regarding the potential of the resources, it is observed that, the wind energy potential of the site is of class 1 type according to the US Department of Energy (DOE) wind mapping. Such wind energy potential, although it may not be sufficient for a large independent wind farm, is a viable preference if incorporated into other energy conversion systems such as PV and others [7]. The data sources for solar energy also confirmed the availability of huge potential of solar energy at the site.

The result of techno-economic feasibility study shows that there are numerous alternative feasible hybrid set ups with 100% contribution by the renewable resources and narrow range of COE (0.310$/kWh to 0.328$/kWh). Accordingly, the community electric demand can be satisfied by a hybrid system power source containing 110 kW PV, one PGE 20/25 wind turbine, a 40 kW converter, 120 Surrette

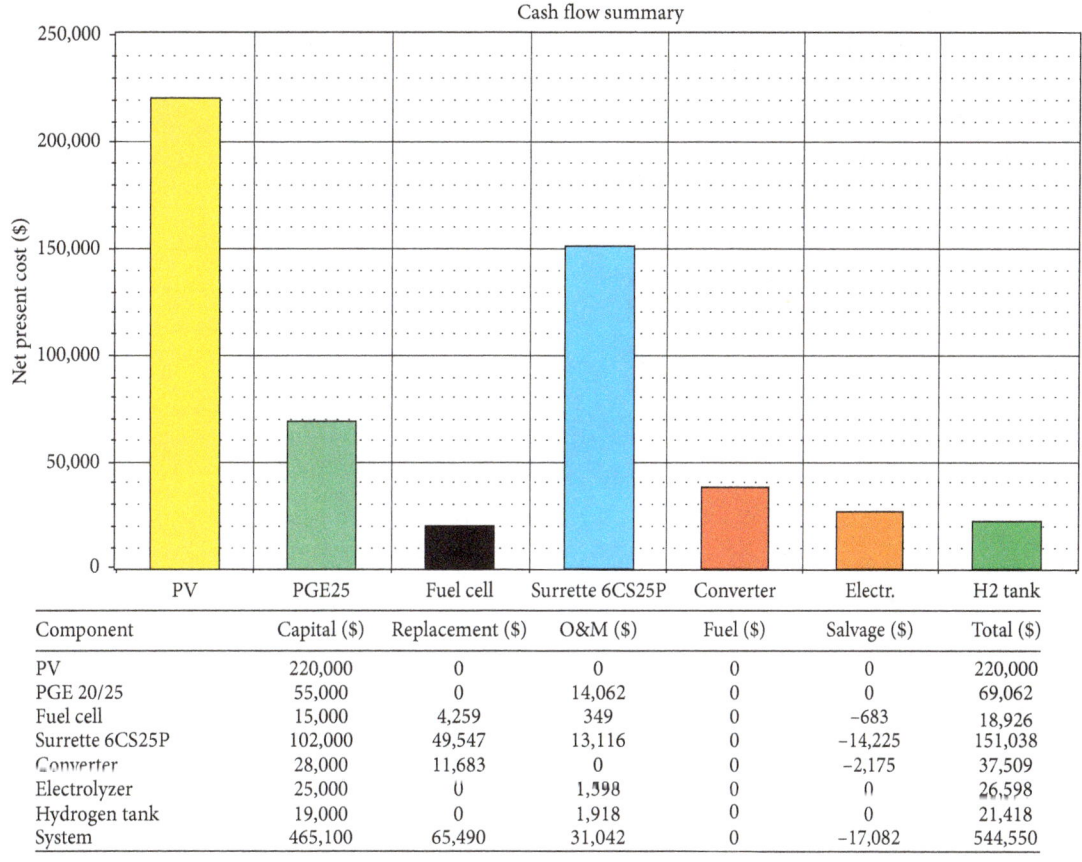

Component	Capital ($)	Replacement ($)	O&M ($)	Fuel ($)	Salvage ($)	Total ($)
PV	220,000	0	0	0	0	220,000
PGE 20/25	55,000	0	14,062	0	0	69,062
Fuel cell	15,000	4,259	349	0	−683	18,926
Surrette 6CS25P	102,000	49,547	13,116	0	−14,225	151,038
Converter	28,000	11,683	0	0	−2,175	37,509
Electrolyzer	25,000	0	1,598	0	0	26,598
Hydrogen tank	19,000	0	1,918	0	0	21,418
System	465,100	65,490	31,042	0	−17,082	544,550

FIGURE 9: Cost share by each component.

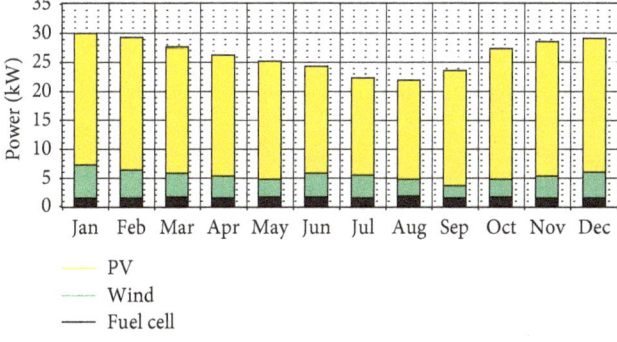

FIGURE 10: Monthly average electricity productions by the optimal hybrid PV-wind-FC power units.

TABLE 3: Electric power production and consumption.

Production			Consumption		
Power unit	kWh/yr	%	Load type	kWh/yr	%
PV array	183,017	80	AC load	129,291	79
Wind turbine	33,461	14	Deferrable load	6,703	4
Fuel cell	13,395	6	Electrolyzer	27,511	17
Total	229,874	100	Total	163,505	100

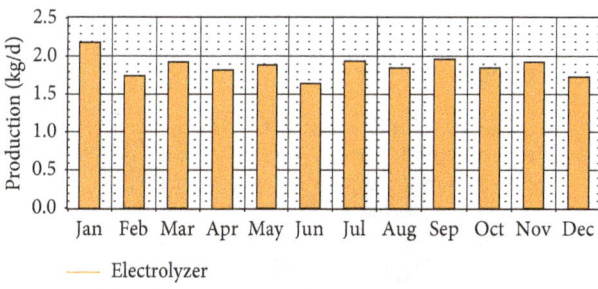

FIGURE 11: Monthly average hydrogen productions by the electrolyzer.

S6CS25P batteries, a 5 kW fuel cell, a 25 kW electrolyzer, and a 15 kg hydrogen tank. This hybrid system needs $465,100 initial capital, $6,215 for operation and maintenance, and $544,550 total NPC over the 25 life-year horizon. With these economics, the COE of the selected hybrid system was found to be 0.313$/kWh.

Neglecting other factors, in terms of COE, the result obtained in this study is comparable and even may be better than what previously studied PV-wind-Genset hybrid set up showed. For instance, in [7, 13], the range of COEs reported are 0.322$/kWh to 0.518$/kWh and 0.302$/kWh to

0.392$/kWh, respectively. But, the renewable fraction in both researches is poor—being too less than unity. Moreover, due to Genset package, there is a considerably high level of emission to the atmosphere such as CO_2, CO, unburned HC,

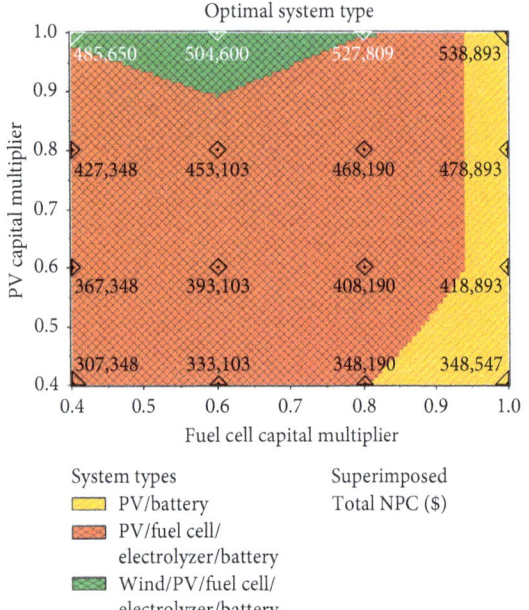

FIGURE 12: Sensitivity of the optimal system type for PV and FC capital multiplier variations.

SO_2, NO_X, and particulate matter (smoke, soot, and liquid droplets) emitted per unit fuel consumed by the generator. Hence, a hybrid PV-wind-fuel cell power system can be considered as a good choice in this regard.

Conflicts of Interest

The authors declare that there are no conflicts of interest regarding the publication of this paper.

Acknowledgments

The authors would like to thank National Meteorological Service of Ethiopia for their kind response to give some valuable solar and wind data.

References

[1] S. Touati, A. Belkaid, R. Benabid, K. Halbaoui, and M. Chelali, "Pre-feasibility design and simulation of hybrid PV/fuel cell energy system for application to desalination plants loads," *Procedia Engineering*, vol. 33, pp. 366–376, 2012.

[2] S. Kumar Kotte, Y. Ravisankar, A. Ahad, and S. Babu, "Dynamic modeling and simulation of a wind fuel cell hybrid energy system for standalone systems with hybrid optimization model for electrical renewable," in *Proceedings of International Colloquiums on Computer Electronics Electrical Mechanical and Civil*, Mulavoor, Kerala, 2011.

[3] P. Ozaveshe and T. C. Jen, "The energy cost analysis of hybrid systems and diesel generators in powering selected base transceiver station locations in Nigeria," *Energies*, vol. 11, no. 3, p. 687, 2018.

[4] O. H. Mohammed, Y. Amirat, M. Benbouzid, and A. A. Elbaset, "Optimal design of a PV/fuel cell hybrid power system for the city of Brest in France," in *Proceedings of 2014 First International Conference on Green Energy (IEEE ICGE 2014)*, pp. 119–123, Sfax, Tunisia, March 2014.

[5] K. Mohamed and M. El Ganaoui, "Feasibility study for the production of electricity using a hybrid PV-wind-generator system in a remote area in Comoros," *International Journal of Research and Reviews in Applied Sciences*, vol. 33, no. 2, 2017.

[6] D. Debnath, A. C. Kumar, and S. Ray, "Optimization and modeling of PV/FC/battery hybrid power plant for standalone application," *International Journal of Engineering Research and Technology*, vol. 1, no. 3, 2012, ISSN: 2278-0181.

[7] G. Bekele, *The study into the potential and feasibility of standalone solar-wind hybrid electric energy supply system for application in Ethiopia*, Ph.D. thesis, KTH Royal Institute of Technology, Stockholm, Sweden, 2009.

[8] M. H. Kebede, "Design of standalone PV system for a typical modern average home in Shewa Robit Town-Ethiopia," *American Journal of Electrical and Electronic Engineering*, vol. 6, no. 2, pp. 72–76, 2018.

[9] K. Gupta, A. Gupta, I. Aziz, A. Sharma, U. Ul Khaliq, and S. Kumar, "Feasibility of solar wind hybrid renewable energy in India," *International Journal of Engineering Research and General Science*, vol. 5, no. 2, 2017, ISSN 2091-2730.

[10] H. Chaouali, H. Othmani, M. Selméne, D. Mezghani, and A. Mami, "Energy management strategy of a PV/fuel cell/supercapacitor hybrid source feeding an off-grid pumping station," *International Journal of Advanced Computer Science and Applications*, vol. 8, no. 8, 2017.

[11] G. Bekele and G. Boneya, "Design of a photovoltaic-wind hybrid power generation system for Ethiopian remote area," *Energy Procedia*, vol. 14, pp. 1760–1765, 2011.

[12] B. Tamrat, "*Comparative analysis of feasibility of solar PV, wind and micro hydropower generation for rural electrification in the selected sites of Ethiopia*," M.Sc. thesis, Addis Ababa Institute of Technology, Addis Ababa, Ethiopia, 2007.

[13] G. Tadesse, "*Feasibility study of small hydro/PV/wind hybrid system for off-grid rural electrification in Ethiopia*," M.Sc. thesis, Addis Ababa Institute of Technology, Addis Ababa, Ethiopia, 2011.

[14] USAID, *Powering Health: Electrification Options for Rural Health Centers*, United States Agency for International Development, Washington, DC, USA, 2005.

[15] M. H. Kebede and G. B. Beyene, "*Dynamic modeling and techno-economic analysis of PV-wind-fuel cell hybrid power system: the case study of Nifasso*," M.Sc. thesis, Addis Ababa Institute of Technology, Addis Ababa, Ethiopia, 2014.

[16] V. Dash and P. Bajpai, "Power management control strategy for a stand-alone solar photovoltaic-fuel cell battery hybrid system," *Sustainable Energy Technologies and Assessments*, vol. 9, pp. 68–80, 2015.

[17] M. H. Kebede, P. Mukilan, and G. B. Beyene, "Dynamic modeling and optimization of self-sustaining solar-wind hybrid street lighting system: the case study of Addis Ababa city," *Journal of Advanced Research in Dynamical and Control Systems*, vol. 9, no. 17, 2017.

[18] NASA, 2012, http://eosweb.larc.nasa.gov/cgibin/sse/retscreen.cgi?email=rets%40nrcan.gc.ca&step=1&lat=9.9777892&lon=39.8342339&submit=Submit.

2

Shunt Active Power Filter based on Proportional Integral and Multi Vector Resonant Controllers for Compensating Nonlinear Loads

Sen Ye[1], **Youbing Zhang**[1], **Luyao Xie**[1] and **Haiqiang Lu**[2]

[1]*College of Information Engineering, Zhejiang University of Technology, Hangzhou, China*
[2]*Jiaxing Heng Chuang Electric Equipment Co., Ltd., Jiaxing, China*

Correspondence should be addressed to Luyao Xie; xieluyao@zjut.edu.cn

Academic Editor: Yandong Chen

The current tracking control strategy determines the compensation performance of shunt active power filter (SAPF). Due to inadequate compensation of the main harmonic by traditional proportional integral (PI) control, a control algorithm based on PI and multi vector resonant (VR) controllers is proposed to control SAPF. The mathematical model of SAPF is built, and basic principle of VR controller is introduced. Under the synchronous reference frame, the proposed control method based on pole zero cancellation is designed, which narrows the order of the control system and improves the system dynamic response and the control accuracy. Then the feasibility of the method is demonstrated by analyzing the closed loop frequency characteristics of the system. Finally, the simulation and experimental results are carried out to verify the performance of the proposed method.

1. Introduction

In recent years, with the intensification of the global energy crisis, the renewable energy and power electric technology have been integrated, which has promoted the development of distributed generators (DGs) [1, 2]. The speed driver of small motor, voltage source converter, and a large number of nonlinear loads including electric arc furnace, high-power rectifier, transducer, and fluorescent are connected to the grid, which will cause harmonic pollution unavoidably [3–5]. The power grid operates in this state for a long run, which will bring great harm to power system and power consumers, mainly manifested in the following respects: (1) The line loss of the system is increasing greatly, and the operating efficiency of the grid is reduced. A lot of 3rd harmonics will also increase the neutral current, leading to the damage of the custom power devices. (2) It makes the motor vibrate and produce noise, which affects the safety operation in production. (3) It makes the relay protection device act by mistake, which interrupts the power supply and makes the extent loss to production. (4) It interferes with measuring instruments and communication systems [6–8].

Therefore, it is urgent to control the power harmonic pollution to improve the power quality.

As a new type of power electronic device that can suppress harmonics dynamically, a three-phase three-wire SAPF with excellent performance has been widely used in the field of power quality. The real-time and accurate compensation for the fast changing harmonic currents is an important guarantee for the operation of SAPF safely and reliably. In the past years, a lot of research and analysis have been done on the detection and tracking control strategy of the harmonic currents. In [9, 10], it transforms the harmonic component to the DC component through multisynchronous rotating coordinate transformation; thus, traditional PI controller can be utilized to track the harmonics with zero steady-state error; however, it is too complicated. The hysteresis current control method presented in [11, 12] with quickly dynamic respond performance is simple and easy to realize without carrier modulation, but the control accuracy is inversely proportional to switching frequency. In addition, the switching frequency is not fixed, which sets a higher request to switching device. In [13, 14], the repetitive control theory was developed to solve harmonic problems, which has the advantages of good robustness, simple structure, and easy implementation. However, dynamic performance is limited if the load changes suddenly. In order to overcome the effect of computational delay and control delay on compensation performance, the dead-beat control in [15, 16] is used to track the reference current, which plays a certain effect, but it highly depends on the system parameters. When the system parameters change, the compensation performance will be greatly affected. In [17, 18], vector resonant(VR) control method is proposed, where proportional controller is to adjust the control bandwidth to improve the dynamic response speed, while the resonant controller is to select frequency for specific frequency signal to improve the control accuracy of current. However, it can not compensate the phase delay of controlled object.

Viewing of the shortcomings of above methods, a current control method based on PI and VR controller is proposed to compensate the selected harmonics accurately. The proportional control can improve the dynamic performance of the current loop and the fundamental current. The VR controller can control a group of positive and negative sequence harmonic currents with zero steady-state error by the idea of zero pole cancellation. Finally, the feasibility and validity of the proposed strategy are confirmed by simulation based on Matlab/Simulink and a prototype of the SAPF using TMS320F28335 as control core.

2. Modeling of SAPF

The structure of two-level three-phase three-wire SAPF used in this paper is shown in Figure 1 [19]. The mathematical model can be described in three-phase ABC static coordinate system as follows:

$$L\frac{di_{Fa}}{dt} + Ri_{Fa} = u_a - e_{Sa},$$

$$L\frac{di_{Fb}}{dt} + Ri_{Fb} = u_b - e_{Sb}, \qquad (1)$$

$$L\frac{di_{Fc}}{dt} + Ri_{Fc} = u_c - e_{Sc},$$

where e_a, e_b, and e_c are the 3-phase source voltage; u_a, u_b, and u_c are the 3-phase output voltage of SAPF; i_{sa}, i_{sb}, and i_{sc} are source current; i_{Fa}, i_{Fb}, and i_{Fc} are the compensation currents of SAPF; $S_a - S_c^*$ are fully controlled switch devices; L is AC-link inductance; C is DC-link capacitance; u_{dc} is DC-link voltage; and R is the equivalent loss of the AC-link inductance and switch device.

By means of Clarke transform and Park transform, the mathematical model under the synchronous reference frame is given as follows:

$$L\frac{di_{Fd}}{dt} + Ri_{Fd} = u_d + \omega_s Li_{Fq} - e_{Sd},$$

$$\qquad (2)$$

$$L\frac{di_{Fq}}{dt} + Ri_{Fq} = u_q - \omega_s Li_{Fd} - e_{Sq}.$$

$e_d, e_q u_d, u_q$ and i_{Fd}, i_{Fq} are d-axis components and q-axis components of the three-phase source voltage, AC-link output voltage, and compensation current of the SAPF

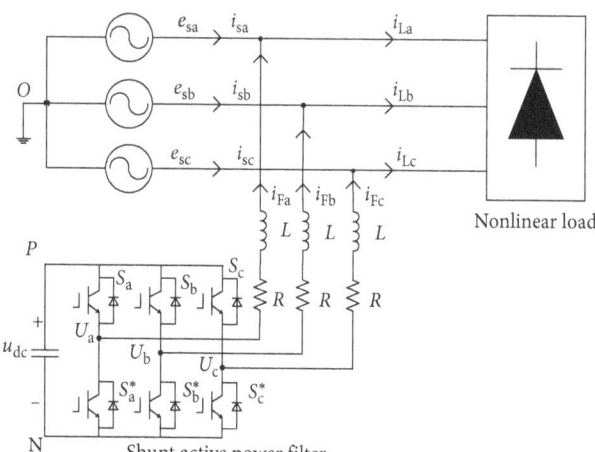

FIGURE 1: Structure of three-phase three-wire SAPF.

under synchronous reference frame, respectively. ω_s represents the voltage angle frequency detected by a software phase-locked loop.

3. Basic Principle of VR Controller

The current controller based on VR control has a higher gain at its resonant frequency, which can control the AC current at the resonant frequency with high precision. The transfer function of the VR controller can be derived as [20]

$$G_{VRh}(s) = \frac{K_{ph}s^2 + K_{ih}s}{s^2 + (h\omega_s)^2}, \qquad (3)$$

where K_{ph} and K_{ih} are proportional coefficient and the integral coefficient, respectively, and $h\omega_s$ is the resonant angle frequency. The absolute value of h represents the order of the harmonic currents, the sign of which represents the rotation direction in sequence component diagram. When h is positive, it represents that harmonic currents are in positive sequence. Otherwise, it represents that harmonic currents are in negative sequence.

The bode diagram of VR controller is shown in Figure 2 for $K_{ph} = 1, K_{ih} = 50, h = 6, \omega_s = 100\pi$. On one hand, the open loop gain of VR controller is infinity at its two resonant points $\pm h\omega_s$, but for other frequencies, the gain is rapidly attenuated. On the contrary, the VR controller is phase-leading about $180°$ at the frequency of $0 \sim h\omega_s$. Once the angular frequency exceeds the resonant frequency, the phase jumps from $180°$ to $0°$. But at negative frequency, it appears on the opposite performance. On the basis of the analysis above, it shows that VR controller has a good control precision for the signal at the resonant frequency.

4. Current Control Strategy of SAPF

4.1. VR Controller for Compensating Harmonic Currents.
For the three-phase uncontrollable rectifier, the main harmonic currents consist of 5th of negative sequence, 7th of positive sequence, 11th of negative sequence, 13th of positive

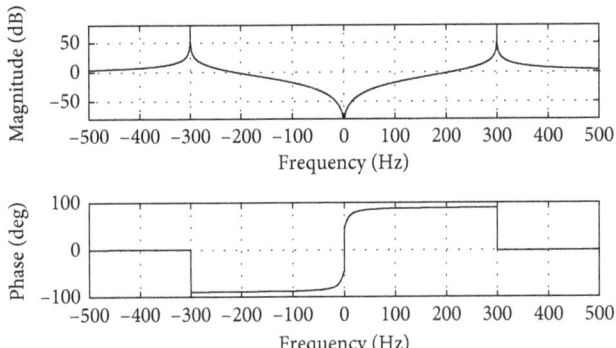

Figure 2: Bode diagram of VR controller.

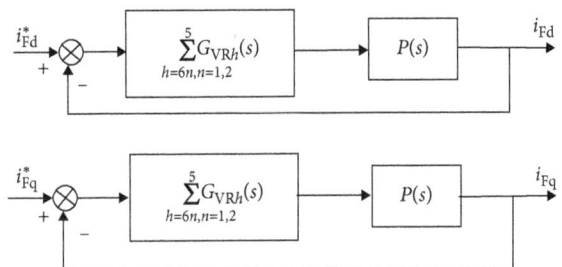

Figure 3: Current control block diagram based on multi-VR controller.

harmonic, and so on, such that $h = \pm 6n + 1, n$ is $1, 2, \ldots$. Since the synchronous reference frame provides a frequency shift of $-50\,\text{Hz}$, the main harmonic current in fundamental frame transforms to 6th of positive and negative sequence harmonic, 12th of positive and negative sequence harmonic, and so on. The harmonic orders become $h = \pm 6n$. Thus the multi-VR controller can compensate the selected harmonic based on formula (3). The equivalent gain of the SAPF K_{PWM} is set to 1. The current control block diagram is shown in Figure 3, where $P(s) = 1/(Ls + R)$ is the transfer function of the controlled object.

Taking the d-axis current as an example, current open loop transfer function of the multi-VR controller is

$$
\begin{aligned}
G_o(s) &= \sum_{h=6n,n=1,2}^{5} G_{\text{VR}h}(s)P(s) \\
&= \sum_{h=6n,n=1,2}^{5} \frac{K_{ph}s^2 + K_{ih}s}{\left(s^2 + (h\omega_s)^2\right)(Ls + R)}.
\end{aligned}
\tag{4}
$$

The basic idea of VR control is to adopt the zero point of the control to offset the poles of the controlled object, so as to compensate for the phase lag of the controlled object, thus improve the control accuracy of the harmonic currents. The zero of the controller can effectively cancel the poles of the controlled object on condition that $K_{ph} = K_{ih} * L/R$, and the simplified open loop transfer function is given as

$$
G_o(s) = \sum_{h=6n,n=1,2}^{5} \frac{K_{ph}s}{L\left(s^2 + (h\omega_s)^2\right)}.
\tag{5}
$$

The closed loop transfer function is

$$
G_c(s) = \frac{G_o(s)}{1 + G_o(s)} = \frac{\sum_{h=6n,n=1,2}^{5}\left(K_{ph}s/L\left(s^2 + (h\omega_s)^2\right)\right)}{1 + \sum_{h=6n,n=1,2}^{5}\left(K_{ph}s/L\left(s^2 + (h\omega_s)^2\right)\right)}.
\tag{6}
$$

The values of $K_{ph}, K_{ih}, L,$ and R in formulae (4)–(6) are both greater than 0. The bode diagram of harmonic current close loop control is depicted in Figure 4, at the resonant frequency, the amplitude gain of the transfer function is 1 without phase delay, which indicates that the VR controller has frequency selection function on a pair of harmonic currents and can be controlled with zero steady-state error.

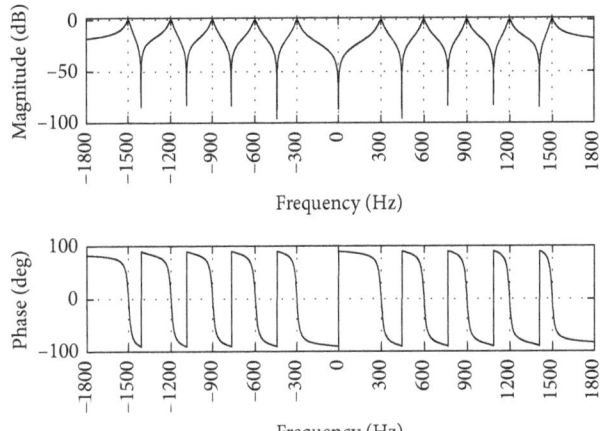

Figure 4: Bode diagram of harmonic current close loop control.

Therefore, the system is stable even when the circuit parameters and VR parameters change.

Whether SAPF can operate stably is affected not only by its own system parameters, but also by external disturbances, and the grid voltage is the most important factor for SAPF. Next, the tracking performance of compensation current is analyzed, when the grid voltage is disturbed. The current control block is shown in Figure 5.

When controlling any order of harmonic, the transfer function of compensation current caused by grid voltage disturbance is

$$
\Phi_{en}(s) = \frac{P(s)}{1 + G_0(s)} = \frac{1}{(Ls + R)\left(1 + \left(K_{ph}s/L\left(s^2 + (h\omega_s)^2\right)\right)\right)}.
\tag{7}
$$

At the resonant frequency, the amplitude gain caused by the grid voltage disturbance tends to 0, that is, the steady-state error is 0, which indicates that the compensation current is affected by the voltage disturbance very little by using the VR controller.

4.2. PI Controller for Compensating Fundamental Current.
The previous section demonstrates that the VR controller with double resonant frequency can compensate the harmonic with zero steady-state error, but from bode diagram of harmonic close loop control, the amplitude gain of the current near 0 Hz is very small, which means that the VR

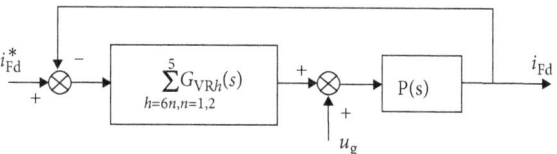

FIGURE 5: Current control block diagram when the system is disturbed by the grid voltage.

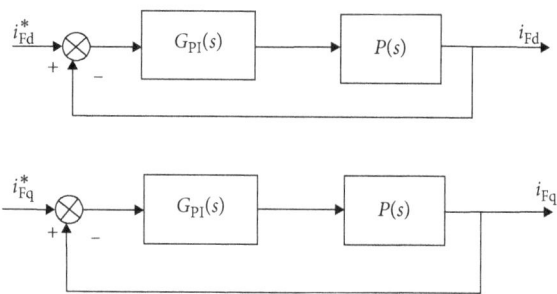

FIGURE 6: Fundamental current control block diagram based on PI controller.

controller is not ideal for the low frequency signal including DC current signal and has the problem of slow dynamic response. However, in addition to controlling the harmonic currents, the SAPF needs to control the fundamental current from voltage controller to compensate for its loss. The DC current obtained fundamental current by the synchronous reference frame. So, it is necessary to ameliorate the VR controller to improve the gain of the low frequency signals. For $h = 0$ in formula (3), the VR controller can be transformed to PI controller, as shown in the following formula:

$$G_{PI}(s) = \frac{K_{p0}s + K_{i0}}{s}. \tag{8}$$

The fundamental current control block diagram based on PI controller is given in Figure 6.

The fundamental current closed loop transfer function based on the PI controller is

$$\Phi_c(s) = \frac{K_{p0}s + K_{i0}}{Ls^2 + (R + K_{p0})s + K_{i0}}. \tag{9}$$

For $K_{p0}/K_{i0} = L/R$, the zero point of the PI controller is used to offset the pole of the controlled object; thus, the phase delay of the controlled object is compensated. That is,

$$\Phi_c(s) = \frac{K_{p0}}{Ls + K_{p0}}. \tag{10}$$

The closed loop transfer function shown in formula (10) is a typical first-order inertia link, and its control bandwidth is $\omega = K_{p0}/L$; that is, the cut-off frequency of the transfer function is $f_c = K_{p0}/2\pi L$. When L is constant, the bigger K_{p0} is, the bigger bandwidth ω is, which represents that the response speed of current loop is faster; when K_{p0} is constant, the bigger L is, the bigger bandwidth ω is, which represents that the response speed of current loop is slower.

The bode diagram of fundamental close loop control is given in Figure 7 for $L = 3$ mH and $K_{p0} = 1, 10, 20$. At the frequency of 0 Hz, the amplitude gain of the fundamental current is 1 without phase delay. However, the gain is rapidly attenuated outside 0 Hz. It indicates that the PI controller has the function of frequency selection for the DC current under the synchronous reference frame, which can achieve high precision control.

4.3. The Influence of Digital Control Delay.
Digital control delay, including computation delay and PWM update link delay, will affect tracking performance of the current loop and even cause instability of the system. Thus, it is necessary to study the effect of control delay on the tracking performance of the current loop. The delay of the digital control is

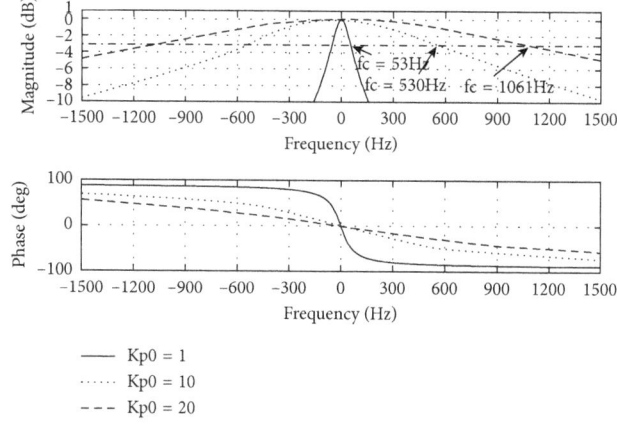

—— Kp0 = 1
······ Kp0 = 10
- - - Kp0 = 20

FIGURE 7: Bode diagram of fundamental close loop control.

generally 1.5 times of the sample period Ts [21], and the transfer function can be expressed as

$$G_d(s) = \frac{1}{1.5T_s * s + 1},$$

$$G_c(s) = \frac{\sum_{h=6n,n=1,2}^{5} \left(K_{ph}s/L\left(s^2 + (h\omega_s)^2\right)(1.5T_s s + 1)\right)}{1 + \sum_{h=6n,n=1,2}^{5} \left(K_{ph}s/L\left(s^2 + (h\omega_s)^2\right)(1.5T_s s + 1)\right)}. \tag{11}$$

Bode diagram of closed loop current considering digital control delay is shown in Figure 8 for $L = 3$mH, $T_s = 10^{-4}$ s, $K_{ph} = 0.4$. Even if affected by the digital control delay, the current loop has 35° of phase margin, which is still stable. In addition, the phase shift is still 0 at every positive and negative sequence harmonic frequency, which indicates that the digital control delay has less influence on the system and the current loop still has high static control accuracy.

4.4. Design of Control System.
The control block diagram of the three-phase three-wire SAPF for the proposed method is shown in Figure 9. The PI control is used to maintain DC-link voltage stability in the external voltage loop. The difference between the reference voltage u_{dc}^* and the actual voltage u_{dc} is input to the PI controller and the output of PI controller as a part of active power loss is added to the active source current component. PI and VR control is used in the internal current loop. The harmonic currents i_{hd}^* and i_{hq}^*

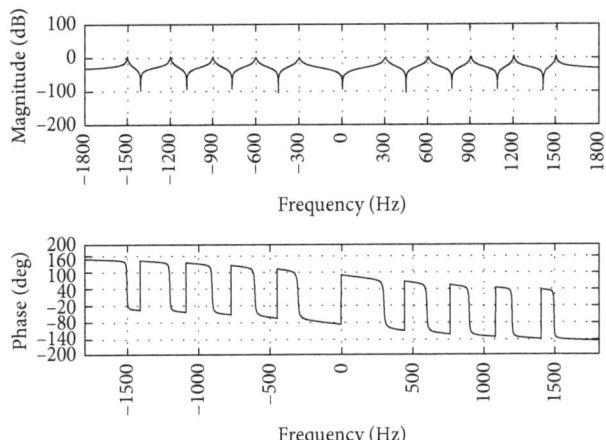

FIGURE 8: Bode diagram of closed loop current considering digital control delay.

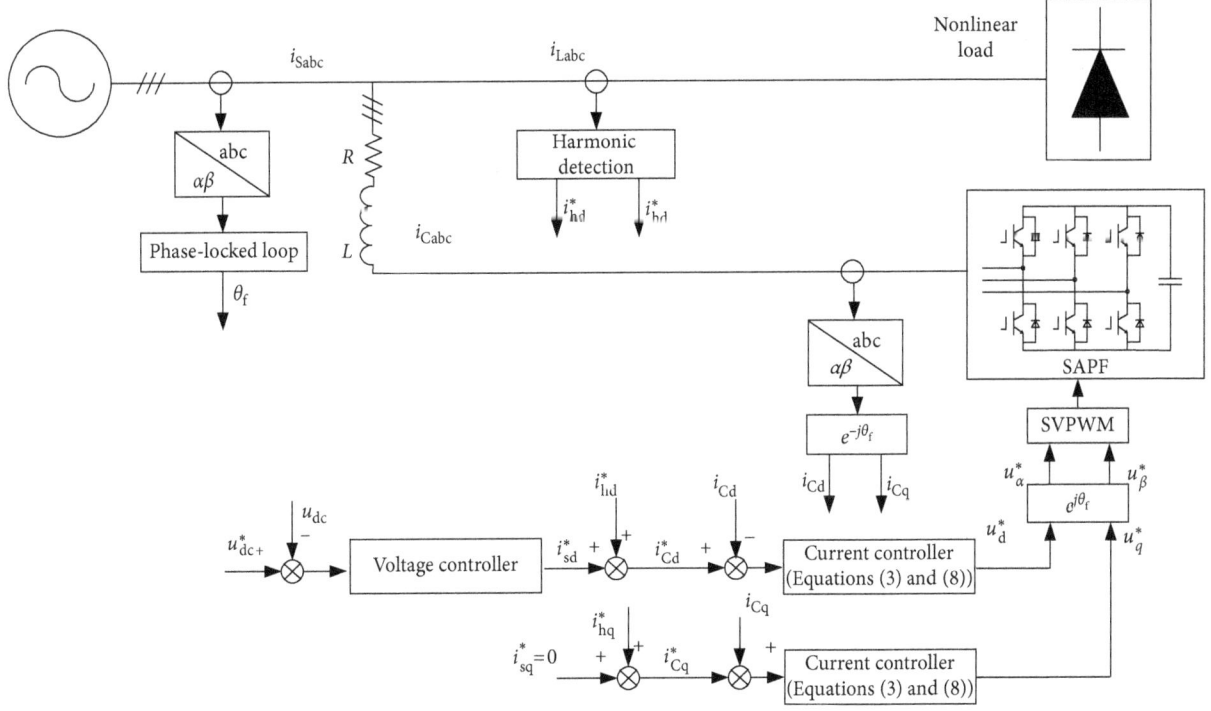

FIGURE 9: Control block diagram of three-phase three-wire SAPF.

extracted from the load current i_{Labc} by instantaneous reactive power theory is added to the output of the PI controller i_{sd}^* and i_{sq}^*. Compensation current i_{Cabc} is turned to i_{cd} and i_{cq} based on synchronous coordinate transformation; then the difference between i_{cd}, i_{cq} and i_{cd}^*, i_{cq}^* is input to the current controller; besides, the control pulse signal required by SAPF is obtained by the space vector pulse width modulation (SVPWM) to compensate the harmonic currents [22].

5. Simulation and Experimental Results

In order to verify the correctness and effectiveness of the proposed current control method based on PI and multi-VR control in this paper, both classical PI controller and PI + multi-VR controller are designed according to the above-mentioned

TABLE 1: Simulation model parameters.

Parameter	Value
Source voltage V_{snom}	220 V, 50 Hz
DC-link voltage V_{dcref}	750 V
DC-link capacitance C	1000 μF
Filter inductance L_{f}	3 mH
Equivalent loss resistance R_{f}	0.3 Ω
Nonlinear load	Three-phase diode bridge rectifier with a 10 Ω DC resistor

method in Matlab/Simulink. The simulation model parameters are shown in Table 1.

Since both of the power supply and the nonlinear load are three-phase symmetry, the A phase is only analyzed in

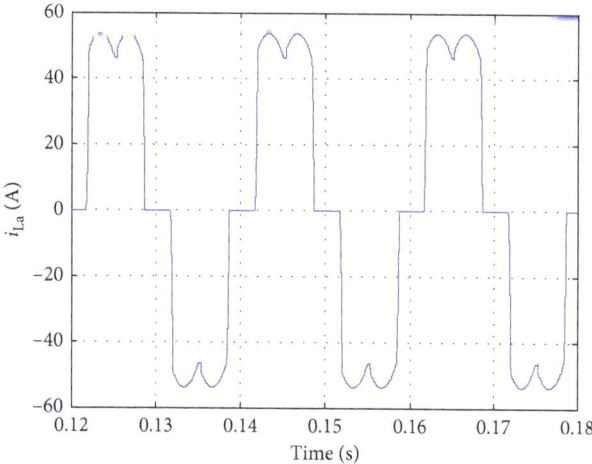

FIGURE 10: Waveform of nonlinear load current.

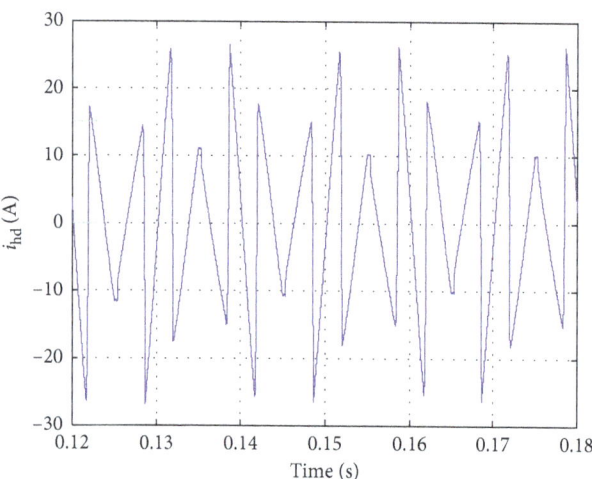

FIGURE 11: Waveform of harmonic currents.

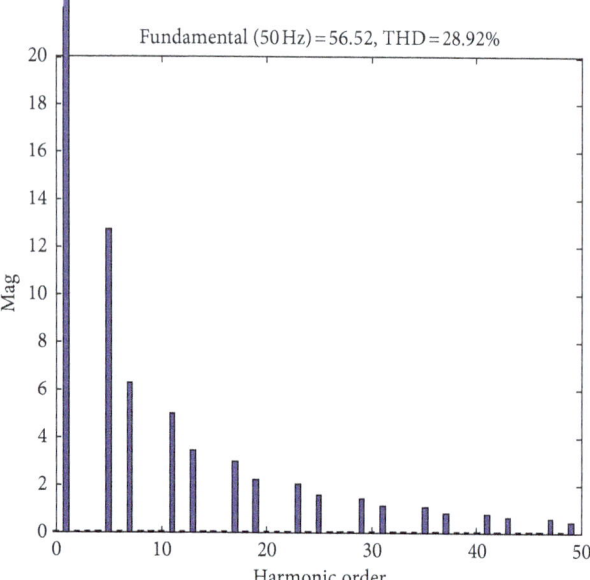

FIGURE 12: FFT analysis of nonlinear load current.

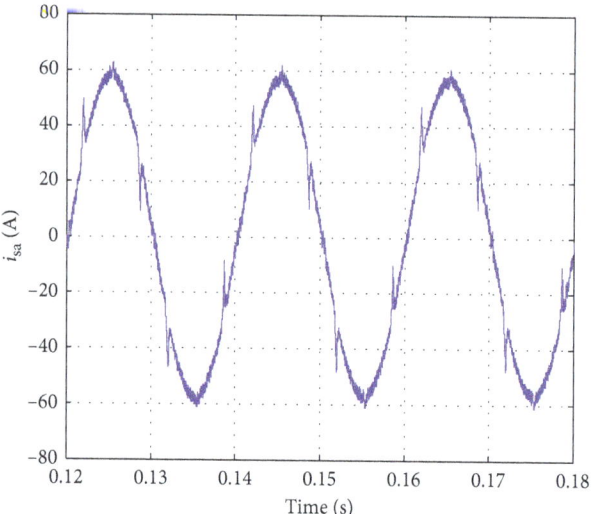

FIGURE 13: Waveform of source current using PI controller.

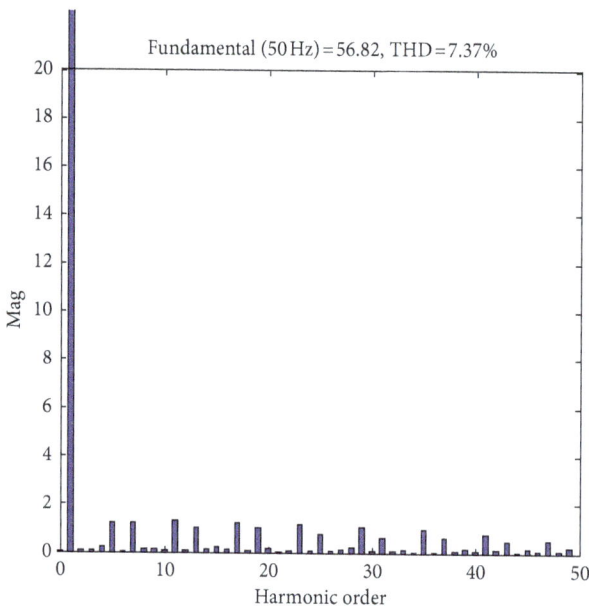

FIGURE 14: FFT analysis of source current using PI controller.

the simulation waveform. The waveform of the load current and harmonic currents are depicted in Figures 10 and 11, and the FFT analysis of the nonlinear load current are shown in Figure 12. From Figure 11, the total harmonic distortion (THD) before compensation is up to 28.92%, and the load current's harmonic spectrum contains harmonics of the order $h = 6n \pm 1, n = 1, 2, \ldots$.

The compensation performance using PI controller is shown in Figures 13 and 14, respectively. The compensation results of the proposed control strategy based on PI and 6th VR controller are shown in Figures 15 and 16, respectively. Comparing the four figures, PI controller can compensate a part of harmonic, which is not too fruitful. After adopting the 6th of the VR controller, the 5th harmonic and 7th harmonic can be compensated effectively. 5th harmonic is reduced to 0.28% from 2.08%. 7th harmonic is reduced to

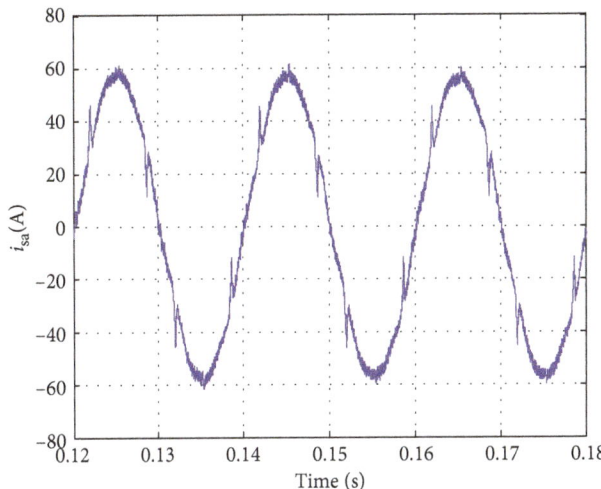

FIGURE 15: Waveform of source current using PI controller and 6th VR controller.

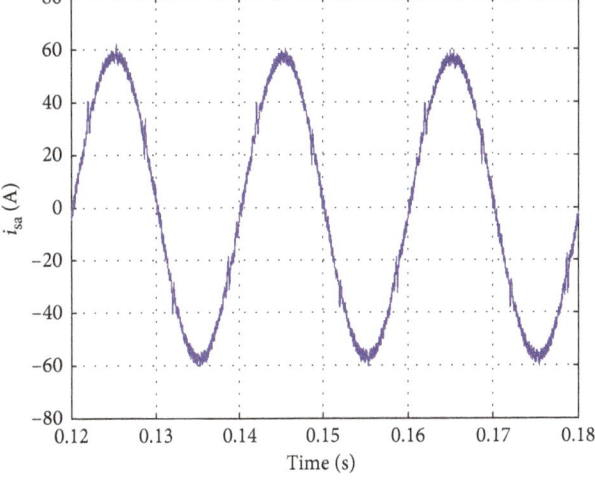

FIGURE 17: Waveform of source current using PI controller and multi-VR controllers.

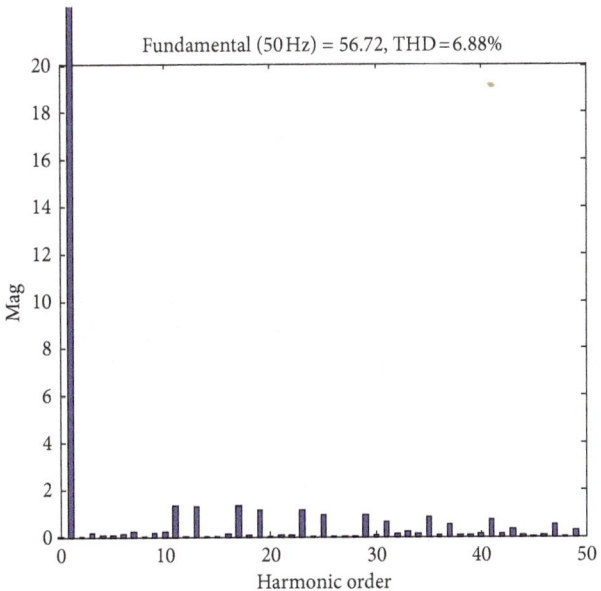

FIGURE 16: FFT analysis of nonlinear load current using PI controller and 6th VR controller.

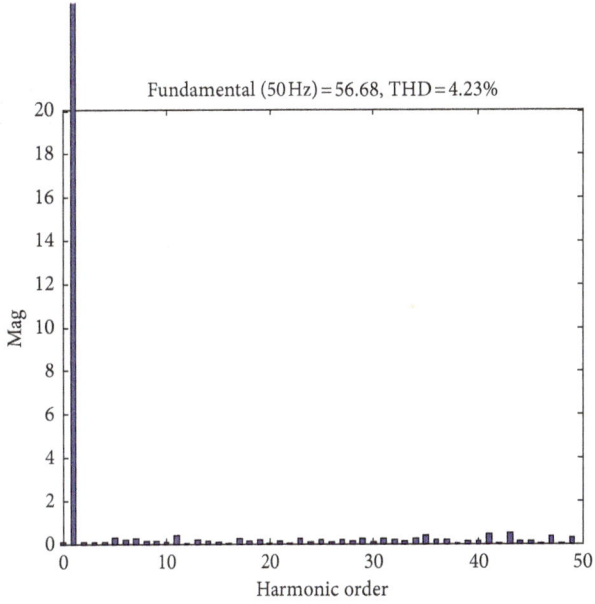

FIGURE 18: FFT analysis of nonlinear load current using PI controller and multi-VR controllers.

0.11% from 2.16%, but distortion rate of other harmonics remains the same. The simulation results show that the 6th resonant controller has good inhibitory effects on 5th harmonic and 7th harmonic, which can further improve the compensation performance for some specific harmonics.

The compensation results using PI controller and multi-VR controller are shown in Figures 17 and 18. When the PI and multi-VR controllers are introduced in the control system, distortion rate of each harmonic decreases remarkably.

The experimental waveforms based on PI + multi-VR controllers proposed in the case of abrupt loading is shown in Figure 19. It is known from Figure 19 that at 0.3 s, a three-phase diode rectifier of 20 kW is added to the system, and the SAPF can still track the abrupt load current quickly.

After about one fundamental period, the system reaches stability. It shows that the proposed control strategy has good dynamic performance.

In order to further verify the effectiveness of the proposed strategy, a SAPF prototype of 15 kVA is built. The main control chip is TMS320F28335 of TI company. PM50RL1B120 produced by Mitsubshi is used as power devices of the main circuit. In the process of prototype testing, the experimental waveform is recorded by the DPO3014 digital fluorescence oscilloscope produced by the Tektronix Inc in the United States, and the harmonic data are recorded by PM3000A Power Quality Analyzer produced by Voltech Inc. Experimental setups of SAPF are shown as Figure 20. The experimental parameters are as follows:

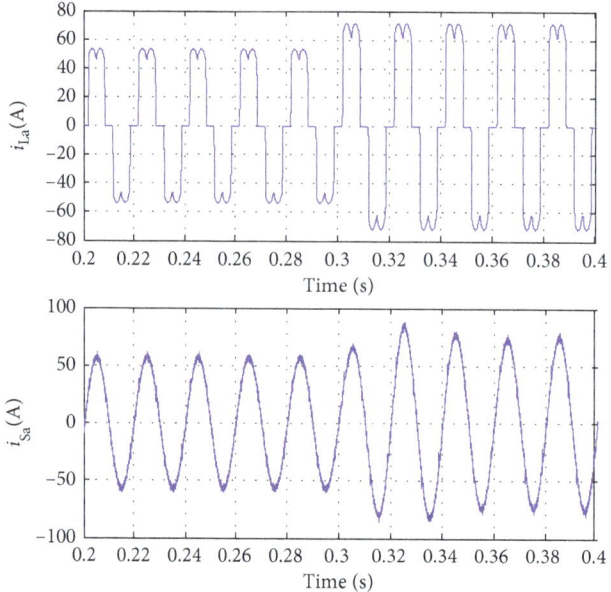

FIGURE 19: Waveforms of load current and source current in the case of abrupt loading.

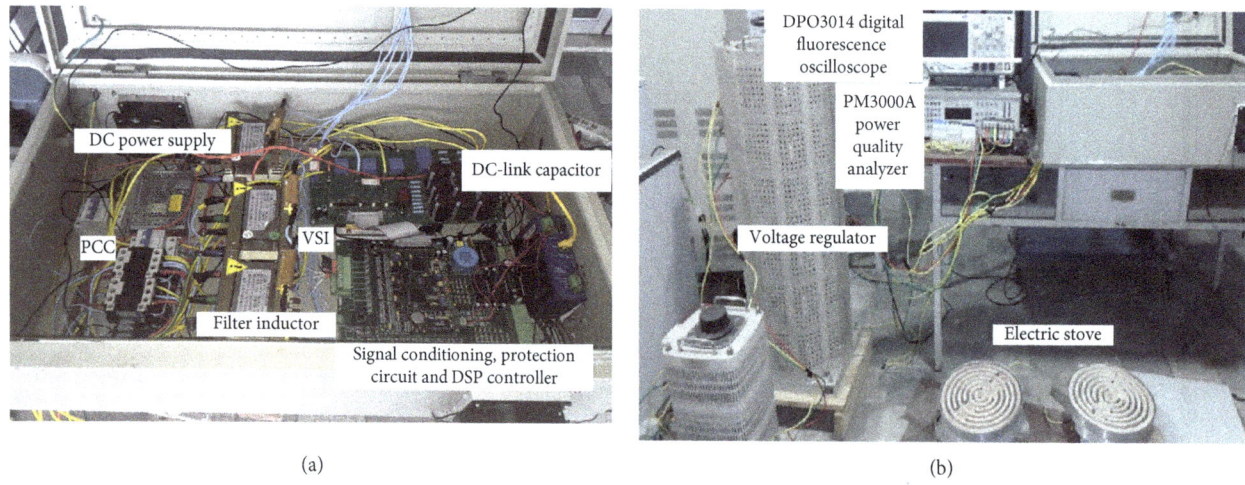

(a)   (b)

FIGURE 20: Experimental setups of SAPF.

(i) Three-phase source voltage: $U_N = 380$ V and $f = 50$ Hz

(ii) Nonlinear load: two-way three-phase diode rectifier with two series two parallel(2S2P) electric stoves of 5 kW in DC link and a three-phase voltage regulator with a rated power of 15 kVA, rated voltage p of 380 V, and output voltage of 220 V in AC link

(iii) Switching frequency of IGBT: 10 kHz

(iv) Voltage controller parameters: ratio coefficient $Kp = 0.1$ and integral coefficient Ki = 10

(v) Current controller parameters: $K_{p0} = 0.5$, $K_i = 160$, $K_{p6} = 0.2$, $K_{i6} = 200$, $K_{p12} = 0.2$, $K_{i12} = 200$, $K_{p18} = 0.5$, $K_{i18} = 200$, and so on

(vi) Other experimental parameters: the parameters including DC-link voltage, DC link capacitance, and

filter inductance are consistent to the simulation parameters

Experimental steady waves using PI controller and PI + multi-VR controllers are shown in Figures 21 and 22. Waveform 2 represents load current, THD of which is 26.1%. Waveform 4 in Figures 21 and 22 represent source current after compensation based on PI controller and PI + multi-VR controllers. Evidently, both of source currents approach nearly sinusoidal by SAPF. However from Table 2, the THD of the source current reduces from 7.34% to 2.86% by using the proposed control strategy. The above results show that the proposed control strategy has good steady performance and control accuracy.

Experimental dynamic results are shown in Figure 23. Before the time of 0.04 s, the SAPF has been invested in the system to finish the harmonic current detection. Besides, voltage controller has been working steadily and DC-link

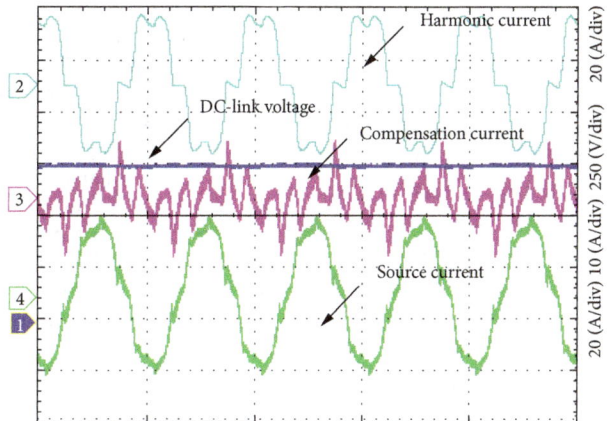

FIGURE 21: Experimental steady waves using PI controller.

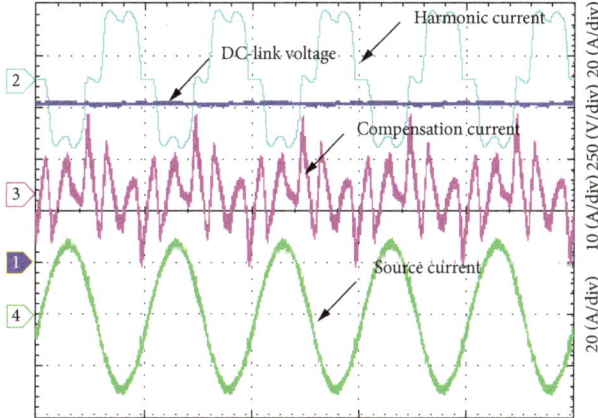

FIGURE 22: Experimental steady waves for proposed scheme.

TABLE 2: Compensation results with two control algorithms.

Order of harmonic		5	7	11	13	17	19	THD (%)
Harmonic distortion before compensation (%)		21.3	9.87	5.60	3.36	1.80	0.92	26.1
Harmonic distortion after compensation (%)	With PI controller	2.16	1.86	1.76	1.80	1.74	1.82	7.34
	With the proposed method	0.72	0.48	0.35	0.2	0.18	0.10	2.86

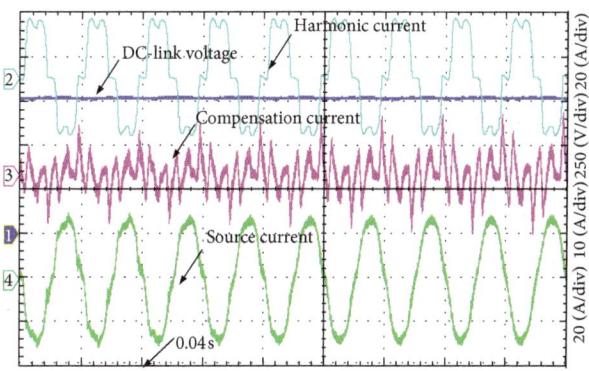

FIGURE 23: Experimental dynamic waveform after adding VR controller.

voltage has been stabilized at 750 V. Moreover, the basic control loop of current controller or PI control has also been run. At the time of 0.04 s, the multi-VR controllers start running. After about 3 fundamental periods, the system enters the new steady state. Experimental results show that the proposed strategy also has good dynamic performance.

6. Conclusion

With development of power electronic technology, a large number of power electronic devices are integrated into the power grid, which causes the harmonic pollution. Through the study on SAPF mathematical model and the principle of VR controller, a current control strategy based on PI and multi-VR controller is proposed in this paper. Through detailed analysis on frequency response characteristic of current closed loop, in that the PI and VR controller can compensate the harmonic currents with zero steady-state error, little phase delay and good dynamic performance are proved. In addition, under the synchronous reference frame, the proposed method is simple enough to compensate the harmonic, which reduces the computation and is better adapted to frequency fluctuation. The simulation and experimental results show that the proposed control strategy is correct and effective, which improves the power quality.

Conflicts of Interest

The authors declare that there are no conflicts of interest regarding the publication of this paper.

Acknowledgments

The authors thank J. Huang and Z. Shen for helpful conversations. This work was partially supported by National Natural Science Foundation of China under grant no. 51507153 and the Zhejiang Provincial Natural Science Foundation of China under grant no. LY16E070005.

References

[1] Q. Zhong, J. Feng, G. Wang, and H. Li, "Feedforward harmonic mitigation strategy for single-phase voltage source converter," *Journal of Electrical and Computer Engineering*, vol. 2018, Article ID 5909346, 10 pages, 2018.

[2] C. Burgos-Mellado, C. Hernandez, R. Cardenas-Dobson et al., "Experimental evaluation of a CPT-based 4-leg active power compensator for distributed generation," *IEEE Journal of Emerging & Selected Topics in Power Electronics*, vol. 5, no. 2, pp. 747–759, 2017.

[3] S. Pettersson, M. Salo, and H. Tuusa, "Applying an LCL-filter to a four-wire active power filter," in *Proceedings of Power Electronics Specialists Conference (PESC'06)*, pp. 1–7, Jeju, South Korea, June 2006.

[4] E. Uz-Logoglu, O. Salor, and M. Ermis, "Real-time detection of interharmonics and harmonics of AC electric arc furnaces on GPU framework," in *Proceedings of IEEE Industry Applications Society Meeting (IAS'17)*, pp. 1–8, Cincinnati, OH, USA, October 2017.

[5] S. Rahmani, A. Hamadi, K. Al-Haddad, and L. A. Dessaint, "A combination of shunt hybrid power filter and thyristor-controlled reactor for power quality," *IEEE Transactions on Industrial Electronics*, vol. 61, no. 5, pp. 2152–2164, 2014.

[6] S. M. M. Gazafrudi, A. T. Langerudy, E. F. Fuchs et al., "Power quality issues in railway electrification: a comprehensive perspective," *IEEE Transactions on Industrial Electronics*, vol. 62, no. 5, pp. 3081–3090, 2015.

[7] M. K. M. Valappil and M. K. Mishra, "Three-leg inverter-based distribution static compensator topology for compensating unbalanced and non-linear loads," *IET Power Electronics*, vol. 8, no. 11, pp. 2076–2084, 2015.

[8] L. Wang, C. S. Lam, M. C. Wong, N.-Y. Dai, K.-W. Lao, and C. K. Wong, "Non-linear adaptive hysteresis band pulse-width modulation control for hybrid active power filters to reduce switching loss," *IET Power Electronics*, vol. 8, no. 11, pp. 2156–2167, 2015.

[9] S. Q. Zhang, D. Ke, B. Xie, and Y. Kang, "Selective harmonic current control based on multiple synchronous rotating co-ordinates," *Proceedings of the CSEE*, vol. 30, no. 3, pp. 55–62, 2010.

[10] A. D. Aquila, M. Liserre, V. G. Monopoli et al., "Overview of PI-based solutions for the control of DC buses of a single-phase H-bridge multilevel active rectifier," *IEEE Transactions on Industry Applications*, vol. 44, no. 3, pp. 857–866, 2008.

[11] Z. Chelli, R. Toufouti, A. Omeiri, and S. Saad, "Hysteresis control for shunt active power filter under unbalanced three phase load conditions," *Journal of Electrical and Computer Engineering*, vol. 2015, Article ID 391040, 9 pages, 2015.

[12] Y.-H. Kim, J.-G. Kim, Y.-H. Ji, C.-Y. Won, and Y.-C. Jung, "Photovoltaic parallel resonant dc-link soft switching inverter using hysteresis current control," in *Proceedings of the Applied Power Electronics Conference and Exposition (APEC'10)*, pp. 2275–2280, Palm Springs, CA, USA, February 2010.

[13] Z. X. Zou, K. Zhou, Z. Wang, and M. Cheng, "Frequency-adaptive fractional-order repetitive control of shunt active power filters," *IEEE Transactions on Industrial Electronics*, vol. 62, no. 3, pp. 1659–1668, 2015.

[14] Y. Wang, H. Zheng, R. Wang et al., "A novel control scheme based on the synchronous frame for APF," *Journal of Power Electronics*, vol. 17, 2017.

[15] W. Jiang, L. Wang, W. Ma et al., "A control method based on current tracking error compensation of deadbeat control for a three-phase active power filter," *Proceedings of the CSEE*, vol. 36, no. 20, pp. 5605–5615, 2016.

[16] S. Buso, T. Caldognetto, and D. I. Brandao, "Dead-beat current controller for voltage-source converters with improved large-signal response," *IEEE Transactions on Industry Applications*, vol. 52, no. 2, pp. 1588–1596, 2016.

[17] L. Herman, I. Papic, and B. Blazic, "A proportional-resonant current controller for selective harmonic compensation in a hybrid active power filter," *IEEE Transactions on Power Delivery*, vol. 29, no. 5, pp. 2055–2065, 2014.

[18] J. B. Hu, Y. K. He, H. S. Wang et al., "Proportional-resonant current control scheme for rotor-side converter of doubly-fed induction generators under unbalanced network voltage conditions," *Proceedings of the CSEE*, vol. 30, no. 6, pp. 48–56, 2010.

[19] B. Singh, S. K. Dube, and S. R. Arya, "An improved control algorithm of DSTATCOM for power quality improvement," *International Journal of Electrical Power & Energy Systems*, vol. 64, no. 64, pp. 493–504, 2015.

[20] H. Yi, F. Zhuo, Y. Zhang et al., "A source-current-detected shunt active power filter control scheme based on vector resonant controller," *IEEE Transactions on Industry Applications*, vol. 50, no. 3, pp. 1953–1965, 2014.

[21] A. Szromba, "Conductance-controlled global compensation type shunt active power filter," *Archives of Electrical Engineering*, vol. 64, no. 2, pp. 259–274, 2015.

[22] X. U. Hailiang, Z. Liao, and H. E. Yikang, "Key points of proportional-resonant controller applied for PWM converters," *Automation of Electric Power Systems*, vol. 39, no. 18, pp. 151–159, 2015.

3

EPS Current Tracking Method Research based on Hybrid Sensitivity H_∞ Control Algorithm

Hairong Wu,[1] **Guangfei Xu,**[1,2] **Jian Wu**⑩**,**[1] **Xue Han,**[3] **Jiwei Feng,**[1]
Shifu Liu,[1] **and Linglong Bu**[2]

[1]*School of Mechanical and Automotive Engineering, Liaocheng University, Liaocheng City, Shandong Province 252000, China*
[2]*Modern Agricultural Demonstration Garden, Liaocheng Academy of Agricultural Sciences, Liaocheng City,*
Shandong Province 252000, China
[3]*Economic & Technology Research Institute, State Grid Shandong Electric Power Company, Jinan City,*
Shandong Province 250000, China

Correspondence should be addressed to Jian Wu; wujian1982@mail.tsinghua.edu.cn

Academic Editor: Ephraim Suhir

For electric power steering system (EPS), road interference, noise of the sensor, and the uncertainty of the steering system may make EPS control effect and the driver's road sense worse. EPS system which takes advantage of good current tracking ability, good anti-interference ability, and good operation stability is becoming more and more important in automotive research. The traditional H_∞ control algorithm can solve the system uncertainty theoretically, but it cannot solve the contradiction between robustness and performance without considering the performance of the system. Therefore, this paper proposes a EPS current tracking method based on the hybrid sensitivity H_∞ control algorithm, which takes the current tracking performance as one of the control objectives, so that the system can maximize the robustness and performance. Firstly, the dynamic model of EPS is established. Then, the two-degree-of-freedom vehicle model and tire model are introduced. The state space equation of the system is constructed on the basis of the system state space with random disturbance signals, the hybrid sensitivity H_∞ controller is designed in the sensitivity index design, and the proposed algorithm can use weighting function to minimize the performance of the current tracking error as well as the robustness of the yaw rate error in response to robustness. Simulation analysis and experimental verification of EPS system are also carried out. The results show that the control method of the hybrid sensitivity H_∞ can better achieve EPS target current tracking, effectively suppress the effect of external interference and noise, improve the system performance and robustness, ensure the driver get good road sense, and improve the system of steering stability.

1. Introduction

Electric power steering (EPS) system is the core technology of the automobile steering system. It is the key component of intelligent driving. It has the advantages of portability, agility, energy saving, environmental protection, and convenient installation. The EPS system is based on the mechanical steering system, and the auxiliary motor is added to control the motor to provide the auxiliary torque to achieve the purpose of power steering. Great progress has been made in EPS system and it has been widely used in automobile industry. The EPS control based on the classical control theory can realize the power steering function, but the EPS

feel is not as well as hydraulic power system (HPS) [1]. It cannot restrain the disturbance caused by nonlinear factors due to the uncertainties of the steering system.

Classical control theory, modern control theory, and advanced control theory have been applied to the study of EPS system one after another. Classical control theory was first applied to EPS. Tan Guangxing et al. [2] proposed a kind of EPS control based on immune fuzzy PID. This kind of control based on fuzzy PID can greatly improve EPS power steering performance and portability. Then, PID control of EPS is further developed by using modern control theory. Zhang jianwei et al. [3] proposed a PID control strategy based on genetic algorithm, which is used to optimize the PID controller parameters by using genetic algorithm so as to achieve better control effect.

The above research does not fundamentally solve the current tracking problem, especially anti-interference and robustness problem. Then, the robustness of EPS system has become the focus of research. In order to improve the anti-interference ability of EPS system, advanced control theory is applied to EPS research. Various anti-interference algorithms have been studied experimentally. Qiu Ming et al. [4] studied the EPS based on the control principle and discussed the uncertainty of the system model and the method of minimizing interference to achieve robust control of the EPS. Wang Qirui et al. [5] studied the standard control problem of EPS and designed the robust controller for the road sense. Zhao Wanzhong et al. [6] studied EPS based on H_2/H_∞ hybrid control using H_∞ to minimize the impact of interference on the output and simultaneously carry out the H_2 optimization. Weng Jingliang et al. [7] studied the robust control strategy of EPS based on handling stability. The stability and tracking problem of the system were transformed into a hybrid sensitivity problem. She Guoqin et al. [8] studied the robust control of EPS and its influence on handling stability by using hybrid sensitivity method. Chen long et al. [9] designed a new μ controller by selecting the related performance indexes based on μ synthesis theory and robust control method. The stability and safety of vehicle are improved effectively. Frédéric Wilhelm et al. [10] considered the effects of friction in EPS and proposed an active compensation control strategy which could estimate the internal friction of the system and compensate for it through the motor input. It is composed of two feedback loops: internal loop for system friction estimation and external loop for minimizing the tracking error. Ying-Chih Hung et al. [11] proposed a wavelet fuzzy neural network using asymmetric membership function (WFNN-AMF) with improved differential evolution (IDE) algorithm to control the EPS. It improved the stability and comfort of the vehicle to great extent. But the performance of EPS is given less attention. Wonhee Kim et al. [12] proposed a lane-keeping system of automated vehicle based on EPS which takes unknown parameters and external disturbance, along with their derivatives into the designment of the augmented observer and nonlinear damping controller. Alaa et al. [13] studied a kind of new EPS control strategy. They, respectively, test the reference target and determine the reference model and finally realize the EPS control by using the two-level synovial control strategy to track the motor corner. Dongwook Lee et al. [14] and Cassio T. Faria et al. [15] all identified the external disturbance, nonlinearity, and uncertainty of EPS system, and the controller design is carried out based on them. Some results have been achieved.

The studies mentioned above about the EPS control achieved considerable effect, but there should be more in-depth study on current tracking performance which is the main indicators of the controller. These studies are based on the simulation level, so they can not verify the real effect of the algorithm due to the inevitable gap between the designed algorithm model and the actual steering system. Therefore, a H_∞ controller based on hybrid sensitivity is designed and the controller is used in the actual steering system.

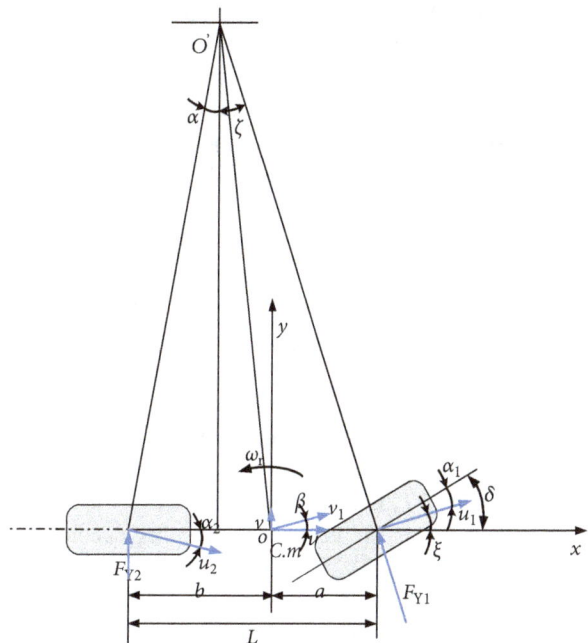

FIGURE 1: Two-degree-of-freedom vehicle model.

This paper is arranged as follows. The second section presents the two-degree-of-freedom vehicle model, EPS steering system model, and others. The third section describes the design of the new hybrid sensitivity H_∞ controller. In the fourth section, the simulation results of the PID and the new hybrid sensitivity H_∞ control strategy are shown. Then, the fifth section is the comparison of experimental results between two control strategies applied on the real experimental bench. Finally, the conclusions are shown in the sixth section.

2. Mathematical Model of EPS System

2.1. Two-Degree-of-Freedom Vehicle Model. In this paper, the steering performance of automobile is studied, so a simplified model of vehicle with 2 degrees of freedom is presented [16] in Figure 1.

The model's motion differential equation is

$$(k_1 + k_2)\beta + \frac{1}{u}(ak_1 - bk_2)\omega_r - k_1\delta = m(\dot{\beta} + \omega_r) \quad (1)$$

$$(ak_1 - bk_2)\beta + \frac{1}{u}(a^2k_1 + b^2k_2)\omega_r - ak_1\delta = I_z\dot{\omega}_r \quad (2)$$

k_1 and k_2 are, respectively, the stiffness of front and rear wheels. β means sideslip angle of vehicle center of mass. u is vehicle centroid velocity. a and b are, respectively, the distance between the car and the center of mass. ω_r means yaw speed for car. δ means front wheel corner. m means mass of vehicle. I_z is the moment of inertia of the car around the z-axis.

2.2. EPS Steering System Model. For the convenience of study, this paper simplifies the front wheel and steering gear to steering shaft [17]. We focus on the situation that the relevent forces are equivalent to the steering axis. On the basis of the reliability of the model, the relation between the input and output variables is established [18]. EPS simplified model is shown in Figure 2.

In Figure 2, T_h means torque input for steering wheel. T_s means measuring torque of torque sensor. T_m means electromagnetic torque of motor. T_r is the steering resistance moment of the road equivalent to the torque on the pinion (gear ratio is N). $N_1 T_m$ is the motor torque acting on the steering shaft torque. N_1 is the drive ratio which is the steering shaft to the motor. θ_h is steering angle. θ_m is the angle of motor. δ_1 is steering shaft angle. J_h is the moment of inertia of steering wheel. J_p is the moment of inertia converted to steering shaft. B_p is the equivalent damping coefficient of system friction.

2.2.1. The Steering Shaft Model. The dynamic analysis of the wheel part above the torque sensor can be

$$(k_1 + k_2)\beta + \frac{1}{u}(ak_1 - bk_2)\omega_r - k_1\delta = m(\dot{\beta} + \omega_r) \quad (3)$$

Dynamic analysis of the steering shaft below the torque sensor is available by

$$K_s(\theta_h - \delta_1) + N_1 T_m - T_r = J_p\ddot{\delta}_1 + B_p\delta_1 \quad (4)$$

K_s is torsion bar stiffness of sensor.

2.2.2. Electrical Machinery Model. The system uses a brushed DC motor. The differential equation can be obtained by Holzer's law of voltage:

$$u_a = L_a\dot{I}_a + R_a I_a + K_b\dot{\theta}_m \quad (5)$$

$$\dot{\theta}_m = N_1\dot{\delta}_1 \quad (6)$$

$$T_m = K_a I_a \quad (7)$$

u_a is the motor terminal voltage. R_a is armature resistance. K_a is torque coefficient of motor. K_b is coefficient of back electromotive force of motor. L_a is inductance coefficient of motor. I_a is motor current.

2.2.3. Steering Resistance Moment Calculation. At the small angle, the tire deformation is approximately linear. The steering resistance moment of the pinion acting on the road through the tire [19] is

$$T_r = \frac{2}{N}dk_1\left(\frac{\delta_1}{N} - a\frac{\omega_r}{u} - \beta\right) \quad (8)$$

d is pneumatic trail.

2.2.4. The State Equation of the System. The state equation of the system can be obtained by formulas (1) ~ (8):

$$\dot{X} = AX + B_1\omega + B_2U \quad (9)$$

Take state variables as $X = \begin{bmatrix} \dot{\theta}_h & \theta_h & I_a & \dot{\delta}_1 & \delta_1 & \beta & \omega_r \end{bmatrix}^T$. The control input is $U = \begin{bmatrix} T_h & u_a \end{bmatrix}^T$. The road signal input is $\omega = s_0\delta(t)$. s_0 is coefficient of interference intensity. $\delta(t)$ is pavement interference noise.

Among them, the coefficient matrices A, B are

$$A = \begin{bmatrix} -\dfrac{B_h}{J_h} & -\dfrac{K_s}{J_h} & 0 & 0 & \dfrac{K_s}{J_h} & 0 & 0 \\ 1 & 0 & 0 & 0 & 0 & 0 & 0 \\ 0 & 0 & -\dfrac{R_a}{L_a} & -\dfrac{N_1 K_b}{L_a} & 0 & 0 & 0 \\ 0 & \dfrac{K_s}{J_p} & \dfrac{N_1 K_a}{J_p} & -\dfrac{B_p}{J_p} & -\dfrac{K_s}{J_p} - \dfrac{2dK_1}{N^2 J_p} & \dfrac{2adK_1}{NuJ_p} & \dfrac{2dK_1}{NJ_p} \\ 0 & 0 & 0 & 1 & 0 & 0 & 0 \\ 0 & 0 & 0 & 0 & -\dfrac{k_1}{Nm} & \dfrac{k_1 + k_2}{m} & \dfrac{ak_1 - bk_2}{mu} - 1 \\ 0 & 0 & 0 & 0 & -\dfrac{ak_1}{NI_Z} & \dfrac{ak_1 - bk_2}{I_Z} & \dfrac{a^2 k_1 + b^2 k_2}{uI_Z} \end{bmatrix} \quad (10)$$

$$B_1 = \begin{bmatrix} 0 & 0 & 0 & 1 & 0 & 0 & 0 \end{bmatrix}^T$$

$$B_2 = \begin{bmatrix} \dfrac{1}{J_h} & 0 & 0 & 0 & 0 & 0 & 0 \\ 0 & 0 & 1 & 0 & 0 & 0 & 0 \end{bmatrix}^T$$

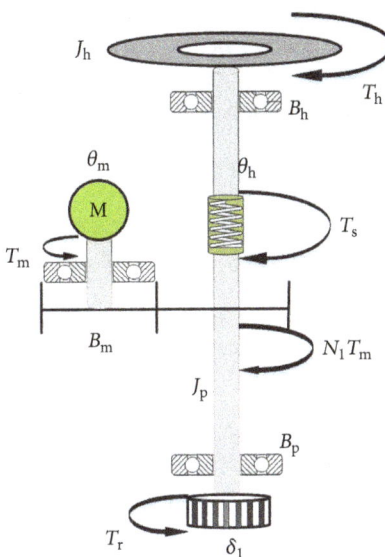

FIGURE 2: Schematic diagram of a simplified dynamic EPS system.

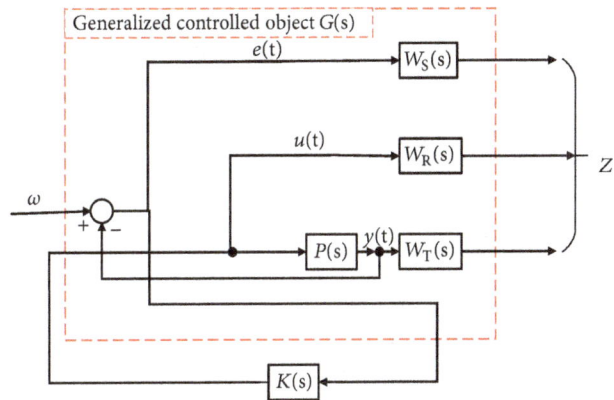

FIGURE 3: Block diagram of hybrid sensitivity design of EPS system.

Take the output variable of the system as $Y = [I_a \ T_s \ \omega_r \ \delta_1 \ T_r]$.

The output equation of the system is

$$Y = CX + DU \tag{11}$$

where

$$C = \begin{bmatrix} 0 & 0 & 1 & 0 & 0 & 0 & 0 \\ 0 & K_s & 0 & 0 & -K_s & 0 & 0 \\ 0 & 0 & 0 & 0 & 0 & 0 & 1 \\ 0 & 0 & 0 & 0 & 1 & 0 & 0 \\ 0 & 0 & 0 & 0 & \dfrac{2dk_1}{N^2} & -\dfrac{2adk_1}{Nu} & \dfrac{2dk_1}{N} \end{bmatrix} \tag{12}$$

$$D = \begin{bmatrix} 0 & 0 & 0 & 0 & 0 \end{bmatrix}^T$$

3. Design of H_∞ Controller Based on EPS

3.1. Hybrid Sensitivity Design Problem. The hybrid sensitivity design structure of EPS system is shown in Figure 3. $P(s)$ is the nominal space model of the system: the state space model of EPS system model. u_1 is the reference input system, including target current, ideal torque, sensor measurement, and ideal yaw rate. u_2 is system control input: the input voltage of the motor. $e(t)$ is the error of the signal, including the deviation of the target current from the actual current, the sensor measurement deviation, and yaw rate deviation. $u(t)$ is the control input, and y is the system output, including the actual current, the sensor torque, and the yaw rate. z is the system output evaluation, mainly referring to the impact of interference on control error (the current tracking performance of sensor, interference suppression, and handling stability), energy control input, and robust stability. W_S is sensitivity weighting factor. W_R is input sensitivity weighting factor, W_T is sensitivity weighted factor, and $K(s)$ is the controller of the system [20].

The sensitivity control function S, the input sensitivity function R, and the complement sensitivity function T are included in the design of the hybrid sensitivity H_∞ control system [21]. Its Laplasse transform [22], respectively, is

$$S(s) = [I + P(s)K(s)]^{-1} \tag{13}$$

$$R(s) = [I + P(s)K(s)]^{-1}K(s) \tag{14}$$

$$T(s) = [I + P(s)K(s)]^{-1}K(s)P(s) \tag{15}$$

I is the unit matrix.

$\|S\|_\infty$ is the infinite norm of the sensitivity function S and it represents the tracking performance of the closed-loop control system to the target. In order to improve the tracking performance of the system, $\|S\|_\infty$ should be as small as possible. $\|T\|_\infty$ is the infinite norm of the complement sensitivity function T and it represents the measure of the perturbation allowed by the closed-loop system. In order to ensure the robustness of the system, $\|T\|_\infty$ should be as small as possible. However, there is a limitation of $S + T = I$ at the same system frequency, so $\|S\|_\infty$ and $\|T\|_\infty$ cannot reach the minimum at the same time [21]. By introducing the sensitivity weighting matrix, W_S, and the complement sensitivity weighting matrix, W_T, the relative performance and stability of the system can be achieved. In addition, in order to avoid the actuator (motor) saturation or overload, the input sensitivity function weighting matrix, W_R, is introduced to limit the overoutput [23].

The hybrid sensitivity design problem is transformed into a controller $K(s)$ about system stability, disturbance rejection, and current tracking problems. The closed-loop control system from ω to z, including the generalized controlled object $G(s)$, is stable and minimal. It can be described as

$$\|T_{z\omega}\|_\infty = \left\| \begin{matrix} W_S S \\ W_R R \\ W_T T \end{matrix} \right\|_\infty \leq 1 \tag{16}$$

The hybrid sensitivity control problem can be reduced to a H_∞ control problem. For generalized space G,

$$G = \begin{bmatrix} G_{11} & G_{12} \\ G_{21} & G_{22} \end{bmatrix} = \begin{bmatrix} W_S & -W_S P \\ 0 & W_R \\ 0 & W_T P \\ 0 & P \end{bmatrix} \tag{17}$$

The solution of the control problem is to find controller K, stabilize G, and make $\|F_l(G, K)\|_\infty < \gamma$.

3.2. The Design of Evaluation Index

3.2.1. Tracking Performance. Due to the influence of sensor noise and other system uncertainties, there is a certain deviation between the actual auxiliary current I_a and the target current value I_a^*. It can be shown as

$$e_1 = I_a^* - I_a \tag{18}$$

In order to reduce the effect of error and improve the current tracking performance, this paper makes a weighted control of the current deviation between the actual current and the target current obtained through the help characteristic curve to obtain better tracking performance.

It can be expressed as

$$\|W_{s1} e_1\|_\infty = \|W_{s1} (I_a^* - I_a)\|_\infty \tag{19}$$

W_{s1} is the current tracking weighting function.

3.2.2. Road Feeling. Road feeling is the reaction of pavement information on the steering wheel. The pavement is measured by the reaction of torque sensor by fixing the steering wheel. We use T_s to represent the measurement of torque sensor. The road surface information is reflected by the steering resistance moment T_r. To get good road sense, we need to control the error between the two.

It can be expressed as

$$\|W_{s2} e_2\|_\infty = \|W_{s2} (T_s - \varphi \dot{T}_r)\|_\infty \tag{20}$$

φ is the road sense coefficient and W_{s2} is the road sense weighting function [24].

3.2.3. Handling Stability. The handling stability is mainly reflected in the changes of yaw rate. The target value of yaw rate is determined by the steady-state yaw rate gain function [20].

$$\omega_r^* = \frac{u/L}{1 + Ku^2} \delta_1 \tag{21}$$

where

$$K = \frac{m}{L^2} \left(\frac{a}{k_2} - \frac{b}{k_1} \right) \tag{22}$$

In order to obtain better handling and stability, the deviation between the actual yaw rate ω_r and the target yaw rate ω_r is controlled as

$$\|W_{S3} e_3\|_\infty = \|W_{S3} (\omega_r^* - \omega_r)\|_\infty \tag{23}$$

W_{S3} is the stability weighting function.

3.2.4. Avoiding Current Overload. In order to avoid the excessive input current of the motor, the function of input W_R is introduced. It makes $\|W_R u_a\|_\infty$ as small as possible to reduce energy consumption and device loss.

3.2.5. The Robust Stability of the System. In order to avoid the system instability due to the performance of the system, the output of the system is weighted by $\|W_T Y\|_\infty$.

W_T is the complementary sensitivity weighting factor.

3.3. Selection of Weighting Function.
The selection of weighting function reflects the requirement of the performance index of the control system and the choice should be based on the specific control target and the control requirement.

In general, the order of a weighting function is not too high. The controller is simple in order to solve it. So we choose the first-order weighting function as follows:

$$W_i(s) = c_i \frac{b_i s + 1}{a_i s + 1} \quad i = S, R, T \tag{24}$$

The choice of weighting function needs iterative computation. In this paper, W_S, W_R, and W_T are obtained by simulation as

$$W_S = \begin{bmatrix} \dfrac{0.01}{200s + 20} & & & & \\ & \dfrac{1}{50s + 100} & & & \\ & & \dfrac{1}{50s + 100} & & \\ & & & 1 & \\ & & & & 1 \end{bmatrix}$$

$$W_R = 0.0001, \tag{25}$$

$$W_T = \begin{bmatrix} \dfrac{s + 50}{0.1s + 5} & & & & \\ & \dfrac{s + 10}{10s + 500} & & & \\ & & \dfrac{1}{5s + 100} & & \\ & & & 1 & \\ & & & & 1 \end{bmatrix}$$

According to the controlled object and the generalized weighting function selected, the controller $K(s)$ is solved by using the MATLAB robust toolbox [25].

4. Simulation Results

In order to verify the control strategy to improve the EPS road sense and effect on vehicle handling performance, this paper builds a EPS simulation framework according to the control model of the system and the hybrid sensitivity H_∞ control model. It is shown in Figure 4.

In the figure, d_s and d_r are, respectively, sensor noise and road interference. ω_r^* is ideal yaw rate. $K_{\omega r} = u/(L(1 + Ku^2))$ is steady-state yaw angle velocity function. $K =$

TABLE 1: Simulation experimental parameters.

Parameter name	Variable name	Unit	Numerical value
Torsional rigidity of motor sensor	K_s	N m/rad	90
Back EMF coefficient	K_b	v s	0.02
Motor torque coefficient	K_a	N m/A	0.02
Steering wheel moment of inertia	J_h	kg m^2	0.04
Equivalent moment of inertia	J_p	kg m^2	0.06
Equivalent damping coefficient of steering shaft	B_h	N m/(rad s^{-1})	0.25
Small gear equivalent damping coefficient	B_p	N m/(rad s^{-1})	0.3
Motor armature resistance	R_a	Ω	0.01
Mass of vehicle	m	kg	1296
The moment of inertia around the plumb shaft of vehicle	I_z	kg m^2	1750
Cornering stiffness of front wheel	k_1	N/rad	95707
Cornering stiffness of rear wheel	k_2	N/rad	84243
The distance from the front wheel to the center of mass	a	m	1.25
The distance from the rear wheel to the center of mass	b	m	1.32
Coefficient of pavement interference intensity	δ_0		0.2
Front wheel trailing distance	d	m	0.1

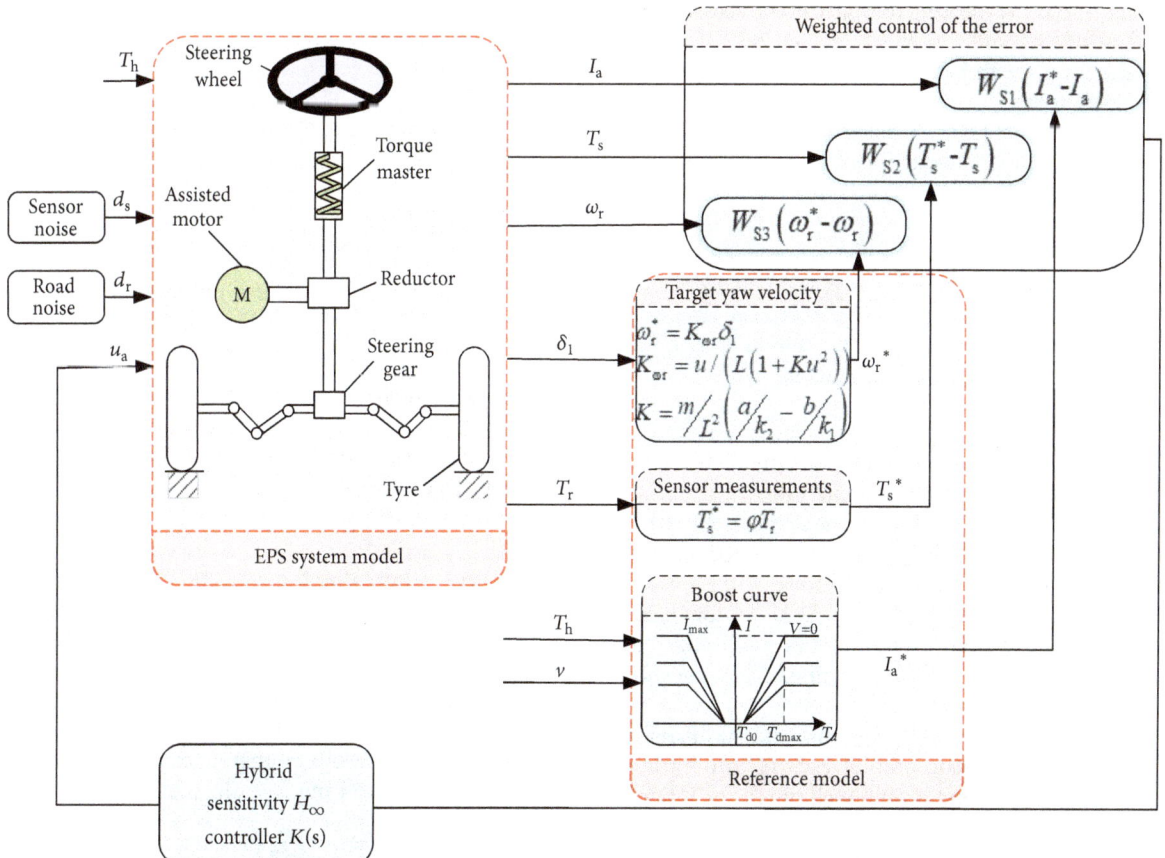

FIGURE 4: EPS hybrid sensitivity H_∞ control system structure diagram.

$(m/L^2)(a/k_2 - b/k_1)$ is stability factor. The ideal yaw rate can be calculated in this way: $\omega_r^* = K_{\omega r}\delta_1$. T_s^* is ideal torque sensor measurement. φ is coefficient of road inductance. The ideal torque sensor has a φ times relationship with the measured resistance torque [24]. It is shown as $T_s^* = \varphi T_r$. I_a^* is ideal current. The steering wheel torque is manipulated by hand, and the ideal target current is calculated by the boost characteristic curve.

The selection of simulation experiment parameters is shown in Table 1.

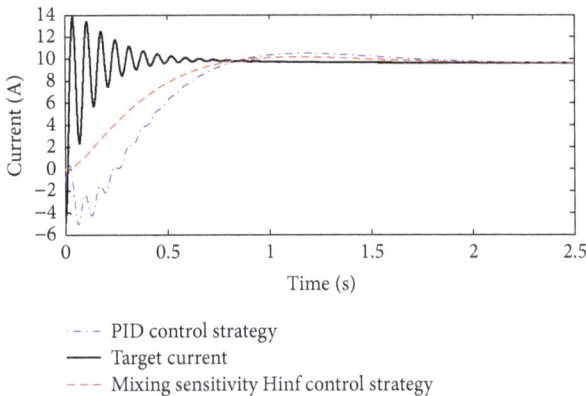

--- PID control strategy
— Target current
--- Mixing sensitivity Hinf control strategy

FIGURE 5: Comparison diagram of current tracking response.

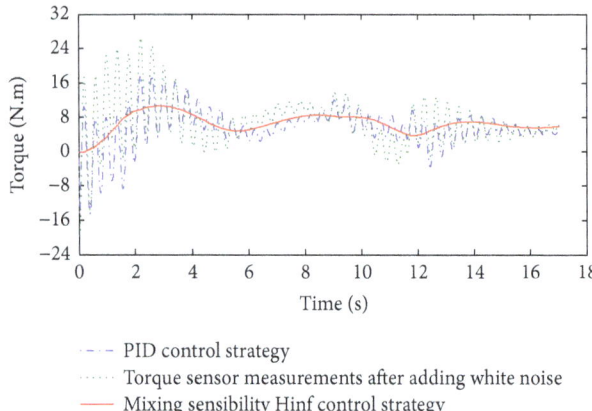

--- PID control strategy
····· Torque sensor measurements after adding white noise
— Mixing sensibility Hinf control strategy

FIGURE 6: Torque sensor measurement response comparison chart.

4.1. Current Tracking Performance. The tracking speed of the actual current of the controlled motor to the target current and the tracking performance determine the response performance of EPS system. The tracking condition of the target current is observed by a step signal of the steering system. Figure 5 is the tracking response of the actual current of the controlled motor to the target current under the PID control strategy and the hybrid sensitivity H_∞ control strategy, respectively.

It can be seen from Figure 5, due to the influence of system stiffness and damping, the step signal will generate about 0.5s oscillation of the target current of the system. When the motor current of PID control is tracking the target, it will generate a certain amount of oscillation in 0-0.5 s and the response time is relatively long. There is a certain overshoot in the time of 0.8-2 s. In addition, the hybrid sensitivity control can eliminate jitter; meanwhile, it can respond more quickly and it has a good effect on tracking current.

4.2. Road Feeling. Road sense is the steering wheel torque that the driver feels during the manipulation of the vehicle. It is represented by the measurement value of the torque sensor to a certain extent. This paper mainly tests the effect of the controller on filtering interference by inputting the step response of the road and adding the noise of the road and the sensor. Set the road step as $\delta(t) = 7\,\mathrm{N}\cdot m$ to simulate. Simulation results of step response under PID control and hybrid sensitivity H_∞ control are obtained. They are shown in Figure 6.

It can be seen from Figure 6 that the measurement value of the torque sensor of PID control can track the target current very well, but it cannot eliminate the influence of jamming noise which makes the driver's road feeling worse. Compared with PID control, torque sensor measurement value controlled by hybrid sensitivity H_∞ control can effectively eliminate the influence of jamming noise, lower overshoot, and shorten the stability time, so that the driver can have a better sense of road in all kinds of interferences and noises.

4.3. Handling Stability. Yaw velocity of the vehicle is the main performance index of the vehicle steering stability.

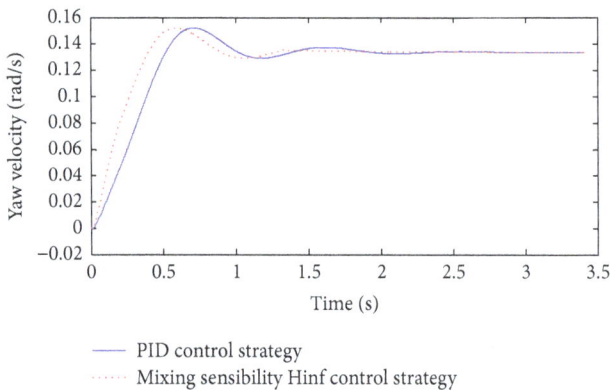

— PID control strategy
····· Mixing sensibility Hinf control strategy

FIGURE 7: Comparison of step response of yaw rate.

The stability of the vehicle under the control strategy can be verified by the change of the angular velocity of the pavement when road step is added. As shown in Figure 7, when the speed is set to 30 m/s, the yaw velocity under the hybrid sensitivity control and the PID control is changed as shown.

Figure 7 shows that the yaw velocity of the PID control is about 7% overshoot in the 0.5-1 s period and a longer stability time is needed, which makes the vehicle's handling stability worse. The design of hybrid sensitivity H_∞ control of yaw velocity response can be very stable and response very quickly. This ensure the vehicle yaw velocity response to the driver's control command be fast and accurate. At the same time, it can improve the handling and stability of vehicle.

5. A Hybrid Sensitivity Robust Controller Is Added to the EPS System Experiment

In the simulation test section, hybrid sensitivity H_∞ controller is verified to a certain extent. In this section, the designed hybrid sensitivity H_∞ controller is embedded into the built EPS test bench [26, 27], and the designed controller mainly investigates the tracking situation of the actual

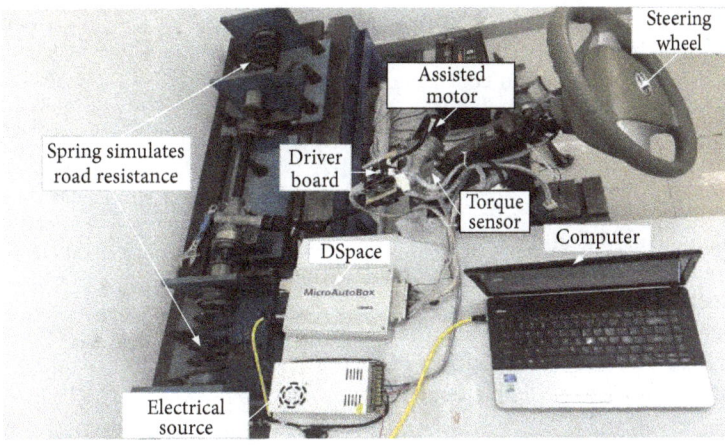

FIGURE 8: EPS system test bench.

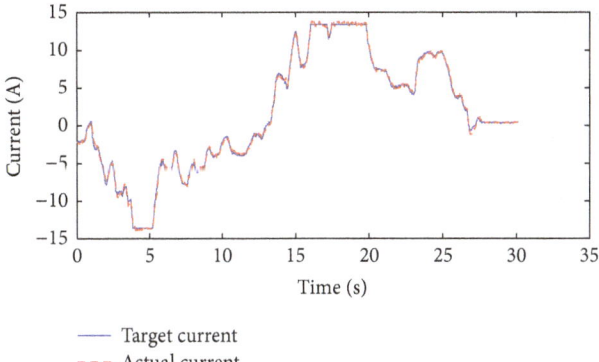

— Target current
--- Actual current

FIGURE 9: Hybrid sensitivity H_∞ control current tracking curve when in turn.

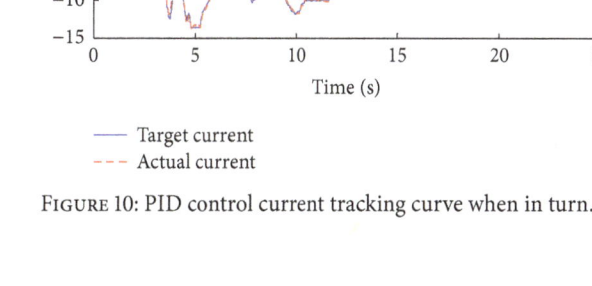

— Target current
--- Actual current

FIGURE 10: PID control current tracking curve when in turn.

current to the target current to verify the performance of the controller [28]. EPS test bench is shown in Figure 8. In situ simulation conditions, the torque is applied to the steering wheel. The tracking data of the actual current to the target current is recorded in real time by DSpace software control desk [29]. The power performance of EPS system under hybrid sensitivity H_∞ controller and PID controller is analyzed and compared. The test results are shown in Figures 9 and 10.

When turning in the same place, it can be seen from Figure 9 that the curve tracking of actual current to target current is almost coincident with the hybrid sensitivity H_∞ controller. It has achieved good tracking performance. As can be seen from Figure 10, the tracking curve of actual current to target current fluctuates under PID control, especially in 13 and 20 seconds, and is not ideal. The experimental results show that the proposed hybrid sensitivity H_∞ control strategy is better than the PID power assisted control strategy. It can ensure both robustness and good performance. Compared with the EPS system under the PID control, the EPS system under the hybrid sensitivity H_∞ control can have faster response speed, better robustness, and better performance.

6. Conclusion

(A) This paper establishes the mathematics model of EPS system and it takes better tracking performance of current, better driver's sense of road, and better handling stability as the control targets. The state space equation of the system is constructed by using modern cybernetics, and a hybrid sensitivity H_∞ controller is designed.

(B) To verify the design of hybrid sensitivity H_∞ controller, the simulation model is built. The simulation results show that the designed controller has good current tracking, which can guarantee that the driver has a good sense of road and can make the vehicle have good maneuvering stability. It can effectively suppress the noise and can improve the system performance while ensuring the robustness and robustness stability of the system compared with the PID controller.

(C) To verify the control effect of the controller, it is embedded into experimental bench experiments. The experimental results show that the design controller can realize the current tracking in the actual system and improve the performance, robustness, and robustness stability of the system compared with PID control.

(D) In this paper, the main factors affecting EPS system are mainly considered. The change of the system stiffness, the perturbation of the motion, and the parameters of the system are not considered. The next step will be to take more overall consideration of more indicators to establish a more eligible control system.

Conflicts of Interest

The authors declare that there are no conflicts of interest regarding the publication of this paper.

Acknowledgments

The authors acknowledge the support from Natural Science Foundation of Shandong Province (ZR2016EEQ06).

References

[1] W. Shufeng, *Automobile Structure*, 285-304, National defense industry press, Beijing, China, 2013.

[2] T. Guangxing, J. Wenguo, G. Yuan, and L. Shan, "EPS control simulation study based on immune fuzzy PID," *Computer Simulation*, vol. 31, no. 9, pp. 170–173, 2014.

[3] Z. Jianwei, X. Shijie, and Q. Baojun, "Research and simulation of control strategy for electric power steering system," *Machine Tool & Hydraulics*, vol. 44, no. 6, pp. 96–100, 2016.

[4] Q. Ming, Y. Jiajun, L. Zhao, and Y. Geng, "Research of electric power steering system based on H_∞ robust control theory," *Journal of Huazhong University of Science and Technology (Natural Science Edition)*, vol. 30, no. 12, pp. 72–73, 2002.

[5] W. Qirui, C. Wuwei, H. Sengren, and Y. Ming, "Research on H_∞ control of automotive electric power steering system," *Automobile Engineering*, vol. 26, no. 5, pp. 609–612, 2004.

[6] Z. Wanzhong, S. Guobiao, L. Yi, and L. Qiang, "Steering sense of electric power assisted system based on hybrid H_2/H_∞ control," *Chinese Journal of Mechanical Engineering*, vol. 45, no. 4, pp. 142–147, 2009.

[7] W. Jingliang and Y. Yunbing, "Robust control strategy of EPS steering system based on handling stability," *Journal of Wuhan University of Science and Technology*, vol. 37, no. 3, pp. 204–209, 2014.

[8] S. Guoqin and Y. Yunbing, "Research on H_∞ robust control of electric power steering system and its influence on handling stability," *Modern Manufacturing Engineering*, vol. 9, pp. 114–120, 2016.

[9] C. Long, Y. Chaochun, and J. Haobin, "Study on the stability of EPS vehicle based on μ integrated robust control," *Automotive Engineering*, vol. 30, no. 8, pp. 705–712, 2008.

[10] F. Wilhelm, T. Tamura, R. Fuchs, and P. Mullhaupt, "Friction compensation control for power steering," *IEEE Transactions on Control Systems Technology*, vol. 24, no. 4, pp. 1354–1367, 2016.

[11] Y.-C. Hung, F.-J. Lin, J.-C. Hwang, J.-K. Chang, and K.-C. Ruan, "Wavelet fuzzy neural network with asymmetric membership function controller for electric power steering system via improved differential evolution," *IEEE Transactions on Power Electronics*, vol. 30, no. 4, pp. 2350–2362, 2015.

[12] W. Kim, Y. S. Son, and C. C. Chung, "Torque overlay-based robust steering wheel angle control of electrical power steering for a lane-keeping system of automated vehicles," *IEEE Transactions on Vehicular Technology*, vol. 10, no. 1109, pp. 1–14, 2015.

[13] A. Marouf, M. Djemai, C. Sentouh, and P. Pudlo, "A new control strategy of an electric-power-assisted steering system," *IEEE Transactions on Vehicular Technology*, vol. 61, no. 8, pp. 3574–3589, 2012.

[14] D. Lee, K.-S. Kim, and S. Kim, "Controller design of an electric power steering system," *IEEE Transactions on Control Systems Technology*, 2018.

[15] C. T. Faria, G. Pulvirenti, and T. Geluk, "Modeling and nonlinear parameter identification of an electric-power steering system," in *Topics in Modal Analysis & Testing*, pp. 127–135, The Society for Experimental Mechanics, 2017.

[16] Y. Zhisheng, *Automobile Theory*, Mechanical Industry Press, Beijing, China, 2000.

[17] C. Yin, Q. Sun, J. Wu, C. Liu, and J. Gao, "Development of electrohydraulic steering control system for tractor automatic navigation," *Journal of Electrical and Computer Engineering*, vol. 2018, Article ID 5617253, 7 pages, 2018.

[18] A. Baier, K. Herbuś, P. Ociepka, and M. Płaczek, "Modal analysis in relation to the casing of an electric power steering system," in *Proceedings of the IOP Conference Series: Materials Science and Engineering*, vol. 227, pp. 1–8, June 2017.

[19] Y. Li, T. Shim, D. Wang, and T. Offerle, "Comparative Study of Rack Force Estimation for Electric Power Assist Steering System," in *Proceedings of the ASME 2017 Dynamic Systems and Control Conference*, vol. 5255, pp. 1–8, Tysons, Virginia, USA, 2017.

[20] W. Z. Zhao, Y. J. Li, C. Y. Wang, Z. Q. Zhang, and C. L. Xu, "Research on control strategy for differential steering system based on H_∞ hybrid sensitivity," *International Journal of Automotive Technology*, vol. 14, no. 6, pp. 913–919, 2013.

[21] S. Tielong, H_∞ *Control theory and application 2*, Tsinghua University press, Beijing, China, 2008.

[22] Oppenheim, *Signals and Systems*, Electronic Industry Press, Beijing, China, 2013.

[23] M. Shengwei, S. Tielong, and L. Kangzhi, *Modern robust control theory and its application*, Tsinghua University press, Beijing, China, 2003.

[24] Z. Huaiquan and G. Jia, "Research on EPS handling stability based on H_∞ hybrid sensitivity control," *Journal of Beijing Institute of Technology*, vol. 32, no. 12, pp. 1252–1257, 2012.

[25] J. G. Michael, *Model predictive control system design and implementation using matlab*, Industrial Control Centre, Glasgow, Scotland, UK, 2015.

[26] H. Zheng and M. Zhao, "Development a HIL test bench for electrically controlled steering system," *SAE Technical Papers*, vol. 01, no. 51, pp. 1–7, 2016.

[27] X. Li and Y. Li, "Research and development of the simulation platform of the nose wheel with digital steering system based on hardware in the loop," in *Proceedings of the 2016 IEEE/CSAA International Conference on Aircraft Utility Systems, AUS 2016*, pp. 729–733, Beijing, China, October 2016.

4

Application of Fuzzy Control in a Photovoltaic Grid-Connected Inverter

Zhaohong Zheng, Tianxia Zhang, and Jiaxiang Xue(ID)

School of Mechanical and Automotive Engineering, South China University of Technology, Guangzhou 510641, China

Correspondence should be addressed to Jiaxiang Xue; mejiaxue@scut.edu.cn

Academic Editor: Andrea Bonfiglio

To realize the maximum power output of a grid-connected inverter, the MPPT (maximum power point tracking) control method is needed. The perturbation and observation (P&O) method can cause the inverter operating point to oscillate near the maximum power. In this paper, the fuzzy control P&O method is proposed, and the fuzzy control algorithm is applied to the disturbance observation method. The simulation results of the P&O method with fuzzy control and the traditional P&O method prove that not only can the new method reduce the power loss caused by inverter oscillation during maximum power point tracking, but also it has the advantage of speed. Inductive loads in the post-grid-connected stage cause grid-connected current distortion. A fuzzy control algorithm is added to the traditional deadbeat grid-connected control method to improve the quality of the system's grid-connected operation. The fuzzy deadbeat control method is verified by experiments, and the harmonic current of the grid-connected current is less than 3%.

1. Introduction

In view of the traditional photovoltaic grid-connected inverter system, light intensity can affect the output power of a photovoltaic solar array to a large extent. Therefore, maximum power point tracking (MPPT) is performed to improve the utilization efficiency of the photovoltaic array and ensure that it maintains maximum power output.

In the first stage of a grid-connected inverter, an MPPT control algorithm mainly includes the constant voltage method, the perturbation and observation (P&O) method, and the conductance increment method. The advantages of simplicity, easy implementation, and rapid MPPT have helped the P&O method to be widely used in an MPPT algorithm. However, the P&O method can easily produce continuous oscillation around the maximum power point; therefore, a nonlinear control method, named fuzzy control, is added based on the traditional P&O method. Fuzzy control can simplify the system design and is particularly useful for a nonlinear, hysteretic, time-varying, and model-incomplete system owing to its excellent robust performance [1, 2]. Bououden et al. added an ant-colony intelligent optimization

algorithm to fuzzy control, which can not only deal with nonlinearity but also reduce the parameter randomization of the algorithm [3].

The control methods in the post-grid-connected stage of full bridge inversion include current instantaneous value control (PI control) algorithm, repetitive control algorithm, deadbeat algorithm, and proportion resonance algorithm. The PI control algorithm is widely used owing to its simplicity and easy implementation. However, these control algorithms can only address parts of the problem. For example, the deadbeat algorithm is widely used owing to its high-speed system response time. The system works steadily when its inverter output is combined with resistive loads, yet when the inductive or capacitive loads are connected to the inverter output of system, as well as when the system is suffering from outside interference, the current and voltage at the load end fail to maintain synchronization and the system lacks stability. Besides, the harmonic rates of current and voltage at both ends of the load increase simultaneously.

Based on analysis of the MPPT fuzzy control of the P&O method in the first stage of a photovoltaic grid-connected inverter, this paper proposes a fuzzy control-based deadbeat

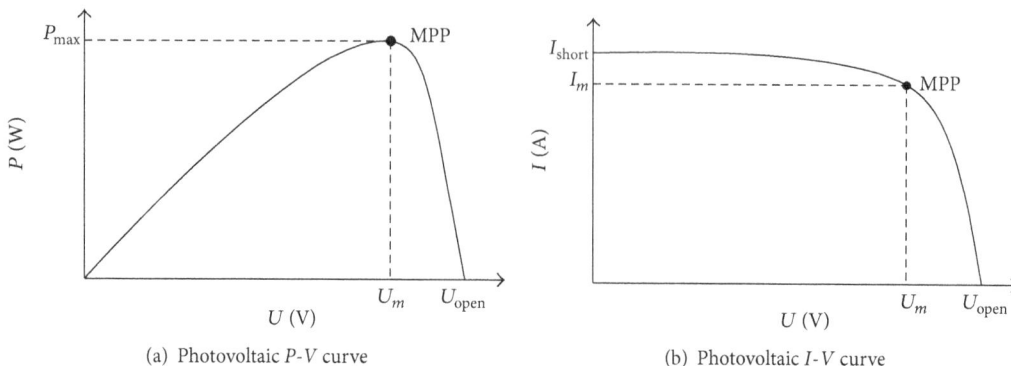

(a) Photovoltaic *P-V* curve   (b) Photovoltaic *I-V* curve

FIGURE 1: Output curve of photovoltaic solar array.

control strategy that can be used in the poststage of the photovoltaic grid-connected inverter, which can not only adapt to the nonlinear load but also reduce the harmonic waves of the inverter output.

2. Design of MPPT Fuzzy Control Algorithm in the First Stage of the Photovoltaic Grid-Connected Inverter

The MPPT of a photovoltaic array needs to be conducted to make the best use of the photovoltaic solar array. The *P-V* and *I-V* curves of a photovoltaic solar cell are shown in Figure 1. where U_{open} indicates the open-circuit voltage of the photovoltaic solar panel and is also its maximum output voltage; I_{short} indicates the short-circuit current of the photovoltaic solar panel and is also its maximum output current; P_{max} indicates the maximum output power of the photovoltaic solar panel in the present case; I_m is the current at the maximum power point; and U_m is the voltage at the maximum power point [4, 5].

2.1. MPPT Control Strategy of P&O Method in the First Grid-Connected Stage. The first-stage MPPT control is conducted using the first-stage interleaving Boost circuit. The output voltage of the photovoltaic solar array used in this study is 200–350 V, and a bus voltage of approximately 400 V can be obtained using the booster circuit. The P&O method is used as the MPPT control method in this study. The disturbance voltages are continuously provided to the output end of the photovoltaic solar array, to calculate the output powers of the two photovoltaic arrays. The output powers are input to PI regulation to produce a pulse width modulation (PWM) control pulse. After passing through the drive circuit, the PWM signal can directly drive the switching element in the boost circuit in order to realize MPPT.

As the name suggests, the P&O method is used to continuously provide the disturbance voltage and calculate the output powers of the two photovoltaic arrays until they are operating around the maximum power point. The operating method is described as follows: "$\delta(n)$" is used as the disturbed value of the photovoltaic arrays' output voltage. $\delta(n) = +\Delta V$ occurs when the perturbation direction is voltage-increasing, while $\delta(n) = -\Delta V$ occurs when the perturbation direction

is voltage-decreasing. The output power $P(n)$ is calculated in accordance with the current and voltage values measured at the output end. In addition, a comparison is made between $P(n)$ and $P(n-1)$. The perturbation direction is correct in the case where $P(n)$ is greater than $P(n-1)$; thus, the perturbation should be continually applied to the photovoltaic array according to the original direction of perturbation voltage, which is $\delta(n) = \delta(n-1)$. The perturbation direction is wrong if the output power decreases; thus, $\delta(n) = -\delta(n-1)$ in the next perturbation. The adjustments should be performed unceasingly according to the perturbation method until the photovoltaic solar array is operating around the maximum power point [6, 7]. The perturbation curve is shown in Figure 2.

The software flow diagram of the P&O method described in this paper is shown in Figure 3. The current output power $P(n)$ of the solar panel is calculated according to the output voltage $V(n)$ and output current $I(n)$ of the solar photovoltaic array sampled by the system. Then, the output power $P(n)$ is subtracted from the last output power $P(n-1)$. If the absolute value of the difference is less than or equal to the set value δ, the output power is equal to the last output power and the system will produce no action output and return to the interrupt subprogram. The purpose here is to prevent the system misjudgments caused by the sampled signal fluctuation of the digital signal processor. If the absolute value of the difference is greater than δ, a comparison between $P(n)$ and $P(n-1)$ is conducted. When $P(n) > P(n-1)$, $V(n) > V(n-1)$, which proves that the perturbation direction is correct and the current operating point is adjustable. Thus, the reference value of voltage $V_{ref}(n)$ can be further increased. When $P(n) > P(n-1)$, $V(n) < V(n-1)$, which proves that the perturbation direction is correct and the current operating point is adjustable. Thus, the reference value of voltage $V_{ref}(n)$ can be further decreased. When $P(n) < P(n-1)$, $V(n) > V(n-1)$, which implies that the perturbation direction is incorrect and the current operating point is located on the right side of the maximum power point. Thus, the direction should be changed and the reference value of voltage $V_{ref}(n)$ should be decreased as well. Yet when $P(n) < P(n-1)$, $V(n) < V(n-1)$, which indicates that the perturbation direction is incorrect and the current operating point is adjustable. Thus, the direction should be

P_m
$P(n)$
$P(n-1)$
I
P-V
$\delta(n)$
$V(n)$ $V(n-1)$
V

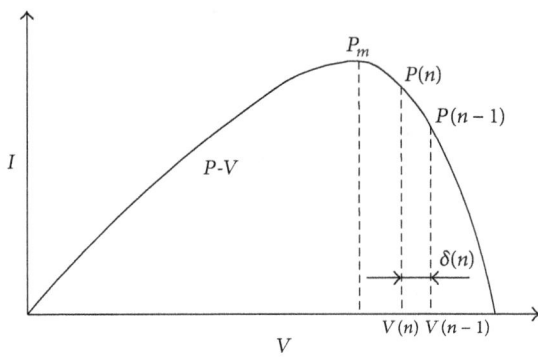

FIGURE 2: P&O curve of solar array.

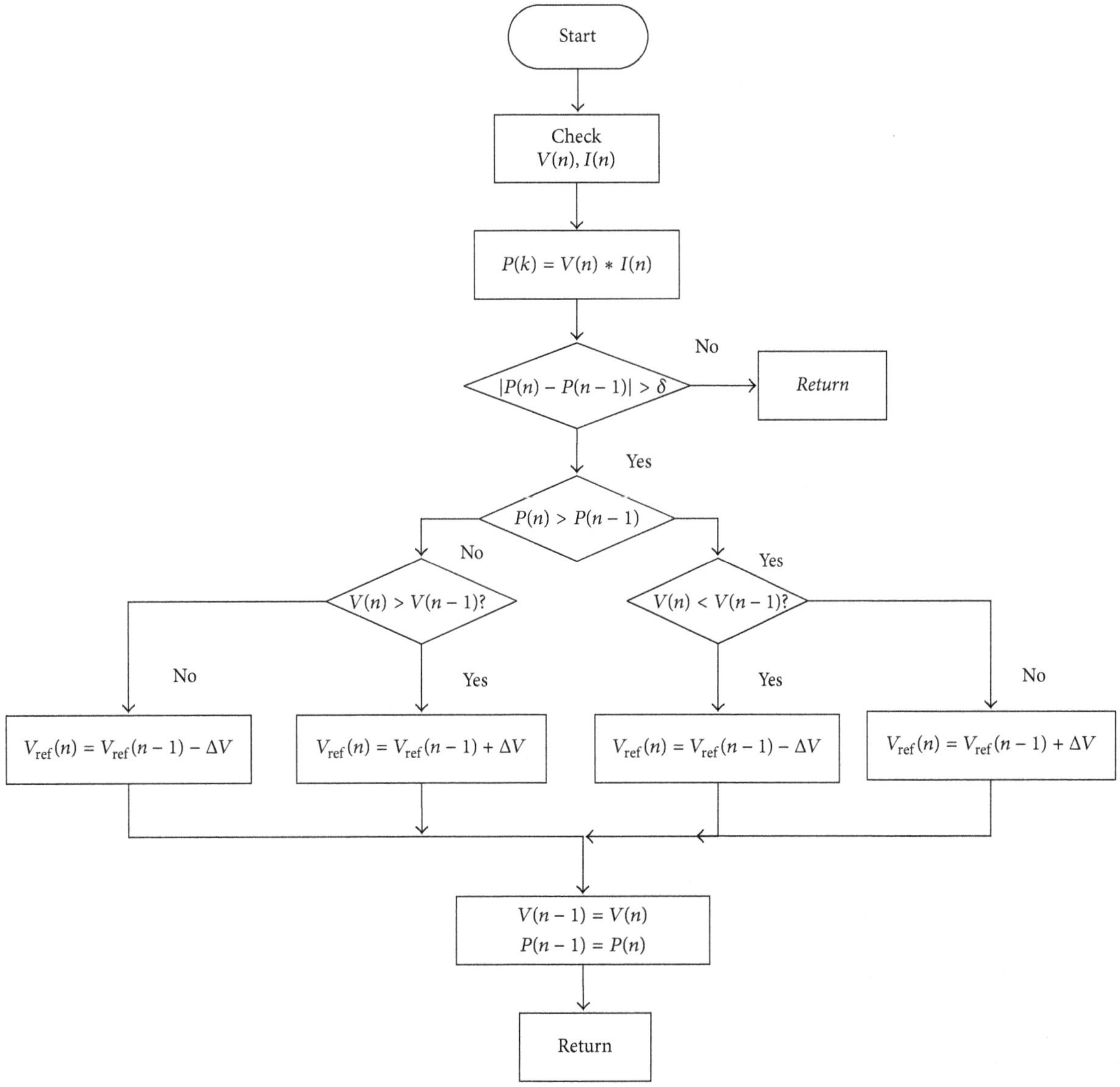

Start
Check
$V(n), I(n)$
$P(k) = V(n) * I(n)$
No
$|P(n) - P(n-1)| > \delta$
Return
Yes
$P(n) > P(n-1)$
No
Yes
$V(n) > V(n-1)$?
$V(n) < V(n-1)$?
No
Yes
Yes
No
$V_{ref}(n) = V_{ref}(n-1) - \Delta V$
$V_{ref}(n) = V_{ref}(n-1) + \Delta V$
$V_{ref}(n) = V_{ref}(n-1) - \Delta V$
$V_{ref}(n) = V_{ref}(n-1) + \Delta V$
$V(n-1) = V(n)$
$P(n-1) = P(n)$
Return

FIGURE 3: Control flow diagram of P&O method.

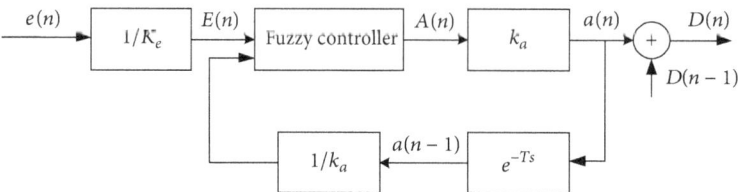

FIGURE 4: Fuzzy controller.

changed and the reference value of voltage $V_{\text{ref}}(n)$ should be increased [8].

2.2. Design of MPPT Fuzzy Control Based on P&O Method. The step size of the traditional P&O method remains unchanged during the process of MPPT, while the fuzzy-controlled P&O method actually improves the traditional P&O method with a fixed step. The control method can adjust the perturbation step according to the real-time output power of the photovoltaic solar cell to ensure that the operating point can be closer to the maximum power point. According to the principle of the P&O method, the output power of the solar cell is used as the objective function, while the duty ratio is used as the control variable. The current step size is adjusted and confirmed based on the variation in the power value and the duty ratio at the last moment. The input of the fuzzy controller at moment n is the variation in the power value in the photovoltaic system at moment n and the step size of the duty ratio [9, 10] at moment $n-1$, while the output at moment n is the step size of the duty ratio at moment n. Thus, the fuzzy controller designing this study is as shown in Figure 4, where k_e and k_a are the quantization factors.

Input. $e(n)$ refers to the actual value of the difference between the output powers at moments n and $n-1$, while $E(n)$ is the corresponding value of the difference in the universe of fuzzy sets. $a(n-1)$ refers to the actual value of the step size at moment $n-1$, while $A(n-1)$ is the corresponding value of the step size in the universe of fuzzy sets.

Output. $a(n)$ is the actual value of the step size at moment n, while $A(n)$ refers to the corresponding value of the step size in the universe of fuzzy sets.

The Mamdani controller is selected in the Matlab fuzzy box and the centroid method is used to solve fuzzification. The fuzzy linguistic variables E and A are defined as five and three subsets, respectively, which are E = {NB, NS, ZE, PS, PB} and A = {N, Z, P}. Here, NB, NS, Z, PS, and PB represent negative large, negative small, positive zero, positive small, and positive large, while N, P, and Z represent negative, positive, and zero fuzzy, respectively. The rule table of MPPT fuzzy control is shown in Table 1.

3. Fuzzy Control Algorithm in the Post-Grid-Connected Stage of Full-Bridge Inversion

Two control models of the poststage full-bridge inversion include controlling of output voltage and controlling of output current. The control strategy of the voltage control

TABLE 1: Rules of MPPT fuzzy control.

$E(n)/A(n-1)$	NB	NS	ZE	PS	PB
N	PB	PM	NS	PM	NB
Z	NM	NS	ZE	PS	PM
P	NB	NM	PS	PM	PB

mode is to consider the entire system as a controlled voltage source and make the inverter output voltage a system control quantity; the control strategy of the current control mode is to consider the entire system as a controlled current source and make the inverter output current a system control quantity.

The control mode of the output voltage is equivalent to a controlled voltage source; therefore, it is easily affected by the power grid voltage. The quality of the inverter output voltage is significantly impacted if the power grid voltage suffers from any abnormality. However, for the control mode of the output current, the controlled output quantity is the inverter output current and the current source is highly resistive to the voltage source; thus, the quality of the output current cannot suffer any impact from the power grid voltage. In short, the control mode of the output current should be used in the grid-connected operation mode, which can improve the quality of the output power as well.

The grid-connected operation mode generally adopts the double closed-loop control algorithm with an outer loop of bus voltage and inner loop of current output. In this paper, a fuzzy control algorithm is added to the PI modulation of the bus-voltage outer ring and the parameters of bus voltage loop are adjusted constantly to make the closed-loop control more precise. Besides, the output quantity of the outer voltage loop is one-unit current. The inner loop of the double closed loop is a current loop, which adopts the deadbeat control algorithm to ensure the synchronization between the output current and power grid voltage.

3.1. Design of Deadbeat Algorithm in the Post-Grid-Connected Stage. The poststage inverter output of the grid-connected operation mode adopts the output current control. The deadbeat control method based on the output current control is used in the grid-connected inverter system described in this paper. The control system is realized using a digital signal processor, which exhibits highly precise AD sampling and rapid internal operation, which is suitable for the deadbeat control.

When the inverter is operating in the grid-connected mode, the poststage inverter circuit is equivalent to the circuit

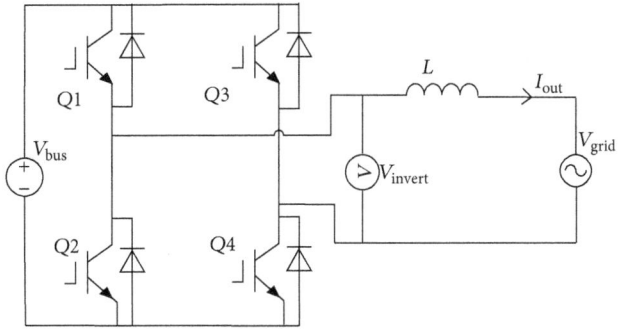

FIGURE 5: Schematic of poststage grid-connected inverter circuit.

diagram shown in Figure 5. The current of the filtering capacitance at the output end and the losses in the circuit should be neglected. The complete bridge is composed of Q1–Q4, where Q1 and Q4 perform the turn-on operations alternatively, while Q2 and Q3 perform the turn-off operations alternatively to transfer power to the grid. L is the filter inductance at the output end, while V_{bus} is the DC bus voltage generated by the first-stage interleaving boost circuit and should be stabilized at about 420 V by the control strategy of a poststage full-bridge inverter. I_{invert} is the inverter output current, V_{invert} is the inverter output voltage, and V_{grid} is the power grid voltage.

The following equation can be obtained according to the output inductance characteristics:

$$L \frac{dI_{out}}{dt} = V_{invert} - V_{grid}.$$ (1)

The above equation can be transformed into the following equation within one control cycle T of the switch tube:

$$V_{inver\text{-}ave}(n) = V_{grid\text{-}ave}(n) + L \frac{I_{out}(n+1) - I_{out}(n)}{T},$$ (2)

where L is the filter inductance at the inverter output end; T is the control cycle of the switch tube; the average value of the voltage at the inverter output end sampled during n sampling period is $V_{inver\text{-}ave}(n)$; the average value of the power grid voltage during n sampling period is $V_{grid\text{-}ave}(n)$; the inverter output current sampled during n sampling period is $I_{out}(n)$; the inverter output current sampled during $(n+1)$ sampling period is $I_{out}(n+1)$; $I_{ref}(n+1)$ is used to replace $I_{out}(n+1)$ in (2) so as to ensure that the current passing through the filter inductance can track the reference current of the inverter output $I_{ref}(n+1)$. Then,

$$V_{inver\text{-}ave}(n) = V_{grid\text{-}ave}(n) + L \frac{I_{ref}(n+1) - I_{out}(n)}{T}.$$ (3)

The average value of the power grid voltage $V_{grid\text{-}ave}(n)$ can be replaced with

$$V_{grid\text{-}ave}(n) = \frac{V_{grid}(n+1) + V_{grid}(n)}{2}$$ (4)

$$V_{grid}(n+1) - V_{grid}(n) = V_{grid}(n) - V_{grid}(n-1).$$

From (4),

$$V_{grid\text{-}ave}(n) = 1.5 V_{grid}(n) - V_{grid}(n-1).$$ (5)

From (3) and (5),

$$\begin{aligned} V_{inver\text{-}ave}(n) = {} & 1.5 V_{grid}(n) - V_{grid}(n-1) \\ & + L \frac{I_{ref}(n+1) - I_{out}(n)}{T}. \end{aligned}$$ (6)

The poststage inverter output voltage of the grid-connected inverter is directly proportional to the first-stage DC bus voltage; thus, the duty ratio of the high-frequency tube during the control cycle is

$$\begin{aligned} D(n) & = \frac{V_{invert\text{-}ave}}{V_{bus}} \\ & = \frac{1.5 V_{grid}(n) - V_{grid}(n-1) + L\left(\left(I_{ref}(n+1) - I_{out}(n)\right)/T\right)}{V_{bus}}, \end{aligned}$$ (7)

where $I_{ref}(n+1)$ is a given reference signal of current at the $(n+1)$ cycle obtained by the combined effect of both PI voltage modulation and poststage phase-locked loop control during the process of full bridge inversion.

V_{bus} is the DC bus voltage, while $V_{grid}(n)$ and $I_{out}(n)$, respectively, are the effective values of power grid voltage and inverter output current sampled during the current cycle. The effective values of power grid voltage and inverter output current sampled during the last cycle can be expressed as $V_{grid}(n-1)$. Therefore, the duty ratio of switch tube D can be obtained through (7) to realize the deadbeat control over the entire circuit. According to (7), the control of the duty ratio of the switch tube can be used to regulate the reference current $I_{ref}(n+1)$. However, the reference signal of the current should maintain its pace with the power grid voltage signal.

3.2. Design of Grid-Connected Fuzzy Control Algorithm Based on Deadbeat Algorithm. The system works steadily when its inverter output is combined with resistive loads, yet when the inductive or capacitive loads are connected to the inverter output of the system, as well as when the system is suffering from outside interference, the current and voltage at the load end fail to maintain synchronization and the system lacks stability. Besides, the harmonic rate of current and voltage

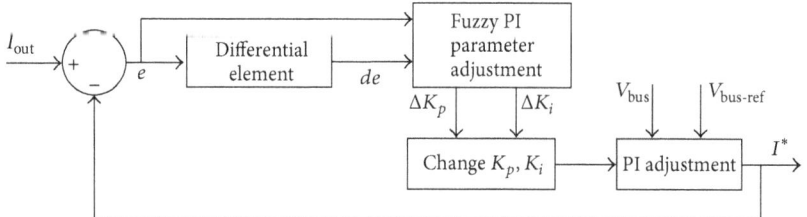

FIGURE 6: Fuzzy controller.

TABLE 2: Rule table of Ki parameter fuzzy control.

$\Delta E/E$	NB	NE	NS	ZO	PS	PE	PB
NB	PB	PB	PB	PB	PE	PS	ZO
NE	PB	PB	PB	PE	PS	ZO	NS
NS	PB	PB	PE	PS	ZO	NS	NE
ZO	PB	PE	PS	ZO	NS	NE	NB
PS	PE	PS	ZO	NS	NE	NB	NB
PE	PS	ZO	NS	NE	NB	NB	NB
PB	ZO	NS	NE	NB	NB	NB	NB

TABLE 3: Rule table of Kp parameter fuzzy control.

$\Delta E/E$	NB	NE	NS	ZO	PS	PE	PB
NB	NB	NB	NB	NB	NE	NS	ZO
NE	NB	NB	NB	NE	NS	ZO	PS
NS	NB	NB	NE	NS	ZO	PS	PE
ZO	NB	NE	NS	ZO	PS	PE	PB
PS	NE	NS	ZO	PS	PE	PB	PB
PE	NS	ZO	PS	PE	PB	PB	PB
PB	ZO	PS	PE	PB	PB	PB	PB

at both ends of the load increases simultaneously. Similar problems exist in the mutual switchover of grid-off and grid-connected operating modes. Owing to the characteristics of fuzzy control, the fuzzy control method can be added to the original unipolar deadbeat control method to improve the stability of the inverter system when the nonlinear load is connected to the output end of the load. Fuzzy control is mainly introduced into the photovoltaic control system to properly modify the PI control parameters, Kp and Ki, which can eventually improve the system stability. The fuzzy controller should be added to the full bridge inverter control to effectively fortify the power factor [11, 12] during the grid-connected process of the inverter.

$e(k)$ and $\Delta e(k)$ are the two inputs of a fuzzy controller, where $e(k)$ refers to the error in the fuzzy controller and $\Delta e(k)$ is the error rate. Their relation can be expressed as

$$e(k) = e(k) - e(k-1)$$

$$\Delta e(k) = \frac{de(k)}{dt} = \frac{e(k) - e(k-1)}{t(k) - t(k-1)}. \tag{8}$$

The double inputs designed in the paper are the current error i_e and the change rate of current error di_e/dt, generated by the combined effect of the actual current at the sampling output end and the predictive current in the next cycle, while the double outputs are ΔKp and ΔKi [13], which are the manipulating variables of Kp and Ki in PI modulation. The fuzzy controller designed for the closed-loop PI modulation of the first-stage bus voltage in the grid-connected system is shown in Figure 6.

The following seven fuzzy variables are added to the fuzzy set: positive large, positive relatively large, positive relatively small, zero, negative relatively small, negative relatively large, and negative large, which are represented as PB, PE, PS, ZO, NS, NE, and NB, respectively.

The program preparation is conducted based on the rule table of fuzzy control, as shown in Tables 2 and 3.

The control strategy chart for the poststage full-bridge inversion of the photovoltaic grid-connected inverter is shown in Figure 7. A comparison between the sampled DC bus voltage V_{bus} and the reference value $V_{bus-ref}$ set in the system is performed via PI modulation. The fuzzy control algorithm is continuously used to adjust Kp and Ki in PI modulation, and the given amplitude of the grid-connected current can be obtained after the controlling operation of PI. The power grid voltage is captured at this

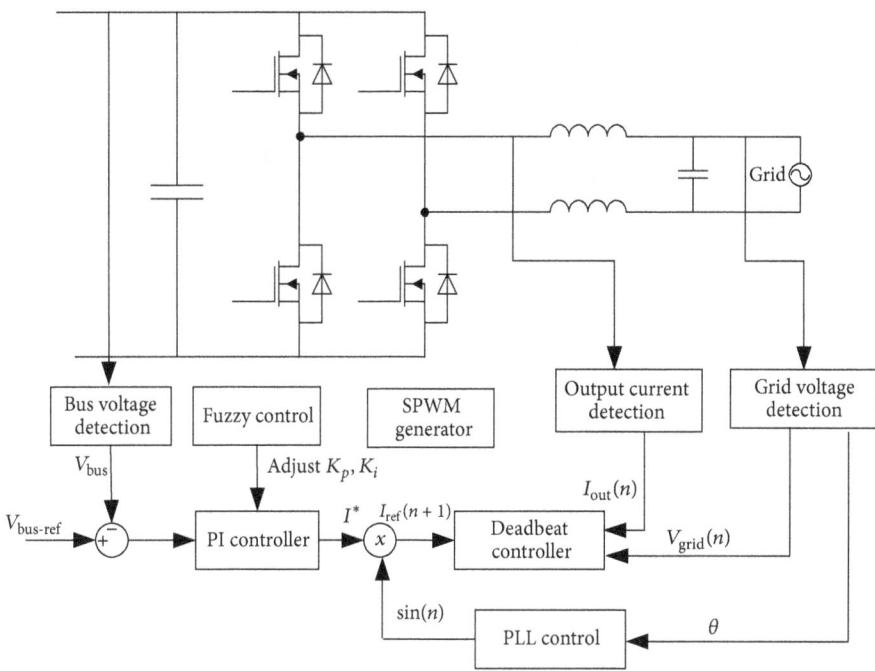

FIGURE 7: Control strategy chart for poststage inversion of grid-connected operating mode.

moment. If the rising edge of the square wave is detected, the capture accessing the digital signal processor is interrupted to generate a unit sinusoidal current signal that is in sync with the power grid voltage. The predicted value of current $I_{ref}(n + 1)$ in the next cycle can be obtained by multiplying the amplitudes of current signal I^* and the sinusoidal signal. Finally, the predicted value of current $I_{ref}(n + 1)$ in the next cycle, the effective value of current $I_{out}(n)$ at the inverting output end, the effective value of power grid voltage $V_{grid}(n)$ sampled in the current cycle, the effective value of power grid voltage $V_{grid}(n - 1)$ sampled in the last cycle, and the effective value of DC bus voltage V_{bus} are inputted into the deadbeat controller to output the PWM signal. After passing through the amplifying and isolating circuits, the square signal can drive the switch tube to produce a current with the same frequency and phase as those of the grid voltage.

4. Simulation and Analysis of Experimental Results

4.1. MPPT Simulation of First-Stage Fuzzy P&O Method.
The light intensity of the photovoltaic array is $G = 1000 \, W/m^2$, the ambient temperature is $T = 25°C$, the maximum power is $P = 270 \, W$, the voltage of the maximum power point is $V_{mpp} = 70 \, V$, the voltage of the open circuit is $V_{OC} = 100 \, V$, the current of the short circuit is $I_{SC} = 5 \, A$, and the current of the maximum power point is $I_{mpp} = 3.8 \, A$. The design power of the boost circuit is 1500 W, $C1 = C2 = 1000 \, \mu F$, $L = 2.0 \, mH$, $R = 592 \, \Omega$, and the operating frequency is 20 kHz. The fuzzy MPPT model is formulated as shown in Figure 8, the power tracking diagram of the simulation output is shown in Figure 9, and the voltage tracking diagram is shown in Figure 10. Thus, the duration of MPPT is shortened to 0.04 s

after adopting the fuzzy control, which is faster than the traditional perturbation method.

4.2. Experiment Design of Poststage Full-Bridge Inverting Fuzzy Deadbeat.
When the output end is composed of loads with different characteristics, the load characteristics can be allocated via an electronic load device and the loads can be resistive, capacitive, inductive, or mixed. The introduction of a fuzzy control algorithm can significantly improve the quality of the grid-connected output current, especially if the output load is not a pure resistant one; the output waveform of the grid-connected output current appears much smoother than that of the current in the original unipolar algorithm, and there is lesser clutter. Meanwhile, the current at the load end is much smoother. The waveforms of the output current and voltage under the grid-connected operating mode are shown in Figure 11, where the first channel is the current at the load end, the second channel is the grid-connected output current, and the third channel is the power grid voltage.

The loads of (a) and (b) and those of (c) and (d) in Figure 11 are the same. (a) and (c) adopt the conventional deadbeat method, while (b) and (d) show the waveform figures of deadbeat control method after the introduction of the fuzzy control algorithm. According to (b) and (d), the waveforms of the grid-connected output current and load current are somewhat smoother and the harmonic wave is shortened.

5. Conclusions

In this study, modeling analysis is conducted for the MPPT of fuzzy control-based P&O method in the first stage of photovoltaic grid connection in Matlab/Simulink. The MPPT

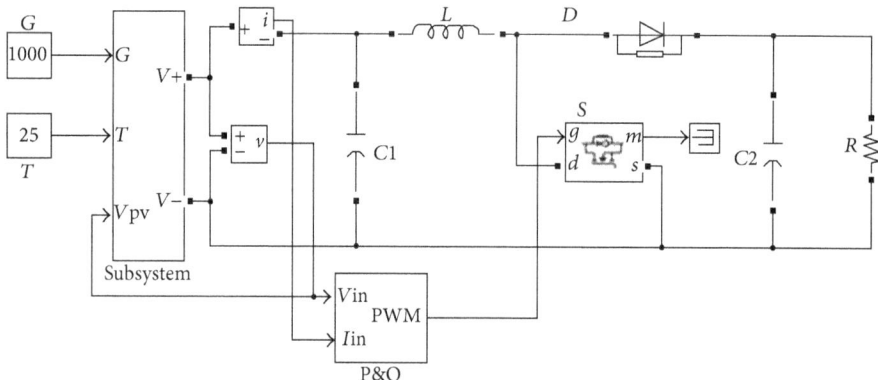

FIGURE 8: Model of fuzzy MPPT control.

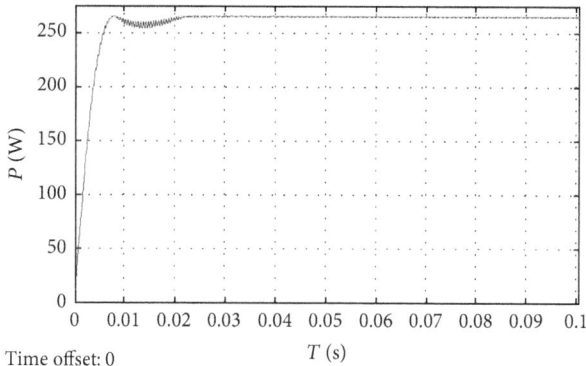

FIGURE 9: Output power of photovoltaic array.

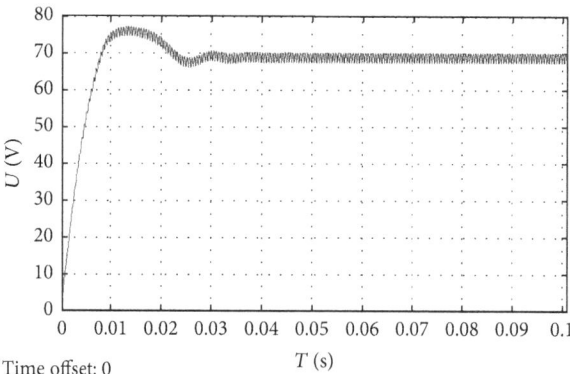

FIGURE 10: Output voltage of photovoltaic array.

of fuzzy control-based P&O method and that of traditional P&O method are analyzed thoroughly by the simulation comparison. According to the simulation results, the MPPT of fuzzy control-based P&O method exhibits rapid response and small steady-state oscillation, which can effectively make up for the shortcomings of the traditional P&O method, improve system efficiency, and reduce power losses to an extreme. In addition, the fuzzy control algorithm is added to the full bridge inversion of the post-grid-connected stage, and the experiment is performed in the formulated experimental platform. Loads with different characteristics are added to the output end of the post-grid-connected stage in order to compare the conventional deadbeat control method and the fuzzy control-based deadbeat control method. The results show that the fuzzy control-based deadbeat control method can enhance the robustness and reduce the harmonic wave when the system relates to nonlinear load.

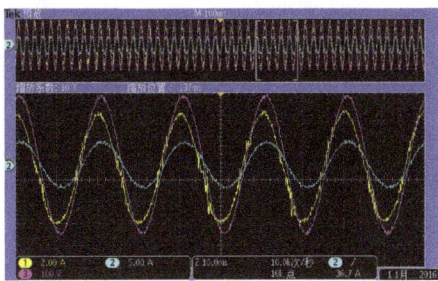

(a) Waveform of conventional deadbeat control with 500 W resistive load + 300 W capacitive load

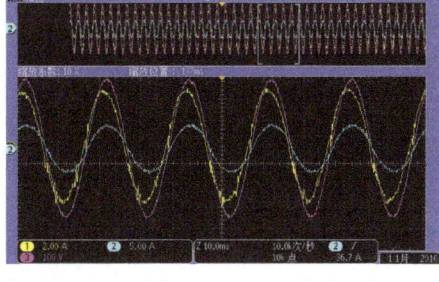

(b) Waveform of fuzzy control method with 500 W resistive load + 300 W capacitive load

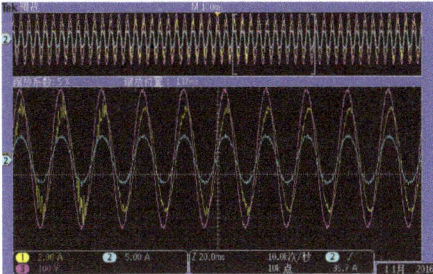

(c) Waveform of conventional deadbeat control with 100 W capacitive load + 150 W inductive load + 500 W resistive load

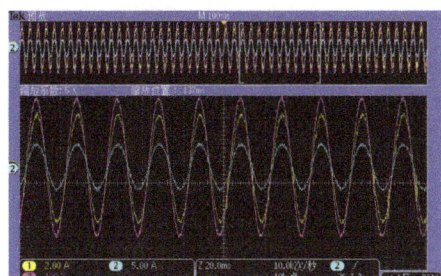

(d) Waveform of fuzzy control method with 100 W capacitive load + 150 W inductive load + 500 W resistive load

FIGURE 11: Grid-connected output current and load current with different load characteristics.

Conflicts of Interest

The authors declare that there are no conflicts of interest regarding the publication of this paper.

Acknowledgments

The authors acknowledge financial support for this research from: Third Batch Innovative Research Team Introduction Program of Dongguan City in 2015 (2017360004004), Special Funding of Collaborative Innovation and Platform Circumstance Construction of Guangdong Province (2016B090918067), and Industry-University-Research of Dongcheng District, Dongguan City, in 2015; Natural Science Foundation of Guangdong Province (2015A030313675).

References

[1] S. Bououden, M. Chadli, S. Filali, and A. El Hajjaji, "Fuzzy model based multivariable predictive control of a variable speed wind turbine: LMI approach," *Journal of Renewable Energy*, vol. 37, no. 1, pp. 434–439, 2012.

[2] H. Dahmani, M. Chadli, A. Rabhi, and A. El Hajjaji, "Road curvature estimation for vehicle lane departure detection using a robust Takagi-Sugeno fuzzy observer," *Vehicle System Dynamics*, vol. 51, no. 5, pp. 581–599, 2013.

[3] S. Bououden, M. Chadli, and H. R. Karimi, "An ant colony optimization-based fuzzy predictive control approach for nonlinear processes," *Information Sciences*, vol. 299, pp. 143–158, 2015.

[4] M. Görig and C. Breyer, "Energy learning curves of PV systems," *Environmental Progress & Sustainable Energy*, vol. 35, no. 3, pp. 914–923, 2016.

[5] E. D. Aranda, J. A. Gómez Galan, M. S. de Cardona, and J. M. Andújar Márquez, "Measuring the I-V curve of PV generators: Analyzing different dc-dc converter topologies," *IEEE Industrial Electronics Magazine*, vol. 3, no. 3, pp. 4–14, 2009.

[6] P. Manimekalai, R. Harikumar, and S. Raghavan, "A hybrid maximum power point tracking (MPPT) with interleaved converter for standalone photo voltaic (PV) power generation system," *Annual Review of Microbiology*, vol. 60, no. 3, pp. 451–475, 2014.

[7] D. Sharma K and G. Purohit, "Advanced perturbation and observation (PO) based maximum power point tracking (MPPT) of a solar photo-voltaic system[C]," in *Proceedings of the IEEE India International Conference on Power Electronics*, p. 1, 2012.

[8] A. Al Nabulsi and R. Dhaouadi, "Efficiency optimization of a dsp-based standalone PV system using fuzzy logic and dual-MPPT control," *IEEE Transactions on Industrial Informatics*, vol. 8, no. 3, pp. 573–584, 2012.

[9] Y.-T. Chen, Y.-C. Jhang, and R.-H. Liang, "A fuzzy-logic based auto-scaling variable step-size MPPT method for PV systems," *Solar Energy*, vol. 126, pp. 53–63, 2016.

[10] M. Abdourraziq A, M. Ouassaid, and M. Maaroufi, "A fuzzy logic MPPT for photovoltaic systems using single sensor," in *Proceedings of the Renewable and Sustainable Energy Conference*, pp. 52–56, IEEE, 2015.

[11] G.-R. Yu, J.-J. Lai, and J.-Y. Liu, "T-S fuzzy control of a single-phase bidirectional inverter," in *Proceedings of the IEEE*

International Conference on Industrial Technology, ICIT 2016, pp. 1462–1467, IEEE, March 2016.

[12] A. Tomova, M. Antchev, M. Petkova, and H. Antchev, "Fuzzy logic hysteresis control of a single-phase on-grid inverter: Computer investigation," *International Journal of Power Electronics and Drive Systems*, vol. 3, no. 2, pp. 179–184, 2013.

[13] L. K. Letting, J. L. Munda, and Y. Hamam, "Optimization of a fuzzy logic controller for PV grid inverter control using S-function based PSO," *Solar Energy*, vol. 86, no. 6, pp. 1689–1700, 2012.

5

Improvement of High-Power Three-Level Explosion-Proof Inverters using Soft Switching Control based on Optimized Power-Loss Algorithm

Shi-Zhou Xu and Feng-You He

Department of Information and Electrical Engineering, China University of Mining and Technology,
No. 1 Daxue Road, Xuzhou, Jiangsu 221116, China

Correspondence should be addressed to Feng-You He; hfy_cumt@263.net

Academic Editor: Ahmed El Wakil

The high-power three-level explosion-proof inverters demand high thermal stability of power devices, and a set of theories and methods is needed to achieve an accurate power-loss calculation of power devices, to establish heat dissipation model, and ultimately to reduce the power loss to improve thermal stability of system. In this paper, the principle of neutral point clamped three-level (NPC3L) inverter is elaborated firstly, and a fourth-order RC equivalent circuit of IGBT is derived, on which basis the power-loss model of IGBT and the optimized maternal power-loss thermal model, using an optimized power-loss algorithm, are established. Secondly, in accordance with the optimized maternal power-loss thermal model, the generic formulas of power-loss calculation are deduced to calculate the power-loss modification values of NPC3L and soft switching three-level (S3L) inverters, which will be the thermal sources during thermal analysis for maternal power-loss thermal models. Finally, the experiment conducted on the 2.1 MW experimental platform shows that S3L inverter has the same excellent output characteristics with NPC3L inverter, reduces the power loss significantly by 213 W in each half-bridge, and decreases the temperature by 10°C, coinciding with the theoretical calculation, which verifies the accuracy of optimized power-loss algorithm and the effectiveness of the improvement.

1. Introduction

In explosion-proof inverters field, the NPC3L inverter is one of the most mature facilities of high-power three-level inverters at present [1]. The high-power explosion-proof inverters have the features of high current, flowing through the main circuit power devices, great power losses, and high reliability requirement. What is more, from the view of applications, there is a serious problem that the power loss of inverter power devices is too great, which will cause a high failure rate of inverter power devices and poor thermal stability of the whole system. In order to improve the existing NPC3L inverters, there are three issues to be addressed. The basal one is the accurate power-loss calculation of power devices, and it is the premise of thermal analysis and converter improvement. The second one is a general power-loss calculation and analysis theory of three-level inverter acting as evaluation criteria to predict the results of improvements.

Finally, a new topology should be introduced to reduce the power loss effectively.

Generally speaking, accurate power-loss calculation can figure out the existing power-loss values of three-level inverters, which will be a thermal source during the thermal analysis of inverter system. The inverter temperature rise is mainly caused by conduction loss and switching loss of power devices, while the conduction and switching characteristics of the power devices are very sensitive to temperature, so calculating the power loss of the device accurately is the foundation to optimize the design of inverters. Currently, there are many researches on power-loss calculation and thermal analysis for single IGBT module and two-level inverters [2–6]. However, three-level and two-level inverter currents are essentially different in the flow paths, and their losses of power devices are of huge difference. The fact that the literature [7–10] did not consider the impact of junction temperature of power devices on power losses is the main reason causing errors between their theoretical calculations and experimental results, where Dieckerhoff et al. [10] considered that the switching power loss of power device has a linear relationship with its withstanding voltage, while this assumption is approximately valid only in ±20% range of the test voltage. A much accurate losses calculation and heat dissipation method was introduced in [11], but it did not take all the thermal sources in consideration, which has an effect on the power devices and thermal analysis. In the literature [12], the transient modeling of loss and thermal dynamics in power semiconductor devices is analyzed, while it needs to improve the model by considering the peripheral circuits. Several soft switching inverter types and control methods are proposed in [13–18], where the S3L inverter in [18] has a much more significant effect on the reduction of power losses. It is the accurate thermal analysis methods of inverter system that can analyze the inverter temperature quantitatively, providing references for inverter improvements [19, 20]. In the existing loss calculation studies of three-level inverters, it lacks a system of theories and methods to provide theoretical support for the improvements. Before a new three-level topology improving the NPC3L inverter, it is necessary to apply a common theoretical calculation and method to anticipate its advantages. The S3L inverter proposed in the literature [18] holds the viewpoint that it can reduce the power loss in terms of the NPC3L inverter under the same conditions, but there is no quantitative experimental temperature to support it and demonstrate its effectiveness of improvement.

For the above reasons, a general power-loss calculation method of three-level inverters was established in this paper based on the optimized power-loss algorithm in Section 2, with which an accurate power-loss calculation and performance evaluation approach of three-level inverters was proposed. To improve NPC3L inverter, the S3L inverter working principle was elaborated in Section 3 and put into the general approach mentioned above. What is more, according to this approach, it is expected in Section 4 that the S3L inverter has the same excellent output characteristics with NPC3L inverter and it can reduce power loss by 213 W, bringing in a 10°C decrease in temperature intuitively. In the same section, the experiment results support the validity of the theoretical prediction. Finally, Section 5 concludes this paper.

2. General Optimized Power-Loss Algorithm Based on NPC3L Inverter

The main circuit topology of NPC3L inverter is shown in Figure 1.

Each leg has four IGBTs, labeled T_{i1}, T_{i2}, T_{i3}, and T_{i4} (where i represents one phase of a, b, and c phases and each IGBT has one antiparallel diode, labeled D_{i1}, D_{i2}, D_{i3}, and D_{i4}, resp.) and two clamping diodes, labeled D_{i5} and D_{i6} [11, 21].

At present, the IGBT device is a power module packaged by one IGBT and a fast recovery antiparallel diode. Therefore,

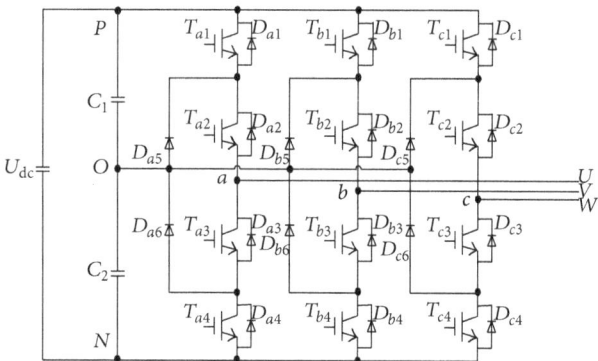

FIGURE 1: The main circuit topology of NPC3L inverter.

its total power loss is composed of these two parts, expressed as follows:

$$P_{\mathrm{mod}} = P_T + P_D. \tag{1}$$

The equivalent structure model of power device and heat sink is shown in Figure 2.

It can be seen from Figure 2 that the whole model consists of four conductive layers, and therefore if we consider the power device and the heat sink as a maternal model, the four conductive layers would be four submodels. On this basis, the thermal resistance and heat capacity of the four submodels can be calculated at first, respectively, and then all the four submodels' thermal resistance and heat capacity constitute the total thermal resistance and heat capacity of the whole model.

The calculation formula of thermal resistance R_{th} is described as follows:

$$R_{\mathrm{th}} = \frac{\Delta T}{P} = \frac{\Delta T}{Q/\Delta t}, \tag{2}$$

where ΔT is the temperature increase of submodel. P and Δt are the heat flow and time period, respectively. Then, the total thermal resistance of maternal model is shown as

$$R_{\mathrm{th\text{-}total}} = R_{\mathrm{th1}} + R_{\mathrm{th2}} + R_{\mathrm{th3}} + R_{\mathrm{th4}}. \tag{3}$$

The calculation formula of thermal capacity C_{th} is delivered as follows [20, 21]:

$$C_{\mathrm{th}} = \frac{Q}{\Delta T}. \tag{4}$$

The total heat capacity of the maternal model can be written as [20, 21]

$$C_{\mathrm{th\text{-}total}} = C_{\mathrm{th1}} + C_{\mathrm{th2}} + C_{\mathrm{th3}} + C_{\mathrm{th4}}. \tag{5}$$

Thus, the maternal model can be replaced alternatively by a fourth-order RC circuit, shown in Figure 3.

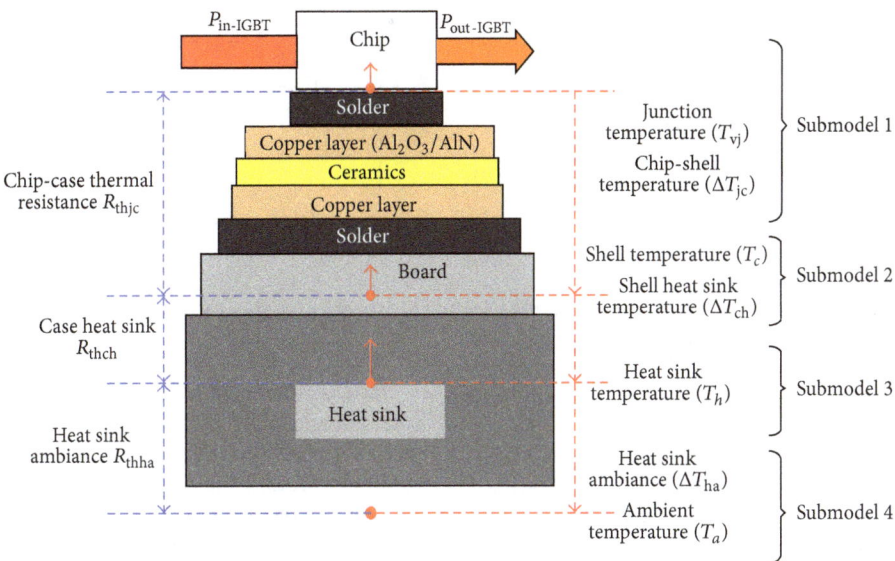

FIGURE 2: Equivalent structure model of power device and heat sink.

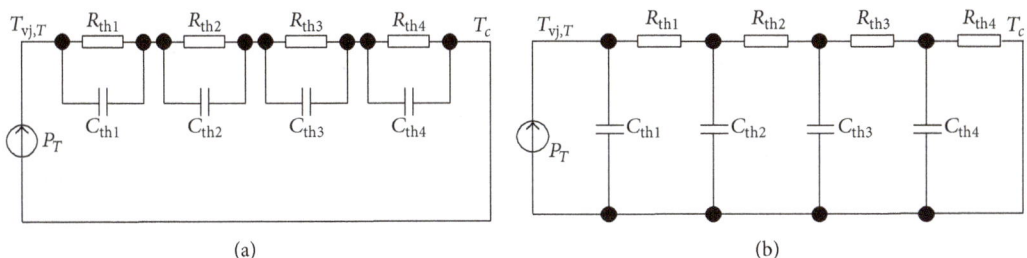

(a)   (b)

FIGURE 3: The fourth-order RC thermal resistance equivalent circuits of IGBT. (a) Partial network; (b) continuous network.

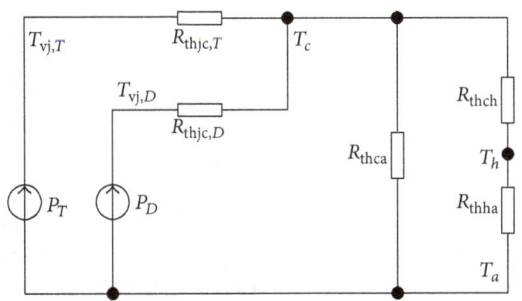

FIGURE 4: The thermal resistance equivalent circuit of IGBT module in steady state.

According to the partial network structure of Figure 3(a), the IGBT thermal resistance can be derived as [20, 21]

$$Z_{\text{thjc},T} = \sum_{i=1}^{n} R_{\text{th}i} \left(1 - e^{-t/\tau_i} \right) \quad (i = 1, 2, 3, 4), \tag{6}$$

where τ_i is the RC time constant of each layer.

The thermal equivalent circuit of IGBT module in steady state is shown in Figure 4.

As a switching device, the IGBT's power loss P_T is primarily composed of conduction loss $P_{\text{con},T}$ and switching loss $P_{\text{sw},T}$; namely [3–5],

$$P_T = P_{\text{con},T} + P_{\text{sw},T}. \tag{7}$$

The conduction resistance, initial saturation voltage, and conduction loss of IGBT can be expressed, respectively, as follows [11]:

$$r_T = r_{T\text{-}25°C} + K_{r,T} \left(T_{\text{vj},T} - 25°C \right),$$

$$v_{0,T} = v_{0,T\text{-}25°C} + K_{v0,T} \left(T_{\text{vj},T} - 25°C \right), \tag{8}$$

$$P_{\text{con},T} = v_{0,T} I + r_T I^2,$$

where $r_{T\text{-}25°C}$ and $v_{0,T\text{-}25°C}$ are the conduction resistance and initial saturation voltage of IGBT with the junction temperature at 25°C; $K_{r,T}$ and $K_{v0,T}$ are the initial saturation voltage and conduction resistance temperature correction factor of IGBT; $T_{\text{vj},T}$ is the junction temperature of IGBT; I is the instantaneous current flowing through the IGBT.

Combine the three-level working principle and optimized IGBT power-loss model, and the average conduction and switching losses formula of T_1 in a modulation voltage period will be as follows [11, 22]:

$$P_{\text{con},T_1}^{\text{npc,spwm}} = f_0 \sum_{k=p}^{q} \left(v_{0,T_1} + r_{T_1} I_L(k) \right) I_L(k) \, \tau(k) \, T_s,$$

$$\tag{9}$$

$$P_{\text{sw},T_1}^{\text{npc,spwm}} = f_0 \sum_{k=p}^{q} E_{\text{sw},T_1} \left(I_L(k) \right),$$

where f_0 is the frequency of modulation voltage; $\tau(k)$ is the kth duty cycle of switching period; $I_L(k)$ is the average load current of the kth switching period; p and q represent the sampling period's beginning and end of T_1 during one modulation period, respectively.

Generally speaking, when the carrier ratio is large enough, the discrete power-loss formula can be transformed into a continuous integral form, and the average conduction losses and switching losses of T_1 can be expressed as [22]

$$P_{\text{con},T_1}^{\text{npc,spwm}} = \frac{1}{2\pi} \int_{\varphi}^{\pi} \left(v_{0,T_1} + r_{T_1} i_L(\alpha) \right) i_L(\alpha) D(\alpha) \, d\alpha$$

$$= \frac{m v_{0,T_1} I_m}{4\pi} \left((\pi - \varphi) \cos\varphi + \sin\varphi \right)$$

$$+ \frac{m r_{T_1} I_m^2}{4\pi} \left(1 + \frac{4}{3} \cos\varphi + \frac{1}{3} \cos 2\varphi \right),$$

$$P_{\text{sw},T_1}^{\text{npc,spwm}} = \frac{1}{2\pi} \int_{\varphi}^{\pi} f_{\text{sw}} E_{\text{sw},T_1} \left(i_L(\alpha) \right) d\alpha$$

$$= \frac{f_{\text{sw}}}{2\pi} \left(A_{\text{sw},T} I_m^2 \frac{1}{2} \left(\pi - \varphi + \frac{1}{2} \sin 2\varphi \right) \right.$$

$$+ B_{\text{sw},T} I_m \left(1 + \cos\varphi \right) + C_{\text{sw},T} \left(\pi - \varphi \right) \right)$$

$$\cdot \left(\frac{U_{\text{dc}}/2}{U_{\text{base}}} \right)^{D_{\text{sw},T}} \left(\frac{T_{\text{vj},T_1}}{T_{\text{base}}} \right)^{K_{\text{sw},T}}.$$

$$\tag{10}$$

In accordance with the same calculation principle as T_1's, the conduction losses and switching losses (or reverse recovery losses) of D_1, T_2, D_2, and D_5 in the same half-bridge leg will be $P_{\text{con},D_1}^{\text{npc,spwm}}$, $P_{\text{rec},D_1}^{\text{npc,spwm}}$, $P_{\text{con},T_2}^{\text{npc,spwm}}$, $P_{\text{sw},T_2}^{\text{npc,spwm}}$, $P_{\text{con},D_2}^{\text{npc,spwm}}$, $P_{\text{rec},D_2}^{\text{npc,spwm}}$, $P_{\text{con},D_5}^{\text{npc,spwm}}$, and $P_{\text{rec},D_5}^{\text{npc,spwm}}$.

Some explanatory notes in expressions (8)~(10) are as follows.

$v_{0,T_x} = v_{0,T\text{-}25°C} + K_{v0,T}(T_{\text{vj},T_x} - 25°C)$ denotes the xth IBGT's initial saturation voltage;

$r_{T_x} = r_{T\text{-}25°C} + K_{r,T}(T_{\text{vj},T_x} - 25°C)$ represents the xth IGBT's conduction resistance;

T_{vj,T_x} means the xth IGBT'S junction temperature;

$v_{0,D_x} = v_{0,D\text{-}25°C} + K_{v0,D}(T_{\text{vj},D_x} - 25°C)$ indicates the xth fast recovery diode's initial saturation voltage;

$r_{D_x} = r_{D\text{-}25°C} + K_{r,D}(T_{\text{vj},D_x} - 25°C)$ stands for the xth fast recovery diode's conduction resistance;

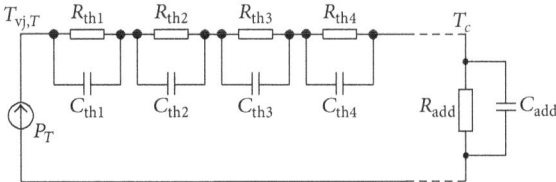

FIGURE 5: The equivalent circuit of optimized power-loss model.

T_{vj,D_x} means the xth fast recovery diode's junction temperature; what is more, $x = 1, 2, 5$ [22].

In accordance with the above step, the power losses of heat sink and IGBT maternal model can be calculated. However, there may be buffer circuit or something like that in the periphery around the power devices of different three-level topologies. That means part of the power difference ($P_{\text{in-IGBT}} - P_{\text{out-IGBT}}$) flowing through the IGBT model is dissipated in the IGBT model and some other part is consumed by the peripheral circuits; namely,

$$P_{\text{in-IGBT}} - P_{\text{out-IGBT}} = P_{\text{IGBT}} + P_{\text{add}}. \tag{11}$$

Hence, it is necessary to optimize the maternal model, and the equivalent circuit of optimized maternal model is shown in Figure 5.

According to the equivalent circuit of optimized maternal model, the total power-loss equation of half-bridge leg can be modified as follows:

$$P_{\text{total}} = P_{T_1} + P_{D_1} + P_{T_2} + P_{D_2} + P_{\text{add}}. \tag{12}$$

The general power-loss calculation of half-bridge leg of three-level inverters using SPWM modulation algorithm in a modulation period will be the one as follows [22, 23]:

$$P_{\text{total}} = P_{\text{con},T_1}^{\text{npc,spwm}} + P_{\text{sw},T_1}^{\text{npc,spwm}} + P_{\text{con},D_1}^{\text{npc,spwm}}$$

$$+ P_{\text{rec},D_1}^{\text{npc,spwm}} + P_{\text{con},T_2}^{\text{npc,spwm}} + P_{\text{sw},T_2}^{\text{npc,spwm}} \tag{13}$$

$$+ P_{\text{con},D_2}^{\text{npc,spwm}} + P_{\text{rec},D_2}^{\text{npc,spwm}} + P_{\text{add}},$$

where the SPWM modulation algorithm can be replaced by the actual algorithm, but the power-loss calculation has the same process based on optimized power-loss algorithm and maternal model, and all we should do is to change the variables consistent with the algorithm we are going to use. In addition, during the analysis of maternal module in Section 4, it is necessary to modify P_{add} by considering precharge current-limiting resistor, balanced resistors, absorption capacitance, DC-link capacitors, and buffer devices of S3L inverter.

3. S3L Inverter Principle

As shown in Figure 6, one full-bridge leg topology of S3L inverter contains four IGBTs ($T_{a1}' \sim T_{a4}'$), four diodes ($D_{a1}' \sim D_{a4}'$), snubber inductor, snubber capacitances C_1' and C_2', and four snubber diodes $D_{aT_1} \sim D_{aT_4}$, where the latter four constitute the snubber circuit [14].

TABLE 1: Switching states of S3L inverter.

Switching state	+	0	−
U'_{load}	$+U_d/2$	0	$-U_d/2$
Conduction	T'_{a1} or D'_{a1}	T'_{a2}, D'_{a2} or T'_{a3}, D'_{a3}	T'_{a4} or D'_{a4}
T'_{a1}	ON	OFF	OFF
T'_{a2}	OFF	ON	ON
T'_{a3}	ON	ON	OFF
T'_{a4}	OFF	OFF	ON

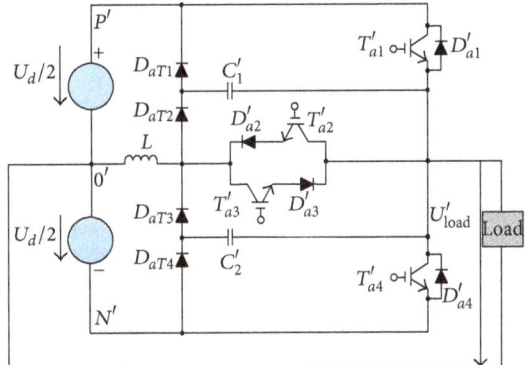

FIGURE 6: One full-bridge leg topology of S3L inverters.

S3L three-level inverter switching state and commutation process are shown in Tables 1 and 2.

For zero load current commutation process, it can be considered as three special cases, specified in Table 3.

Each of these commutation processes is slightly different, and therefore only the $T'_{a1} \rightarrow D'_{a3}, T'_{a3}$ was chosen to describe the working details as an example. In order to facilitate the analysis, the load current in the commutation process is supposed to be constant substantially and its path is marked in red.

Before the commutation process begins, T'_{a1} carries the positive load current I_{Load}, and T'_{a3} is switched on (but does not carry current, because of diode D'_{a3}); T'_{a2} and T'_{a4} are switched off. The output terminal is connected to the positive terminal of the input DC voltage. The capacitor C'_1 is discharged; the capacitor C'_2 is charged to $-U_d$. The current in the snubber inductor is zero (Figure 7(a)) [14].

The $T'_{a1} \rightarrow D'_{a3}, T'_{a3}$ commutation process starts as soon as T'_{a1} is switched off, when T'_{a2} is switched off, and what is more T'_{a3} and T'_{a4} remain switched on and off, respectively. In accordance with the different current path and IGBT action sequences, the whole process can be divided into two periods.

(1) $t_0 \leq t < t_1$ Period. Two current loops are generated during this stage. One of them is the oscillating current loop constituted by $C'_2, T'_{a2}, D'_{a2}, L, U_d/2$, and D_{aT4}; the other is the load current loop generated by the load current flowing through C'_2, load, midpoint 0, $U_d/2$, and D_{aT_4}. As shown in Figure 7(b), the two current paths overlap each other. It is noteworthy that the current flow decreases rapidly to zero, and the rising slope of the voltage both ends is limited to a small amplitude, so that the power loss is correspondingly

small. At this time, the switching-off process of T'_{a1} is the so-called soft switching.

(2) $t_1 \leq t < t_2$ Period. The first period of commutation process comes to an end, when C'_2 discharges and D'_{a4} starts conducting. At the same time, the current flowing through inductor L reduces to 0, and what is more T'_{a3} and D'_{a3} start to conduct as soon as D'_{a2} switches off. Since the voltage applied to inductor L is the constant $U_d/2$, the current flowing through L increases linearly with time. In contrast, the current flowing through D'_{a4} decreases linearly with time (as shown in Figure 7(c)). At the same time when the current flowing through D'_{a4} decreases to 0, the current flowing through the snubber inductor is equivalent to the load current, and then the whole commutation process comes to an end. D'_{a4} is blocked; T'_{a3} and D'_{a3} are carrying the load current and C'_2 is discharged simultaneously (Figure 7(d)).

The red lines represent the current path of $T'_{a1} \rightarrow D'_{a3}, T'_{a3}$ commutation process in S3L inverter during different periods.

It can be seen from the figures that the ratios of the current flowing through $S'_{a2}, D'_{a2}, S'_{a3}$, and D'_{a3} are limited within a limited range. Meanwhile, the switching process of S'_{a3} is soft switching and its power loss is small as well. Similarly, the ratios of currents flowing through D'_{a4} and D'_{a1} are limited in a certain range. Therefore, a substantial reduction of charging energy is realized during the reverse recovery and the power loss of charging is reduced with it as well.

The rest of commutation processes in Table 3 work in a similar way, which is soft switching type and has nothing to do with the influences caused by the amplitude and angle ($0° \sim 360°$) of load current, so it will not be detailed, respectively.

4. Simulation and Experiment

Based on the theories and algorithms above, the experiment was conducted on a 2.1 MW experimental platform (Figures 8 and 9), which includes two 2.1 MW motors and both of them are controlled running under the same conditions by NPC three-level inverter and S3L inverter, respectively, to carry out the comparing experiment of improvement effectiveness.

It can be seen by analyzing the waveforms in Figures 10 and 11 that the output waveforms of the two three-level inverters, which use the same SPWM modulation algorithm and control parameters, are almost consistent in waveform distortion and harmonic content, when the peak value of output phase voltage is $V_{\text{dc}}/\sqrt{3}$. Therefore, it is considered that the S3L inverter has the same output characteristics with NPC three-level inverter with the same modulation conditions, and S3L inverter has a much higher harmonic content in output voltage and slightly smaller total distortion rate; however, its low-order harmonics account for a much bigger proportion. It can be summarized by analyzing Figures 12 and 13 that S3L inverter has the same excellent output waveforms with NPC3L inverter and its output waveforms of current are smoothing and approximate sine curve.

It can be seen by analyzing Figures 14(a)~14(f) that the IGBT (T'_{a2}) current surge of S3L inverter is only two-thirds

TABLE 2: Commutation processes of S3L inverter.

	Load current is positive			Load current is negative	
Commutation	Allowed	Involved	Commutation	Allowed	Involved
$T'_{a1} \rightarrow D'_{a3}, T'_{a3}$	YES	C'_2	$D'_{a1} \rightarrow D'_{a2}, T'_{a2}$	YES	C'_2
$D'_{a3}, T'_{a3} \rightarrow T'_{a1}$	YES	C'_2	$D'_{a2}, T'_{a2} \rightarrow D'_{a1}$	YES	C'_2
$D'_{a3}, T'_{a3} \rightarrow D'_{a4}$	YES	C'_1	$D'_{a2}, T'_{a2} \rightarrow T'_{a4}$	YES	C'_1
$D'_{a4} \rightarrow D'_{a3}, T'_{a3}$	YES	C'_1	$T'_{a4} \rightarrow D'_{a2}, T'_{a2}$	YES	C'_1
$T'_{a1} \rightarrow D'_{a4}$	NO	—	$D'_{a1} \rightarrow T'_{a4}$	NO	—
$D'_{a4} \rightarrow T'_{a1}$	NO	—	$T'_{a4} \rightarrow D'_{a1}$	NO	—

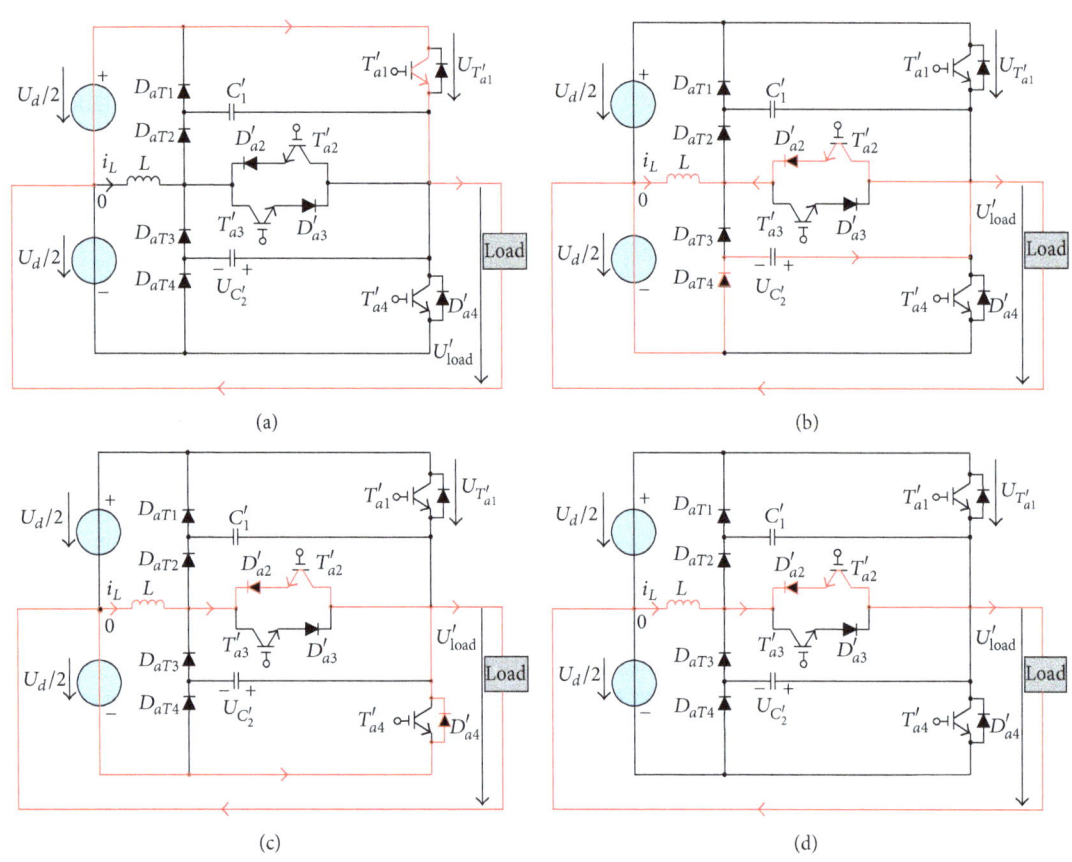

FIGURE 7: Commutation process of $T'_{a1} \rightarrow D'_{a3}, T'_{a3}$. (a) Before commutation; (b) $t_0 \leq t < t_1$ period; (c) $t_1 \leq t < t_2$ period; (d) after commutation.

TABLE 3: Commutation processes with zero load current.

Commutation	Allowed	Involved
$T'_{a1} \rightarrow D'_{a3}, T'_{a3}$	YES	C'_2
$D'_{a3}, T'_{a3} \rightarrow T'_{a1}$	YES	C'_2
$T'_{a1} \rightarrow D'_{a4}$	NO	—
$D'_{a1} \rightarrow D'_{a2}, T'_{a2}$	YES	C'_1
$D'_{a2}, T'_{a2} \rightarrow D'_{a1}$	YES	C'_1
$D'_{a1} \rightarrow T'_{a4}$	YES	—

FIGURE 8: The 2.1 MW dragging platform-motor part.

of NPC3L inverter (T_{a2}) at the switching-on instant; the IGBT (T'_{a2}) voltage surge of S3L inverter is only half of NPC3L inverter (T_{a2}) at the switching-off instant; overall, S3L inverters have much lower switching-on and switching-off voltage and current surges than NPC3L inverters.

NPC3L explosion-proof inverter (1 MW)

S3L explosion-proof inverter (1 MW)

FIGURE 9: The 2.1 MW dragging platform-inverter part.

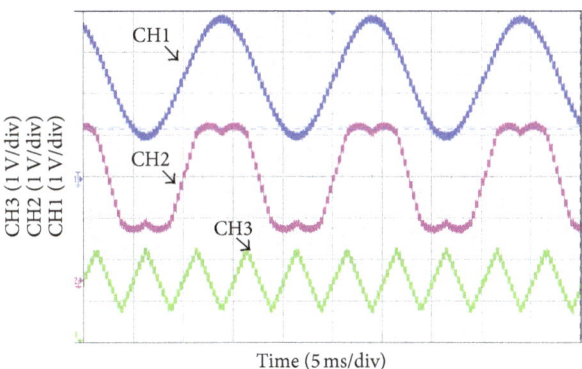

(a) The modulation waveform at $m = 1.154$

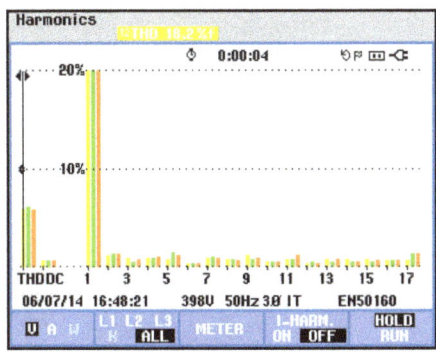

(b) NPC harmonic content

(c) S3L harmonic content

FIGURE 10: The comparison of phase-voltage harmonic characteristics with $m = 1.154$.

It is pointed out in the optimized power-loss calculation algorithm and maternal module concept that it is necessary to calculate the power losses of other devices except power devices on the same bridge to modify the total loss when performing the system thermal analysis. In this paper the modification aspects include the power losses of precharge current-limiting resistor, balanced resistors, absorption capacitance, DC-link capacitors, and buffer devices of S3L inverter. By the way, in any other cases the modification can be calculated in accordance with the actual conditions. These devices are usually fixed in the explosion-proof inverter housing and some of them are working all the way releasing some power as constant thermal sources, which will elevate the ambient temperature of the whole cabinet and affect the thermal flow in the cabinet. All of this above will finally influence the power-loss calculation and generate an error between theoretical calculation and actual value. The conclusion can be drawn by analyzing Figures 15(a)~15(d) that the power loss of IGBT T_{a1} and its antiparallel diode D_{a1} in NPC three-level inverters is much bigger than T'_{a1} and its antiparallel diode D'_{a1} in S3L inverter with different modulations and load impedance angles. According to the optimized power-loss calculation algorithm, the additional power loss of each power device module was calculated and completed

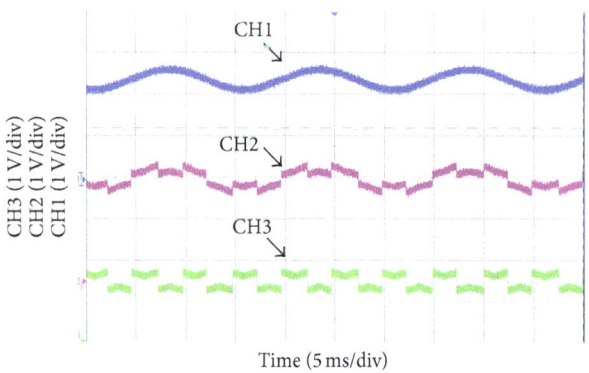

(a) The modulation waveform at $m = 0.1$

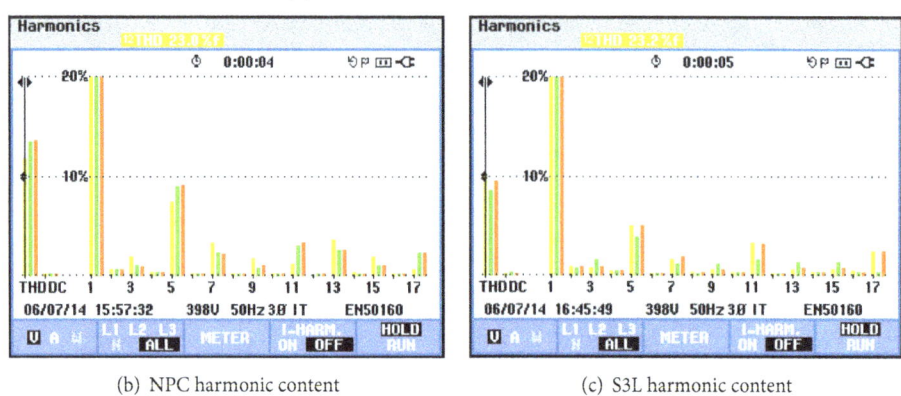

(b) NPC harmonic content

(c) S3L harmonic content

FIGURE 11: The comparison of phase-voltage harmonic characteristics with $m = 0.1$.

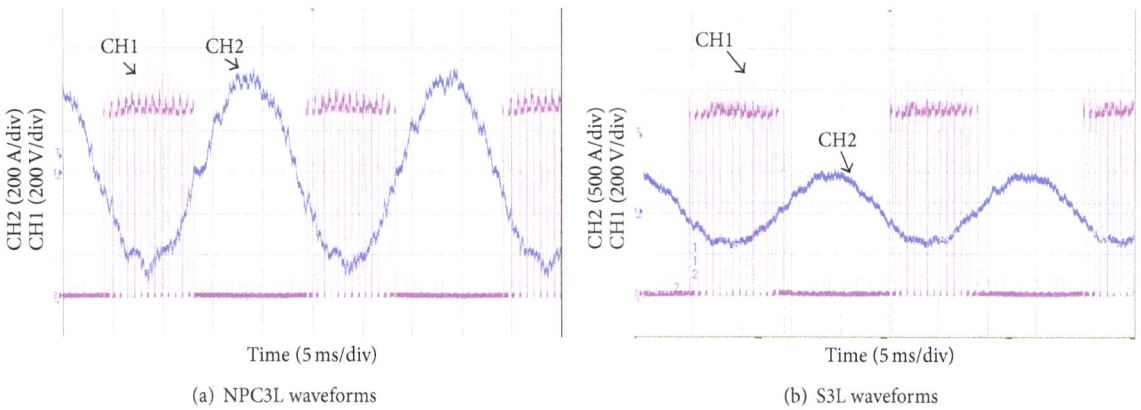

(a) NPC3L waveforms

(b) S3L waveforms

FIGURE 12: The comparison of voltage and current in the same IGBT of two inverters.

the modification of maternal model power loss; in summary, the half-bridge total power loss of S3L inverter maternal model is much smaller than NPC's inverter after the modification by 213 W, on which basis the total power loss of inverters can be obtained.

The calculated power-loss value before modification and after modification was put into the thermal model of maternal model built by ANSYS ICEPAK, respectively, as a thermal resource value, and the thermal analysis results of two inverters were presented in Figures 16 and 17. It can be derived by analyzing the Figures 17(a) and 17(b) that the heat sink temperature of S3L inverter is about 8°C lower than that

of NPC three-level inverter running in the same cooling systems and operating under the same conditions, while the same value in Figures 16(a) and 16(b) before modification is 3°C. It is easy to find that the substrate temperature of S3L inverter is 10°C lower than that of NPC inverter approximately running in the same cooling systems and operating under the same conditions by analyzing Figures 17(c) and 17(d). But there is a 4°C decline in Figures 16(c) and 16(d) before modification. Overall, the analysis shows that the power devices' temperature of S3L inverter has a 9°C advantage over NPC inverter under modification and a 3.5°C advantage without modification.

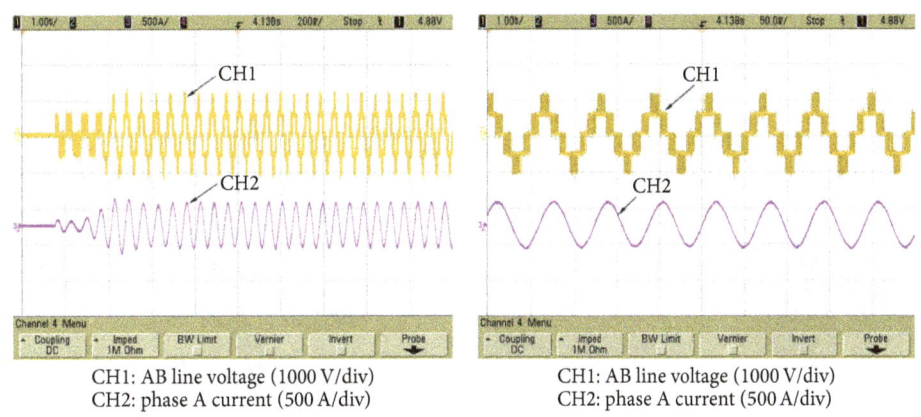

CH1: AB line voltage (1000 V/div)
CH2: phase A current (500 A/div)

(a) The AB line voltage and phase A current

CH1: AB line voltage (1000 V/div)
CH2: phase A current (500 A/div)

(b) Details of AB line voltage and phase A current

FIGURE 13: The output voltage and current waveforms of S3L inverter.

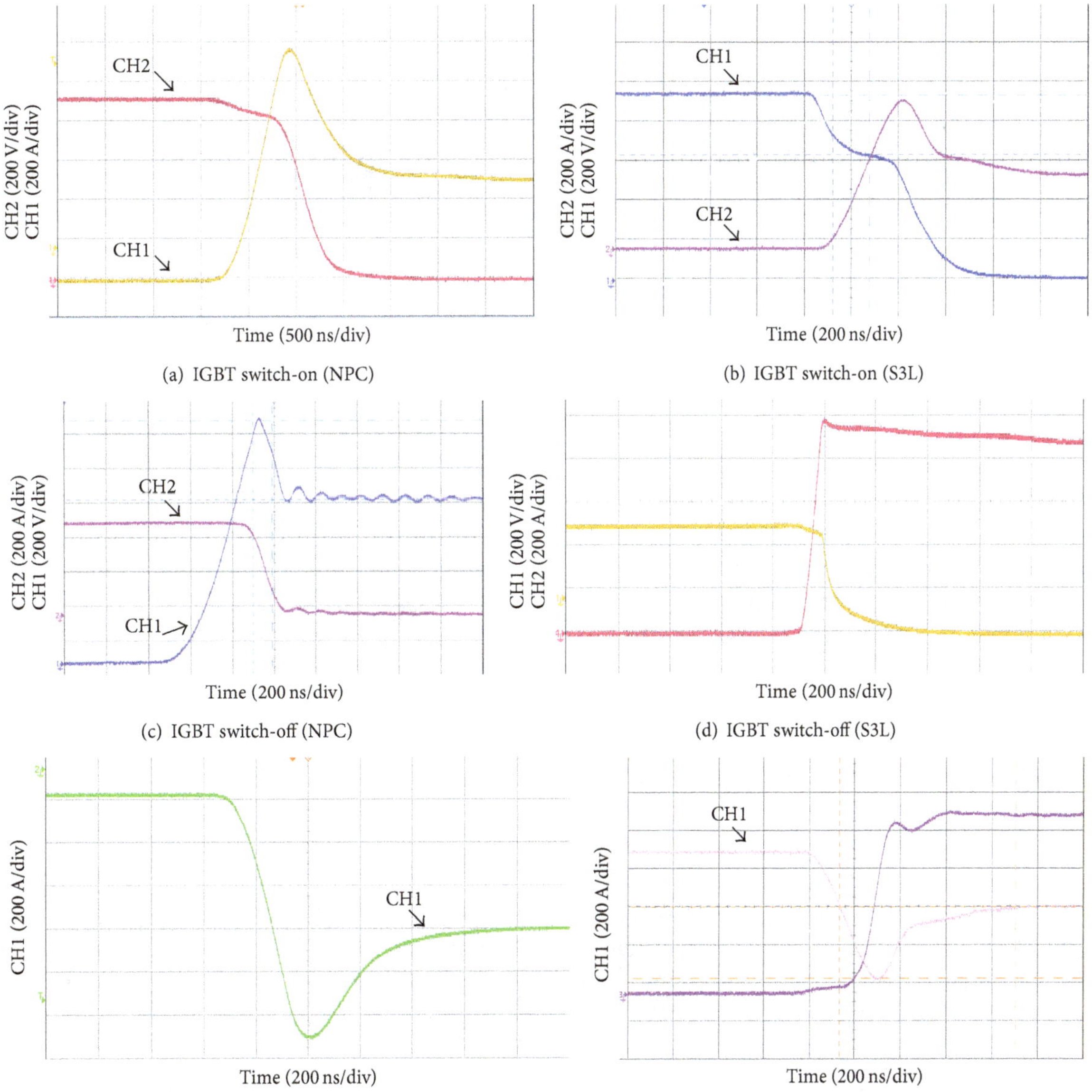

(a) IGBT switch-on (NPC)

(b) IGBT switch-on (S3L)

(c) IGBT switch-off (NPC)

(d) IGBT switch-off (S3L)

(e) NPC diode reverse recovery

(f) S3L diode reverse recovery

FIGURE 14: The comparative experiments of IGBT and diode's characteristics.

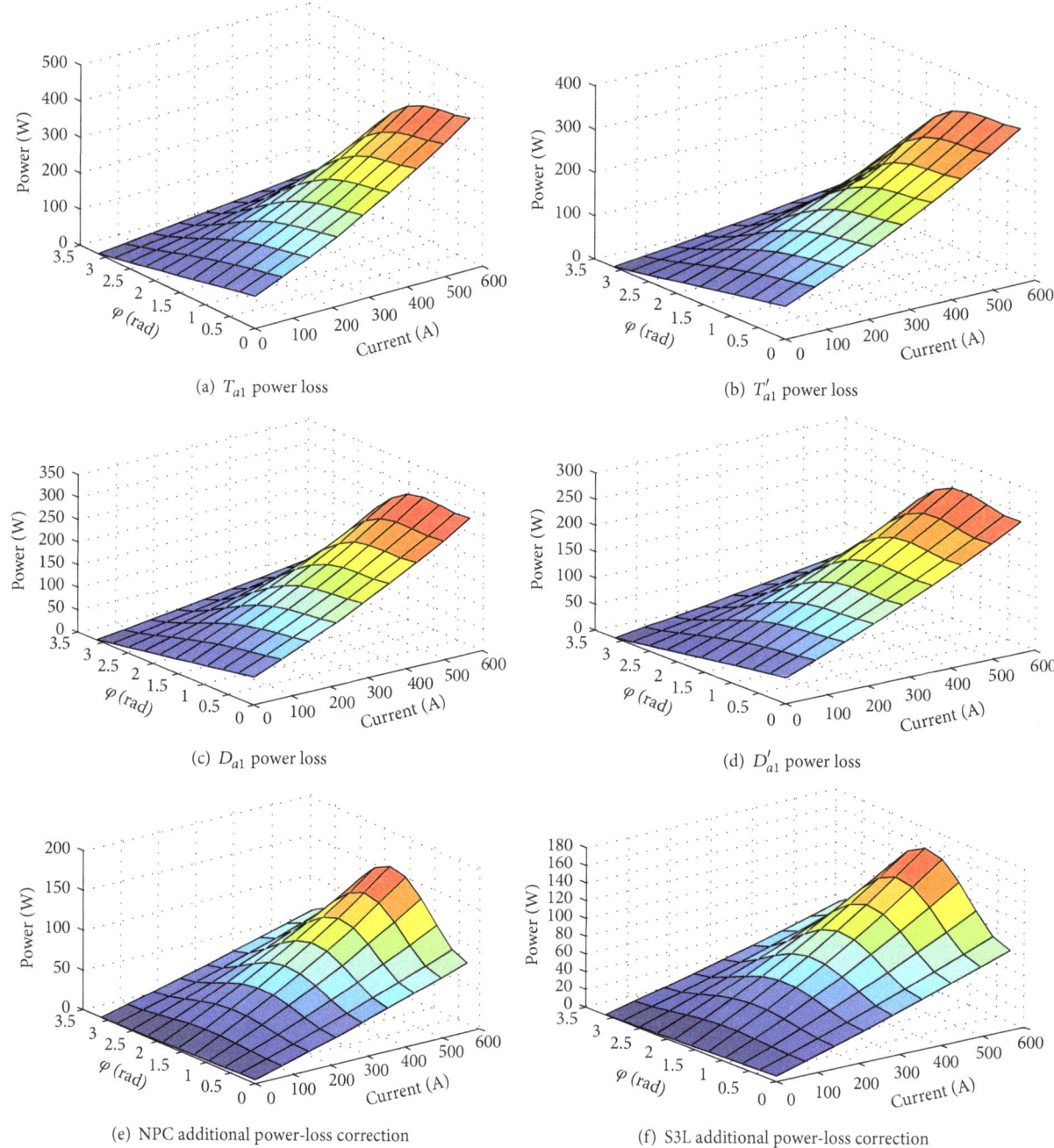

(a) T_{a1} power loss

(b) T'_{a1} power loss

(c) D_{a1} power loss

(d) D'_{a1} power loss

(e) NPC additional power-loss correction

(f) S3L additional power-loss correction

FIGURE 15: The half-bridge power devices' power-loss calculation of the two inverters based on optimized power-loss algorithm. (a) T_{a1} power loss; (b) T'_{a1} power loss; (c) D_{a1} power loss; (d) D'_{a1} power loss; (e) NPC additional power-loss correction; (f) S3L additional power-loss correction.

By observing the experiment results in Figures 18(a) and 18(b) the fact that the heat sink surface temperature of S3L inverter is about 6°C lower than that of NPC three-level inverter in average under the same conditions can be obtained, which is in line with the theoretical analysis expectation; by observing the experiment results in Figures 18(c) and 18(d) the fact that the IGBT substrate temperature of S3L inverter is lower than NPC three-level inverter by 11°C can be figured out, which is consistent with the theoretical analysis results. Generally speaking, the experimental results

show that the power device temperature of S3L inverter is lower than that of NPC three-level inverter by 10°C approximately, which is much closer to the theoretical analysis result 9°C with modification and only has 1°C error under these experiment conditions.

In summary, the maternal model based on the optimized power-loss algorithm has a much higher thermal analysis accuracy in the improvement process of three-level inverters, which offers a 1°C error between theoretical calculation and experiment value in this paper, and can be used as

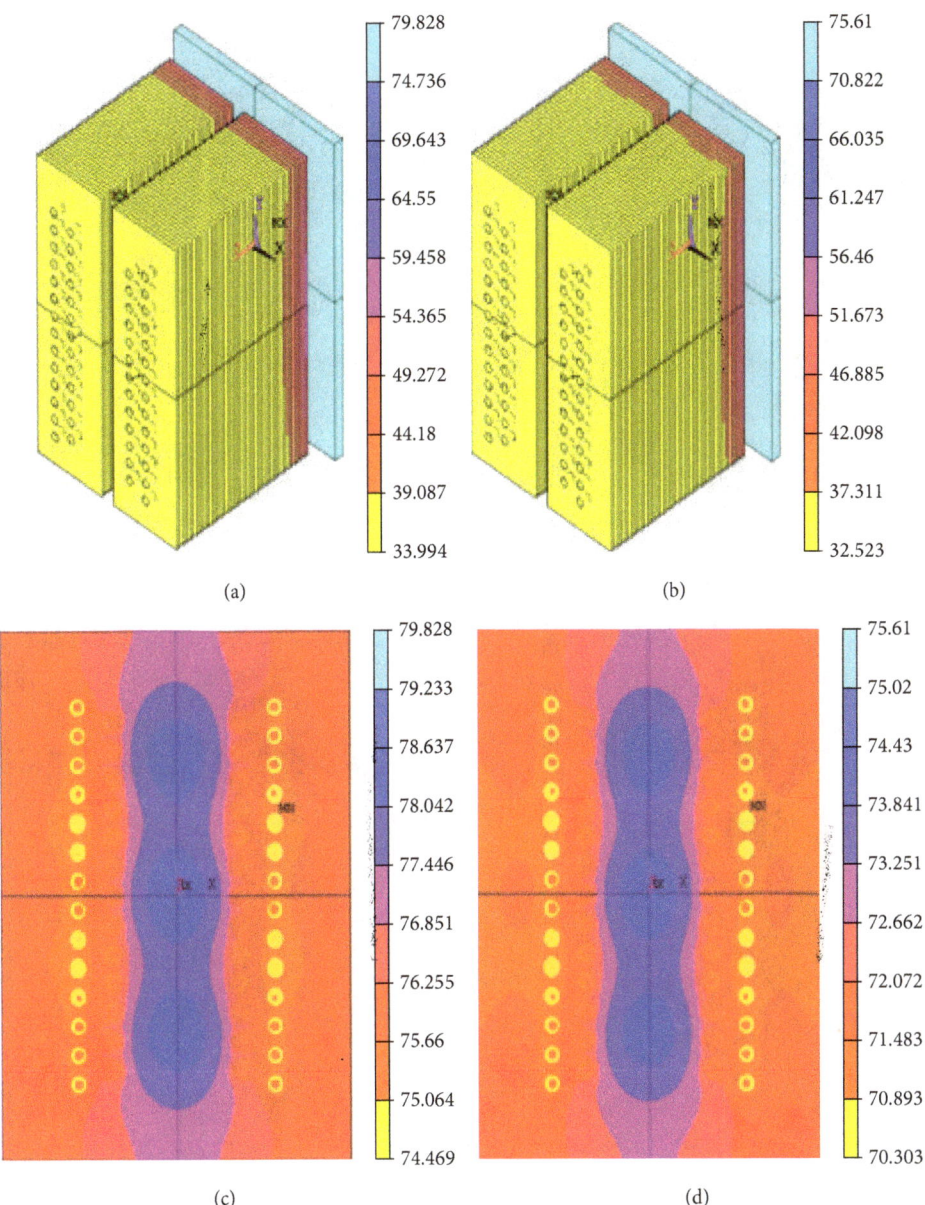

FIGURE 16: The temperature distribution 3D map of maternal thermal models before modification. (a) One heat sink temperature chart of NPC3L inverter; (b) one heat sink temperature chart of S3L inverter; (c) one substrate temperature chart of NPC3L inverter; (d) one substrate temperature chart of S3L inverter.

a tool to support the accurate power-loss calculation and thermal analysis; using the S3L inverter based on the soft switching control to improve the NPC three-level inverter can get a good result that S3L inverter has the same excellent output characteristic with NPC three-level inverter and has a great advantage in reducing power loss, with a 213 W decline in each half-bridge and 10°C decline on power device temperature. The thermal stability of three-level inverters can be enhanced by this improvement.

5. Conclusions

The optimized maternal power-loss thermal models of NPC three-level inverter and S3L inverter were established based on the optimized power-loss algorithm, and a set of general optimized power-loss calculation formulas was derived to modify the total power loss and figure out the modification power-loss values. Then, these values were considered as thermal sources to analyze the maternal thermal models. The three-level inverter can be improved by comparing and analyzing power-loss modification values and experiment results. Based on this principle and methods, the fact that, under the same conditions, the power-loss modification value of S3L inverter is smaller than that of NPC three-level inverter by 213 W and has a 9°C advantage is obtained, which is only 1°C smaller than the experiment result. Experimental results validate the proposed theoretical calculation and analysis and prove the effectiveness of the improvement.

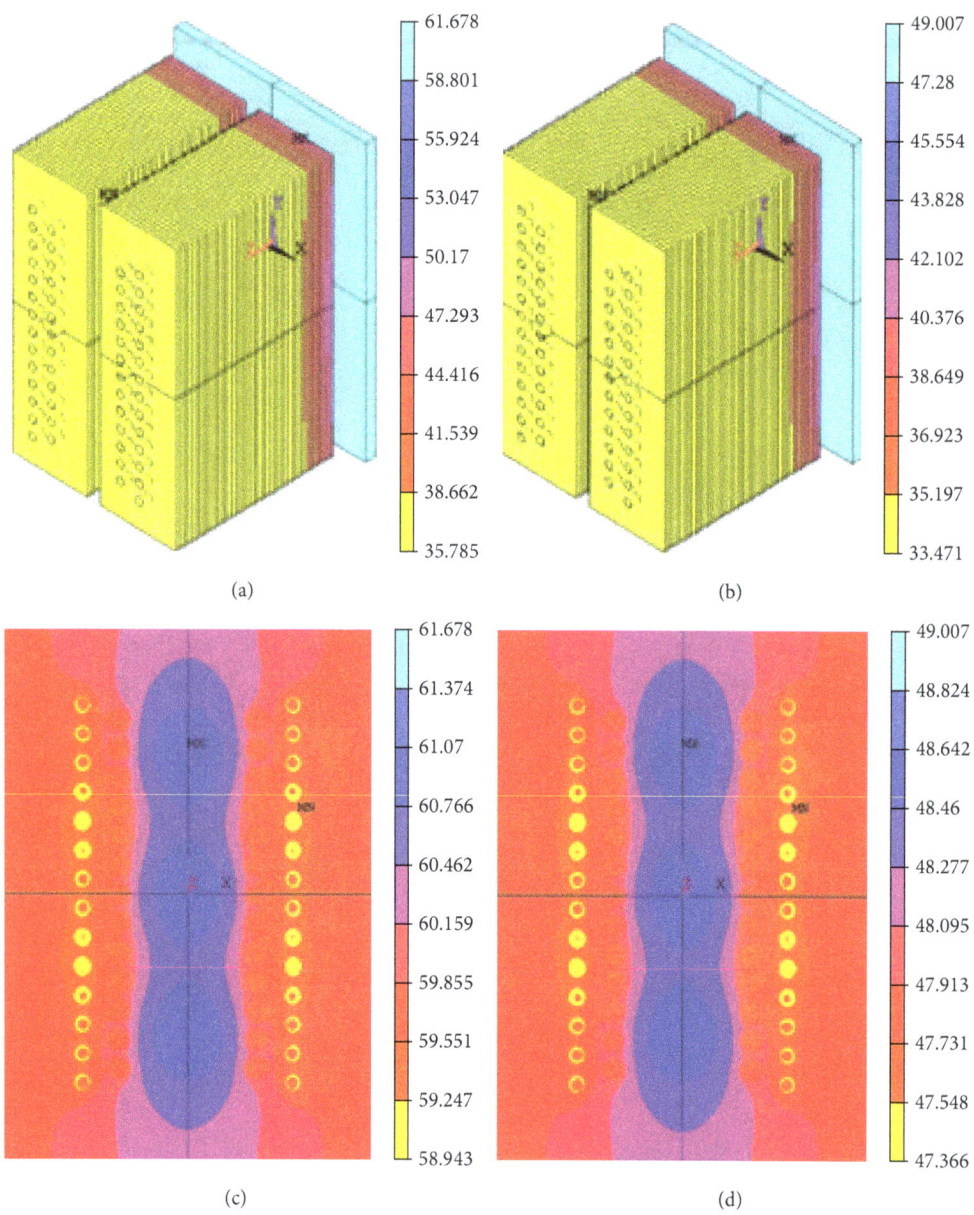

FIGURE 17: The temperature distribution 3D map of maternal thermal models after modification. (a) One heat sink temperature chart of NPC3L inverter; (b) one heat sink temperature chart of S3L inverter; (c) one substrate temperature chart of NPC3L inverter; (d) one substrate temperature chart of S3L inverter.

Appendix

The parameters of double-fed induction motor used in this experiment are as follows:

  rated power: 2100 KW,

  rated speed: 1000 rpm,

  number of pole-pairs: 3,

  efficiency at full load: 97%,

  network voltage: 690 V,

  network frequency: 50 Hz,

  stator current: 1662 A,

  rotor current: 743 A,

  coupling: stator Δ; rotor Y,

  rotor open voltage: 1710 V,

  inertia: 94 kgm^2,

  stator maximum short-circuit current: 4250 A,

  rotor maximum short-circuit current: 2900 A,

  stator resistance $R1$: 0.006013 Ω,

  stator leakage reactance $X1$: 0.045062 Ω,

  rotor resistance (equivalent) $R2$: 0.004193 Ω,

  rotor leakage reactance (equivalent) $X2$: 0.2298 Ω.

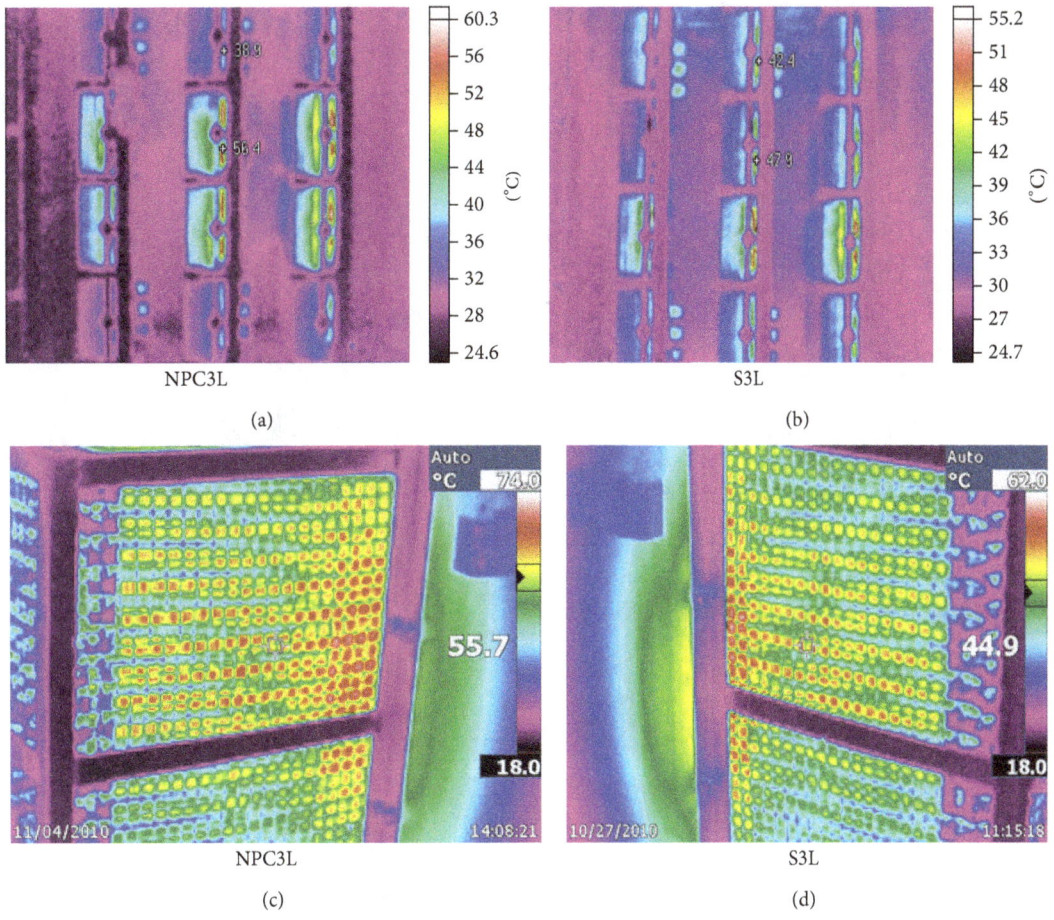

FIGURE 18: The experimental temperature results of IGBT's surface and substrate. (a) IGBT surface temperature of NPC3L inverter; (b) IGBT surface temperature of S3L inverter; (c) IGBT substrate temperature of NPC3L inverter; (d) IGBT substrate temperature of S3L inverter.

Conflict of Interests

The authors declare that there is no conflict of interests regarding the publication of this paper.

Acknowledgments

The authors would like to thank 2014 Jiangsu Province Natural Science Foundation (BK20140204), the Research and Innovation Program of Postgraduates in Jiangsu Province (CXZZ13_0930), and the Fundamental Research Funds for the Central Universities (2012LWB73).

References

[1] S. Bernet, "Recent developments of high power converters for industry and traction applications," *IEEE Transactions on Power Electronics*, vol. 15, no. 6, pp. 1102–1117, 2000.

[2] P. Mao, S.-J. Xie, and Z.-G. Xu, "Switching transients model and loss analysis of IGBT module," *Proceedings of the Chinese Society of Electrical Engineering*, vol. 15, article 007, 2010.

[3] A. D. Rajapakse, A. M. Gole, and P. L. Wilson, "Electromagnetic transients simulation models for accurate representation of switching losses and thermal performance in power electronic systems," *IEEE Transactions on Power Delivery*, vol. 20, no. 1, pp. 319–327, 2005.

[4] J. Hu, J. Li, J. Zou, and J. Tan, "Losses calculation of IGBT module and heat dissipation system design of inverters," *Transactions of China Electrotechnical Society*, vol. 24, no. 3, pp. 159–163, 2009.

[5] M. H. Bierhoff and F. W. Fuchs, "Semiconductor losses in voltage source and current source IGBT converters based on analytical derivation," in *Proceedings of the IEEE 35th Annual Power Electronics Specialists Conference (PESC '04)*, vol. 4, pp. 2836–2842, IEEE, June 2004.

[6] F. Krismer and J. W. Kolar, "Accurate power loss model derivation of a high-current dual active bridge converter for an automotive application," *IEEE Transactions on Industrial Electronics*, vol. 57, no. 3, pp. 881–891, 2010.

[7] Q. Chen, Q. Wang, W. Jiang, and C. Hu, "Analysis of switching losses in diode-clamped three-level converter," *Transactions of China Electrotechnical Society*, vol. 23, no. 2, pp. 68–75, 2008.

[8] J. Wang, Q. Chen, W. Jiang, and C. Hu, "Analysis of conduction losses in neutral-point-clamped three-level inverter," *Transactions of China Electrotechnical Society*, vol. 3, p. 12, 2007.

[9] T. J. Kim, D. W. Kang, Y. H. Lee, and D. S. Hyun, "The analysis of conduction and switching losses in multi-level inverter system," in *Proceedings of the IEEE 32nd Annual Power Electronics Specialists Conference (PESC '01)*, pp. 1363–1368, June 2001.

[10] S. Dieckerhoff, S. Bernet, and D. Krug, "Power loss-oriented evaluation of high voltage IGBTs and multilevel converters in transformerless traction applications," *IEEE Transactions on Power Electronics*, vol. 20, no. 6, pp. 1328–1336, 2005.

[11] W. Jing, G. Tan, and Z. Ye, "Losses calculation and heat dissipation analysis of high-power three-level converters," *Transactions of China Electrotechnical Society*, vol. 26, no. 2, pp. 134–140, 2011.

[12] K. Ma, Y. Yang, and F. Blaabjerg, "Transient modelling of loss and thermal dynamics in power semiconductor devices," in *Proceedings of the IEEE Energy Conversion Congress and Exposition (ECCE '14)*, pp. 5495–5501, IEEE, 2014.

[13] U. R. Prasanna and A. K. Rathore, "Analysis, design, and experimental results of a novel soft-switching snubberless current-fed half-bridge front-end converter-based pv inverter," *IEEE Transactions on Power Electronics*, vol. 28, no. 7, pp. 3219–3230, 2013.

[14] X. Ruan, L. Zhou, and Y. Yan, "Soft-switching PWM three-level converters," *IEEE Transactions on Power Electronics*, vol. 16, no. 5, pp. 612–622, 2001.

[15] R. Li and D. Xu, "A zero-voltage switching three-phase inverter," *IEEE Transactions on Power Electronics*, vol. 29, no. 3, pp. 1200–1210, 2014.

[16] P. Köllensperger, R. U. Lenke, S. Schröder, and R. W. de Doncker, "Design of a flexible control platform for soft-switching multilevel inverters," *IEEE Transactions on Power Electronics*, vol. 22, no. 5, pp. 1778–1785, 2007.

[17] G. Ortiz, H. Uemura, D. Bortis, J. W. Kolar, and O. Apeldoorn, "Modeling of soft-switching losses of IGBTs in high-power high-efficiency dual-active-bridge DC/DC converters," *IEEE Transactions on Electron Devices*, vol. 60, no. 2, pp. 587–597, 2013.

[18] M. W. Gekeler, "Soft switching three level inverter (S3L inverter)," in *Proceedings of the 15th European Conference on Power Electronics and Applications (EPE '13)*, pp. 1–10, IEEE, September 2013.

[19] S. Munk-Nielsen, L. N. Tutelea, and U. Jaeger, "Simulation with ideal switch models combined with measured loss data provides a good estimate of power loss," in *Proceedings of the Conference Record of the IEEE Industry Applications Conference*, vol. 5, pp. 2915–2922, Rome, Italy, October 2000.

[20] O. S. Senturk, L. Helle, S. Munk-Nielsen, P. Rodriguez, and R. Teodorescu, "Converter structure-based power loss and static thermal modeling of the press-pack IGBT three-level ANPC VSC applied to multi-MW wind turbines," *IEEE Transactions on Industry Applications*, vol. 47, no. 6, pp. 2505–2515, 2011.

[21] D. A. B. Zambra, C. Rech, F. A. S. Gonçalves, and J. R. Pinheiro, "Power losses analysis and cooling system design of three topologies of multilevel inverters," in *Proceedings of the 39th IEEE Annual Power Electronics Specialists Conference (PESC '08)*, pp. 4290–4295, IEEE, June 2008.

[22] W. Jing, *Three-level inverter power loss power devices research [Doctoral dissertation]*, China University of Mining and Technology, Xuzhou, China, 2011.

[23] D. Floricau, E. Floricau, and G. Gateau, "Three-level Active NPC converter: PWM strategies and loss distribution," in *Proceedings of the 34th Annual Conference of the IEEE Industrial Electronics Society (IECON '08)*, pp. 3333–3338, November 2008.

6

Hybrid Natural and Forced Active Balancing Control of Battery Packs State of Charge based on Partnership for a New Generation of Vehicles

Yewen Wei [ID],[1,2] Shuailong Dai [ID],[1,2] Jiayu Wang,[1] Zhifei Shan,[1] and Jie Min[1]

[1]*College of Electrical Engineering & New Energy, China Three Gorges University, Yichang, China*
[2]*Hubei Province Collaborative Innovation Center for New Energy Microgrid, CTGU, Yichang, China*

Correspondence should be addressed to Shuailong Dai; daishuailong@126.com

Academic Editor: Pascal Venet

Battery packs are widely used in electric vehicles, and their state-of-charge is one of the essential issues that affect the performances, whilst the balance between parallel and series cell of the battery pack always has an obvious effect. To enhance the working performance of the lithium-based power battery pack, a hybrid natural and forced active balancing control (HNFABC) strategy is proposed and adopted to the balancing circuit that is proposed in this work. These converters, which are advantageous in natural balancing and forced equalization, accelerate the balance speed of natural equilibrium in the final stage and protect the battery from being repeatedly charged and discharged. Simulation and experimental results show that HNFABC is not only simpler than other traditional balance control strategies but also faster in the equalization process. The idea of combining natural equilibrium and forced equilibrium can be inspired to be used in some related industries.

1. Introduction

With the gradual deterioration of the global natural environment, new energy electric vehicles have attracted wide attention and are gradually applied around the world due to their environmental friendliness and can be used as controllable loads in close cooperation with power grids and charging stations [1–3]. In these electric vehicles, batteries are generally needed for energy and power storage. However, the inconsistency of individual cells in the battery pack leads to a reduction in overall charge and discharge performance [4–7]. Therefore, various balanced topologies and control strategies have been proposed to reduce battery inconsistencies.

In [8], to reduce the switching loss and overcome the overvoltage problem, a zero-voltage switch and zero-current switch circuit topology based on DCM operation was proposed. However, the interleaving technique proposed requires a sixteen-channel PWM control signal fulfilling the balancing requirement. In [9], Lee et al. proposed a modular

equilibrium idea, in which the system consists of N batteries and M equalization modules. The circuit structure is complex, and it is necessary to adjust the state of twelve switching tubes to achieve the primary energy balance. In [10], the circuit is based on a time-sharing flyback converter. Each cell shares an equalization module in the control gap of the low-power microcontroller. However, transformer equalization increases the weight and volume of the equalization circuit to a certain extent. In [11], a new type of switching circuit that does not require voltage detection was proposed. In order to use the single-charge equalizer of the multiwinding transformer, the energy balancing between the cells is achieved by the magnetization energy of the multiwinding transformer, which causes a large amount of magnetic loss inevitably. In [12], the idea of energy sharing was used to adaptively adjust the charge and discharge rates of all batteries while maintaining the DC bus voltage, which solved the unbalancing problems of SOC between battery pack cells. Using a low-power DC, power converter may take a long time to achieve equilibrium. In [13], this paper aimed at solving this problem by using a buck chopper and an adaptive unscented Kalman algorithm for estimating the state-of-charge (SOC) of batteries. The scheme in the work presents a simple circuit structure, and shows a faster response compared with traditional methods, while a four-cell balancing experiment is provided. However, this technology is difficult to be applied when the number of battery cells increases. As closed-loop, online SOC estimation methods, Kalman filter (KF) and other observer-based algorithms have gained growing attention from both academia and industry. Nowadays, extended Kalman filtering (EKF) [14] and sigma-point Kalman filtering, i.e., central difference Kalman filter (CDKF) or unscented Kalman filter (UKF) [15], based on equivalent circuit models (ECMs), were applied. To make physical meanings of model parameters and more insights into internal electrochemical reactions, various electrochemical models were also combined with these filters, e.g., single particle model (SPM) [16], single particle model with electrolyte dynamics (SPMe) [17], and electrode average model [18]. Other KF variants were also studied, including robust EKF (REKF) [19], adaptive EKF (AEKF) [20], and adaptive UKF (AUKF) [21]. Besides KFs, smooth variable structure filter (SVSF) [22], sliding-mode observer (SMO) [23], backstepping PDE observer [24], nonlinear geometric observer [25], and Luenberger-type observers [26] were applied as well. Based on a reduced electrochemical model, the optimization-based moving horizon estimation (MHE) framework has been systematically assessed for advanced battery condition monitoring [27]. MHE and mMHE are more precise than EKF/UKF; however, it has slower computation.

In this work, a new balancing strategy called hybrid natural and forced active balancing control (HNFABC) is put forward, the operating principle of which is mainly based on the energy transmission of unbalanced current flowing from overcharged batteries to overdischarged batteries. In this case, energy naturally flows to a low-energy battery without any measurement sensor equipment. Additionally, there is no need to use a sensor to measure the voltage of batteries by utilizing the proposed scheme. Forced balancing works by precisely controlling the drive switch to transfer energy from inconsistent battery packs to all other battery packs. The proposed natural equalization control can achieve active equalization by utilizing the energy difference of the battery itself without external forced control. Forced equalization can solve the problem of slow equilibrium in the final stage of natural equilibrium, but the battery will be repeatedly charged and discharged. The finally proposed hybrid natural forced equalization control can combine the advantages of the two kinds of control above and realize the rapid equalization while the battery is only in a state of charging or discharging. In this paper, the HNFABC means that energy is naturally transferred from overcharged cells to overdischarged cells with forced energy transfer donors and acceptors during equilibrium process. To further verify the feasibility of the proposed strategy, the equalization circuit simulation system is built in PSIM. Whilst an unscented Kalman filter is used in a Matlab/Simulink model of battery packs based on the partnership of a new generation of

vehicles (PNGV). Finally, experiments are provided to verify the effectiveness and superiority of the proposed technology.

This paper is organized in the following ways. The design and operational principle of the complementary equalization topology and its control algorithm are explained in Section 2. Simulation and experimental results are presented in Section 3. The conclusions and key research content for future work are provided in Section 4.

2. Operating Principle and Design

Replacing the diodes in the traditional buck-boost circuit with MOSFET transistors enables bidirectional power transfer and continuous current mode [28–32]. The system flow of the equalization technology is shown in Figure 1. This improved buck-boost circuit can transfer energy from batteries to any other batteries. Each equalization module is connected inductively between two battery packs. The structure consists of n cells in series, which is applicable to the battery pack equalization circuit, is as shown in Figure 2. The access of the equalization module divides the battery components into upper and lower parts, which can transfer energy in the upper and lower parts, thereby achieving equalization of the battery pack. The circuit topology proposed in this paper can also individually adjust the energy of any battery to be transferred to the rest of other batteries in the battery pack. Compared with the traditional equalization circuit, this topology can flexibly realize battery energy balance and is easy to expand. Each equalization circuit divides the battery components into two parts, whose energy can be primarily transferred to each other. Furthermore, combining two equalization circuits enables any single battery energy to be transferred to the remaining batteries in the battery pack by controlling the two switching tubes.

2.1. Natural Active Balancing Control. The transistors in each equalizer block in the circuit operate in a complementary drive mode with a duty cycle a set as a function of input voltage and output voltage:

$$\alpha_i = \frac{U_{\mathrm{H}i}}{(U_{\mathrm{L}i} + U_{\mathrm{H}i})},$$

$$\alpha_i U_{\mathrm{L}i} - R_{\mathrm{INeq}} \alpha_i I_{\mathrm{Lm}} = (1 - \alpha_i) U_{\mathrm{H}i} + R_{\mathrm{OUTeq}} (1 - \alpha_i) I_{\mathrm{Lm}},$$

$$I_{\mathrm{Lm}} = \frac{\alpha_i U_{\mathrm{L}i} - (1 - \alpha_i) U_{\mathrm{H}i}}{R_{\mathrm{INeq}} \alpha_i + R_{\mathrm{OUTeq}} (1 - \alpha_i)},$$

$$(1)$$

where subscript i is the number of cells on the input leg, $i \in \mathrm{int}\ [1; N-1]$, $U_{\mathrm{H}i}$ is the input cell voltage, $U_{\mathrm{L}i}$ is the output cell voltage, R_{INeq} stands for the equivalent resistance of the input circuit, R_{OUTeq} stands for the equivalent resistance of the output circuit, and I_{Lm} means the balancing current that natural energy transfer from any overcharged cells to any undercharged cells is possible which are produced by any unbalanced voltage across each converter inductor leg. This fixed duty cycle equalization mode produces an average voltage on one side of the inductor that is

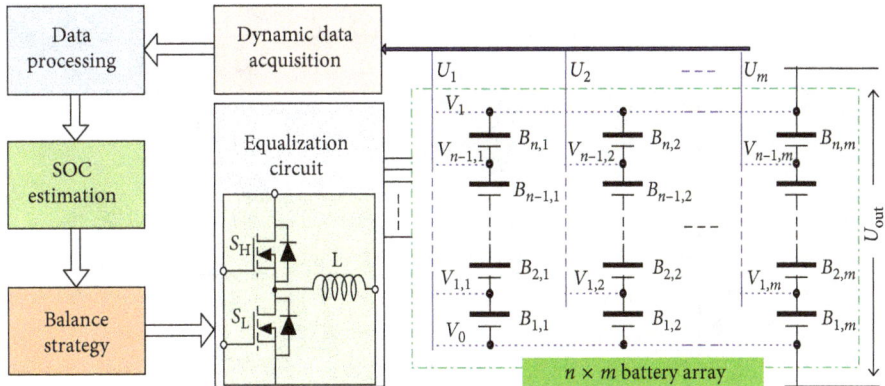

FIGURE 1: Equalization system block diagram.

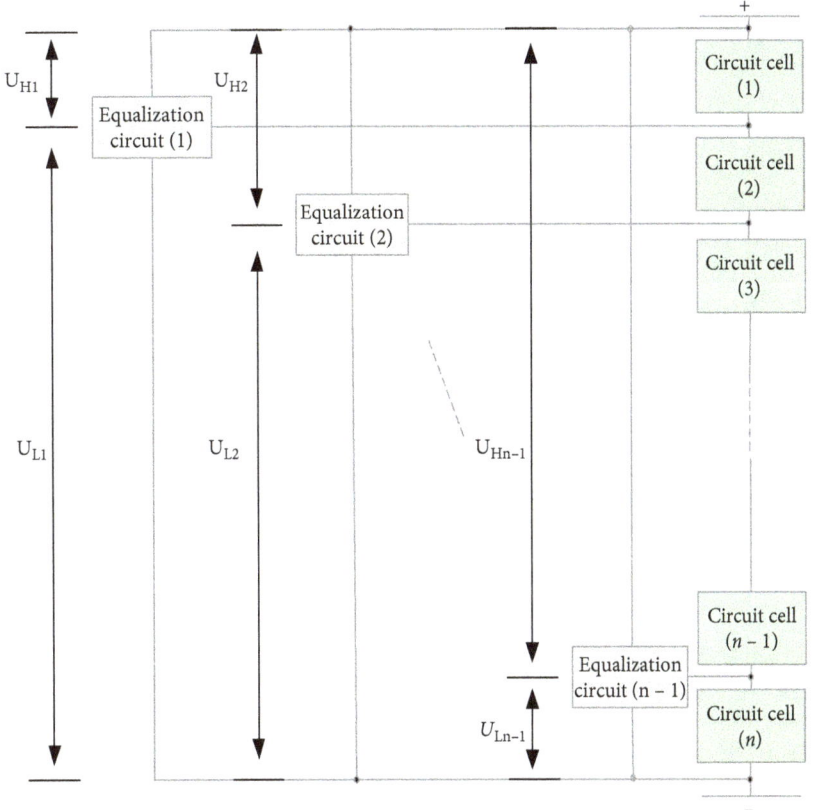

FIGURE 2: Equalization control circuit topology.

equal to a portion of the battery pack voltage. If the energy of the two cells separated by the inductance is balanced, the potential generated at both ends of the inductor is the same; that is, the energy absorbed by the inductor from a part of the circuit is equal to the energy released from the other part of the circuit. If the energy of the two parts of the battery pack is not balanced, the inductor will transfer the energy of the battery pack whose energy is higher than the set value to the battery pack below the set value. The duty ratio of the switching tubes in each equalization module in the equalization circuit is given in Table 1. Further, under natural active balancing control strategy, these switches are operated in a complementary mode.

TABLE 1: Duty cycle rate of equalization circuit.

Switch	Duty cycle					
	EC(1)	EC(2)	...	EC(i)	...	EC($n-1$)
S_H	$n-1$	$n-2$	...	$n-i$	...	1
S_L	1	2	...	i	...	$n-1$

2.2. Forced Active Balancing Control. To reduce the energy circulation caused by natural active balancing control, a novel forced active equalization control is proposed. Current flow direction is indicated by the red arrow. As shown in Figure 3, the topology can easily achieve the balance control of the front battery and the rear battery: turn on the switch

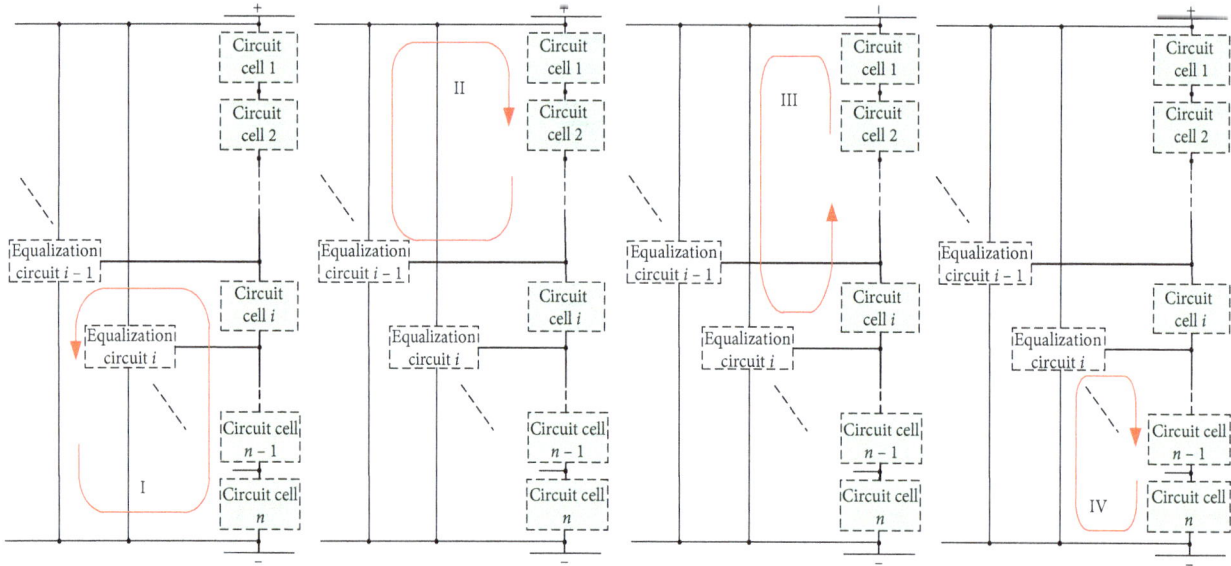

FIGURE 3: Energy transfer path schematic. (a) Mode I, (b) mode II, (c) mode III, and (d) mode IV.

$S_{\text{L}i-1}$, the current flow is shown in mode I; then, turn off the switch $S_{\text{L}i-1}$, the current flow is shown in mode II. Further, if the energy of an overcharged battery in the series battery pack is transferred to all other batteries, the above steps need to be performed, and the switch $S_{\text{H}i}$ needs to be controlled to turn on. The current flow is as shown in mode III; then, the switch $S_{\text{H}i}$ is turned off, current flow is shown in mode IV. The status of each switch from mode I to mode IV is shown in Table 2.

2.3. Hybrid Naturally and Forced Active Balancing Control.
Further, the series equilibrium of N batteries is mathematically verified. The principle is as follows: supposing that the energy of B_i is more than the average value of other batteries, it should be transferred to the other batteries. The inductor absorbs energy given out from each battery between B_i and B_n is $\Delta\varepsilon_1$ under mode I. The inductor releases energy to the rest batteries from B_1 to B_{i-1} under mode II. The inductor that absorbs energy given out from each battery between B_1 and B_i is $\Delta\varepsilon_2$ under mode III. The inductor releases energy to the rest batteries from B_{i+1} to B_n under mode IV. The difference between the energy of the battery B_i and the average energy of the battery pack is defined as Q_i^{extra}. For the amount of energy that is transferred from the single cell in the battery pack, formula (4) and formula (5) are listed:

$$\frac{(n-i+1)\Delta\varepsilon_1}{(i-1)-\Delta\varepsilon_2} = \frac{Q_i^{\text{extra}}}{n}, \tag{2}$$

$$\frac{-\Delta\varepsilon_1 + i\Delta\varepsilon_2}{(n-i)} = \frac{Q_i^{\text{extra}}}{n}. \tag{3}$$

The simultaneous equations are solved as follows:

TABLE 2: Switching state during equalization.

Switch	Mode I	Mode II	Mode III	Mode IV
$S_{\text{H}(i-1)}$	Off	Off	Off	Off
$S_{\text{L}(i-1)}$	On	Off	Off	Off
$S_{\text{H}(i)}$	Off	Off	On	Off
$S_{\text{L}(i)}$	Off	Off	Off	Off

$$\Delta\varepsilon_1 = \frac{(i-1)Q_i^{\text{extra}}}{n},$$
$$\Delta\varepsilon_2 = \frac{(n-i)Q_i^{\text{extra}}}{n}. \tag{4}$$

This process can be realized only by controlling two switch tubes. The circuit simplifies the design, facilitates the controlling strategy and is easy to expand for different numbers of series battery packs. Because the structure of the N-phase circuit is completely the same, the difference is that the inductor is connected to the battery pack at one end, so the circuit will be further simplified as shown in Figure 4.

3. Battery Model and Equilibrium Simulation

3.1. Battery Parameters Module.
PNGV model is shown in Figure 5 [33–35]. Compared with the Thevenin model and R_{int} model, this PNGV model has higher precision and describes the transient response process of the battery better. In the model, U_{OC} is the ideal voltage source and represents the open-circuit voltage of the battery; R_0 is the internal resistance of the battery; R_{P} is the polarization resistance; C_{P} is the polarization capacitance; I_{P} is the current floated on the polarization resistance; when describing the load current, C_{B} is the capacitance that describes the changing voltage, which, accumulates in the open circuit.

Available for circuit diagrams for C_{B} and C_{P}:

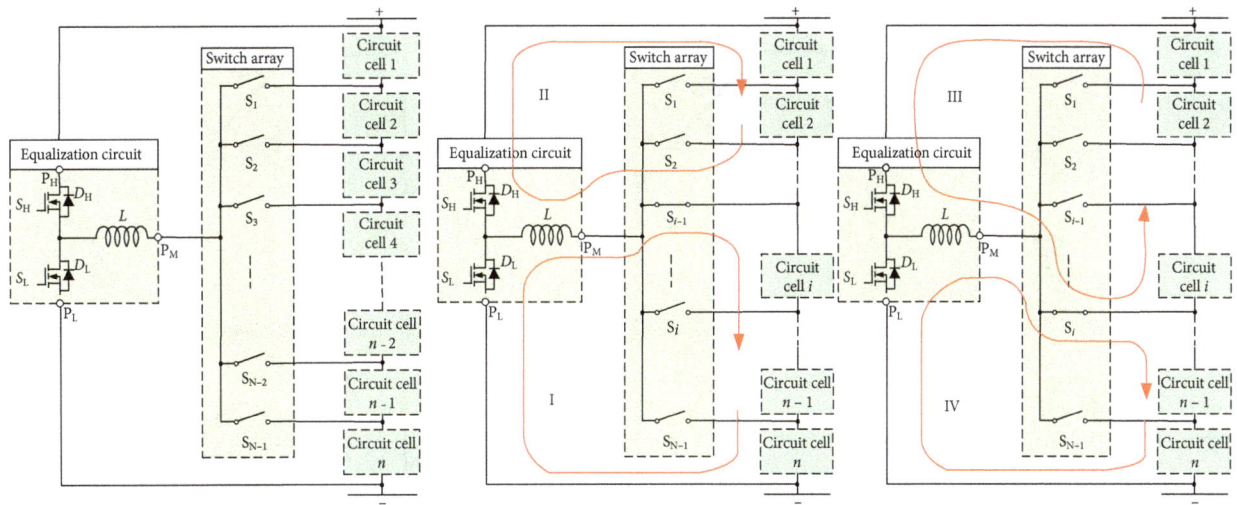

FIGURE 4: Simplified equalization circuit and its working principle. (a) Simplified equalization circuit, (b) when switch S_{i-1} is closed, and (c) when switch S_i is closed.

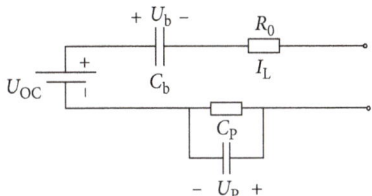

FIGURE 5: PNGV equivalent circuit model.

$$C_b \frac{dU_b}{dt} = I_L,$$

$$C_p \frac{dU_p}{dt} = I_L - \frac{U_P}{R_P}. \tag{5}$$

According to Kirchhoff's voltage law, the open circuit voltage U_{OC} can be as follows:

$$U_{OC} = U_b + U_P + I_L R_0 + U_L. \tag{6}$$

The state equation for establishing the PNGV model with the two capacitor voltages U_B and U_p in the model is

$$\begin{cases} \begin{bmatrix} U_b \\ U_P \end{bmatrix} = \begin{pmatrix} 0 & 0 \\ 0 & -\dfrac{1}{C_p R_p} \end{pmatrix} = \begin{bmatrix} U_b \\ U_P \end{bmatrix} \begin{bmatrix} \dfrac{1}{C_b} \\ \dfrac{1}{C_P} \end{bmatrix} [I_L], \\ \\ UL = [-1,\ -1] \begin{bmatrix} U_b \\ U_P \end{bmatrix} + [-R_0][I_L] + [U_{OC}]. \end{cases} \tag{7}$$

The state equation uses the two capacitor voltages of the PNGV model as the state variables and the battery terminal voltage as the output variable. The SOC is not a measure that can be measured specifically. The electrical quantity that can be measured in the model is only the port voltage U_{OC} and the port current I_L. Therefore, the existing measurement value must be used to select an appropriate SOC estimation method.

3.2. The Analysis of the Simulation Results. The natural equalization control strategy and HNFABC strategy of the circuits above are verified, respectively. In this section, a software simulation, including a PNGV model and an experimental system that contains four cells of lithium-ion batteries, is built.

In the experiment, the energy of the B_2 was set higher than other batteries in the battery pack ($U_{B2} = 4$ V, U_{B1}, U_{B3}, and $U_{B4} = 3.6$ V, $i = 2$). Specifically, the process of transferring the excess energy of the B_2 to other batteries was simulated and verified. For the equalization circuit that consisted of four batteries, the equalization process will undergo four modes as shown in Figure 3. VP1 to VP4 stand for the voltage of each battery. The equalization circuit composed of two batteries will be equipped with an equalization module. The equalization circuit composed of three batteries will be equipped with two equalization modules. The equalization circuit composed of n batteries has an equalization module number of $n - 1$. The simulation experiments of the four-cell battery are built under the natural equalization and HNFABC strategy, respectively. The dynamic voltage waveform is shown in Figures 6 and 7.

Figure 8 is the inductor current of the EC1 equalization module and the EC2 equalization module. The function of the equalization module EC1 is to achieve the balance between the B_1 and the battery 234. The function of the equalization module EC2 is to achieve the balance between the battery and the B_3 or B_4. As is shown in Figure 6(a), when the equalization circuit operates, the switch S_{L1} is turned on first, transferring the energy of B_2, B_3, and B_4 to inductor L_1, and the inductor L_1 transfers energy to B_1. As is shown in Figure 6(b), when the equalization circuit is working, the switch S_{H2} is turned on first, transferring the energy of the B_1 and B_2 to the inductor L_2, and then, the inductor L_2 transfers the energy to B_3 and B_4.

3.3. The Analysis of the Experiment Results. The experimental platform was built in the laboratory as shown in Figure 9.

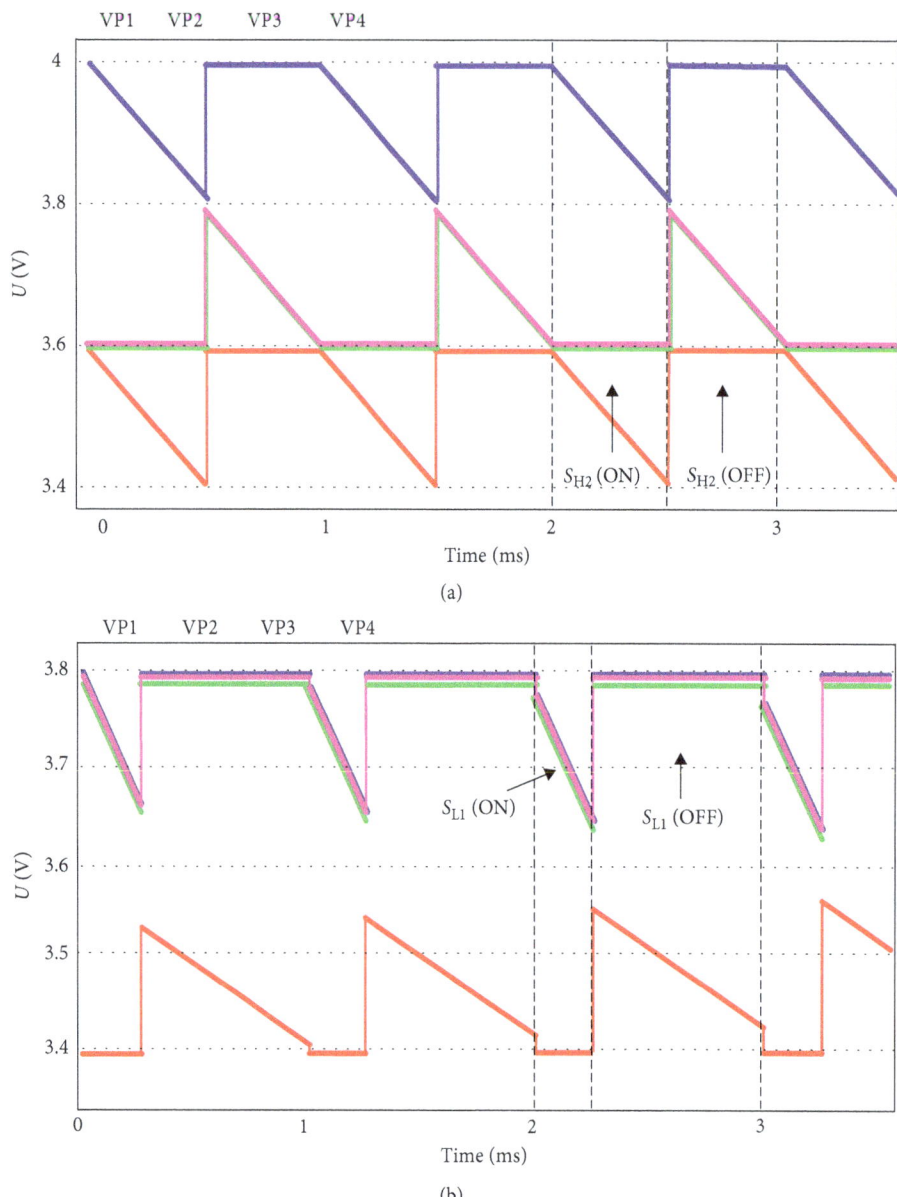

FIGURE 6: Waveforms for four battery cells energy equalization. (a) Battery voltage in mode I and mode II and (b) battery voltage in mode III and mode IV.

Simulation and experimental verifications have been completed, respectively. The system parameters of the experiment are shown in Table 3.

For simulating and comparing the battery balance in the idle state ($B_H > B_L$), Figure 10(a) shows the simulation results in a dynamic equalization process and Figure 10(b) shows the experiment results. Basically, it can be seen from the figure that the experimental waveform and the simulation results are consistent with the theoretical design.

3.4. Analysis of the Results. The feasibility of the equalization circuit and its control strategy described in this paper is verified, after comparing the simulation with experiment results. Based on the above four-cell experimental circuit, the same parameters were used for simulation. Taking a battery

equalization unit as an example, to compare its dynamic balance effect, Figure 8 shows the working waveforms when the same battery operates in idle states, including switch status and current, where Figure 8 is the actual results of the simulation and experiments, respectively. The waveforms of transforming current in experimental results are basically consistent with which in the simulation process, through comparison and observation.

Further, during the equalization process, the equalization voltage of the four-cell battery, under traditional control and HNFABC, is drawn into a line graph as Figure 11 shows. In the experiment, the battery voltage was measured once a minute and was more intuitively plotted with line graphs. Compared with the traditional control strategy, it is obvious that the HNFABC strategy can achieve a faster balance.

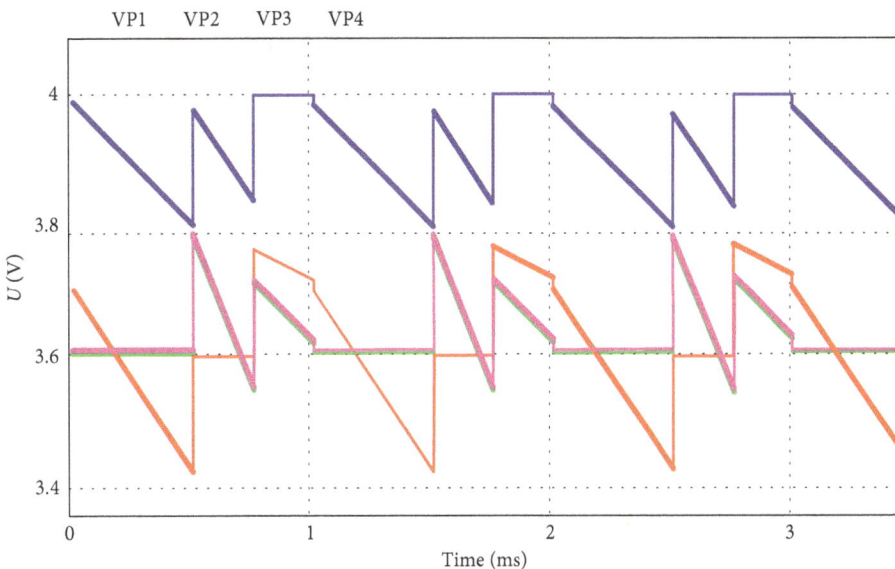

FIGURE 7: Battery pack voltage under HNFABC.

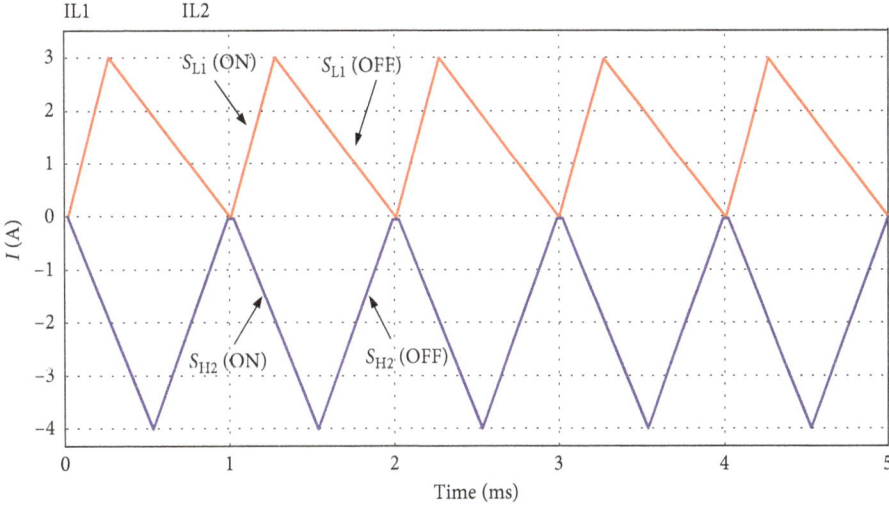

FIGURE 8: Current flow on inductor in HNFABC.

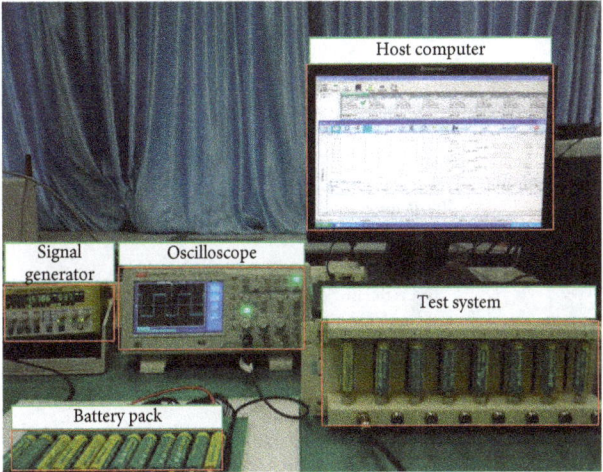

FIGURE 9: Experiment platform of the equalization circuit.

TABLE 3: Simulation and experimental specifications.

Item	Description
Control chip	STM32F103
Battery parameter	4.2 V, 1 AH
Inductance value	100 μH
Switch model	IRF530
Operating frequency	10 kHz
Duty cycle of S_{H2}	50.0%
Duty cycle of S_{L1}	33.3%
Initial state of B_1	3.6 V
Initial state of B_2	4.0 V
Initial state of B_3	3.6 V
Initial state of B_4	3.6 V

In the experiment, the energy of one battery is higher than that of other battery cells. The distinctive of HNFABC strategy will be more obvious than the traditional strategy,

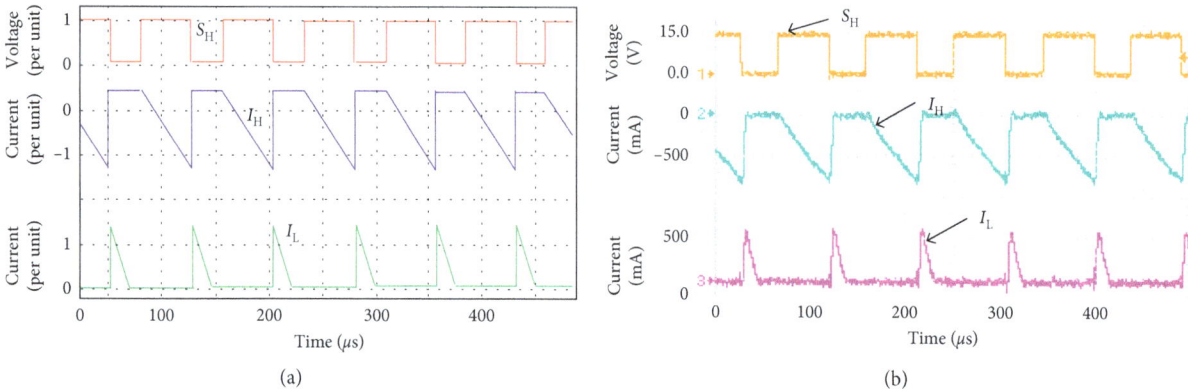

(a)   (b)

FIGURE 10: The dynamic characteristics of the battery equalization process: (a) simulation and (b) experiment.

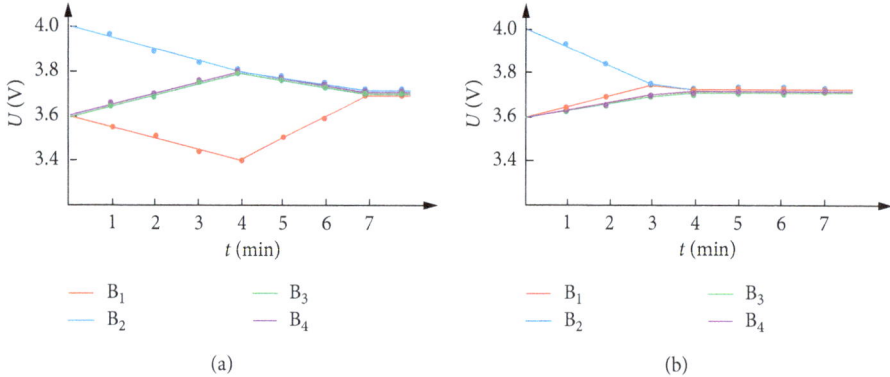

(a)   (b)

FIGURE 11: Equalization time for experiments: (a) traditional control and (b) HNFABC.

however, considering the battery's power distribution is rather complicated in actual circumstances.

The characteristics of the equalization circuit are characterized by loss, time, complexity, and extended battery life. After a comprehensive comparison of the above three equalization methods, the overall advantage of the HNFABC strategy is obvious. As is shown in Figure 12, especially in terms of loss, the energy loss is reduced by more than 50% compared to the resistance dissipation equalization method.

4. Conclusions

This paper proposed the concept of HNFABC strategy for a fast-balancing battery cells system based on the PNGV model. The experimental results have demonstrated that the complementary equalization circuit topology with the HNFABC strategy achieves battery balancing, prolongs the life of the battery pack, and reduces the energy loss in the balancing progress. The main advantage of this structure is that, in order to transfer the energy that belongs to one battery to all others, only two switches are needed to control. The simulation results also proved that compared to traditional balancing control, the new strategy effectively utilizes switches to change and implement the controlling effect. The waveform and experiment results have indicated that the proposed control strategy is effective and feasible. In practical applications, the number of batteries with

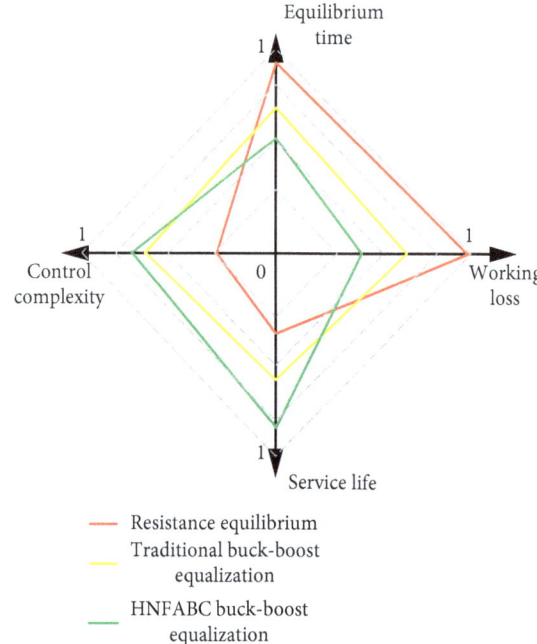

FIGURE 12: Multidimensional comparison equalization strategy (p.u.).

abnormal energy in the battery pack is arbitrary. This design only transfers the energy of one abnormal battery to the remaining batteries in the battery pack. Future research will be more focused on the actual circumstances, to design and to propose a multicell abnormal energy balance control method that is more suitable for engineering practice.

Conflicts of Interest

The authors declare that they have no conflicts of interest.

Authors' Contributions

Shuailong Dai proposed a novel battery pack equalization system, aiming to achieve the balancing process under hybrid natural and forced active balancing control and set up the experimental platform and conducted the experiments. Jiayu Wang and Zhifei Shan designed and implemented the balancing circuit. Jie Min drafted the manuscript. Yewen Wei finalized and polished the manuscript.

Acknowledgments

This research work was supported by the Excellent Young and Middle-aged Science and Technology Innovation Team of Hubei Education Department T201504 and National Innovation and Entrepreneurship Training Program for College Students 201711075009.

References

[1] K. Yang and A. Walid, "Outage storage tradeoff in frequency regulation for smart grid with renewables," *IEEE Transactions on Smart Grid*, vol. 4, no. 1, pp. 245–252, 2013.

[2] T. Masuta and A. Yokoyama, "Supplementary load frequency control by use of a number of both electric vehicles and heat pump water heaters," *IEEE Transactions on Smart Grid*, vol. 3, no. 3, pp. 1253–1262, 2012.

[3] J. Escudero-Garzas, A. Garcia-Armada, and G. Seco-Granados, "Fair design of plug-in electric vehicles aggregator for V2G regulation," *IEEE Transactions on Vehicular Technology*, vol. 61, no. 8, pp. 3406–3419, 2012.

[4] A. Khaligh and Z. Li, "Battery ultracapacitor fuel cell and hybrid energy storage systems for electric hybrid electric fuel cell and plug-in hybrid electric vehicles: state of the art," *IEEE Transactions on Vehicular Technology*, vol. 59, no. 6, pp. 2806–2814, 2010.

[5] J. C. Gómez and M. M. Morcos, "Impact of EV battery chargers on the power quality of distribution systems," *IEEE Transactions on Power Delivery*, vol. 18, no. 3, pp. 975–981, 2003.

[6] S. Han, S. Han, and K. Sezaki, "Development of an optimal vehicle-to-grid aggregator for frequency regulation," *IEEE Transactions on Smart Grid*, vol. 1, no. 1, pp. 65–72, 2010.

[7] E. Inoa and J. Wang, "PHEV charging strategies for maximized energy saving," *IEEE Transactions on Vehicular Technology*, vol. 60, no. 7, pp. 2978–2986, 2011.

[8] L. Ni, D. J. Patterson, and J. L. Hudgins, "High power current sensorless bidirectional 16-phase interleaved DC-DC converter for hybrid vehicle application," *IEEE Transactions on Power Electronics*, vol. 27, no. 3, pp. 1141–1151, 2012.

[9] S. W. Lee, K. M. Lee, Y. G. Choi, and B. Kang, "Modularized design of active charge equalizer for Li-ion battery pack," *IEEE Transactions on Industrial Electronics*, vol. 65, no. 11, pp. 8697–8706, 2018.

[10] A. M. Imtiaz and F. H. Khan, "Time shared a flyback converter based regenerative cell balancing technique for series connected Li-Ion battery strings," *IEEE Transaction on Power Electronics*, vol. 28, no. 12, pp. 5960–5975, 2013.

[11] C. S. Lim, K. J. Lee, N. J. Ku, D. S. Hyun, and R. Y. Kim, "A modularized equalization method based on magnetizing energy for a series-connected lithium-ion battery string," *IEEE Transactions on Power Electronics*, vol. 29, no. 4, pp. 1791–1799, 2014.

[12] W. Huang and J. A. A. Qahouq, "Energy sharing control scheme for state-of-charge balancing of distributed battery energy storage system," *IEEE Transactions on Industrial Electronics*, vol. 62, no. 5, pp. 2764–2776, 2015.

[13] Y. Wei, Y. Li, B. Cao, B. Zhu, and G. Liu, "Research on power equalization of lithium-ion batteries with less-loss buck chopper," *Transactions of China Electrotechnical Society*, vol. 33, no. 11, pp. 2575–2583, 2018.

[14] G. L. Plett, "Extended kalman filtering for battery management systems of LiPB-based hev battery packs: part 3. state and parameter estimation," *Journal of Power Sources*, vol. 134, no. 2, pp. 277–292, 2004.

[15] G. L. Plett, "Sigma-point kalman filtering for battery management systems of LiPB-based hev battery packs: Part 2: simultaneous state and parameter estimation," *Journal of Power Sources*, vol. 161, no. 2, pp. 1369–1384, 2006.

[16] S. Santhanagopalan and R. E. White, "Online estimation of the state of charge of a lithium ion cell," *Journal of Power Sources*, vol. 161, no. 2, pp. 1346–1355, 2006.

[17] S. Santhanagopalan and R. E. White, "State of charge estimation using an unscented filter for high power lithium ion cells," *International Journal of Energy Research*, vol. 34, no. 2, pp. 152–163, 2010.

[18] D. D. Domenico, A. Stefanopoulou, and G. Fiengo, "Lithium-ion battery state of charge and critical surface charge estimation using an electrochemical model-based extended kalman filter," *Journal of Dynamic Systems, Measurement, and Control*, vol. 132, no. 6, article 061302, 2010.

[19] X. Hu, F. Sun, and Y. Zou, "Comparison between two model-based algorithms for li-ion battery soc estimation in electric vehicles," *Simulation Modelling Practice and Theory*, vol. 34, pp. 1–11, 2013.

[20] J. Han, D. Kim, and M. Sunwoo, "State-of-charge estimation of lead-acid batteries using an adaptive extended kalman filter," *Journal of Power Sources*, vol. 188, no. 2, pp. 606–612, 2009.

[21] F. Sun, X. Hu, Y. Zou, and S. Li, "Adaptive unscented kalman filtering for state of charge estimation of a lithium-ion battery for electric vehicles," *Energy*, vol. 36, no. 5, pp. 3531–3540, 2011.

[22] T. Kim, Y. Wang, H. Fang et al., "Model-based condition monitoring for lithium-ion batteries," *Journal of Power Sources*, vol. 295, pp. 16–27, 2015.

[23] I.-S. Kim, "Nonlinear state of charge estimator for hybrid electric vehicle battery," *IEEE Transactions on Power Electronics*, vol. 23, no. 4, pp. 2027–2034, 2008.

[24] S. Moura, N. Chaturvedi, and M. Krstic, "PDE estimation techniques for advanced battery management systems-part i: SOC estimation," in *Proceedings of American Control Conference (ACC), 2012*, pp. 559–565, Montreal, Canada, June 2012.

[25] Y. Wang, H. Fang, Z. Sahinoglu, T. Wada, and S. Hara, "Adaptive estimation of the state of charge for lithium-ion batteries: nonlinear geometric observer approach," *IEEE Transactions on Control Systems Technology*, vol. 23, no. 3, pp. 948–962, 2015.

[26] S. Dey, B. Ayalew, and P. Pisu, "Nonlinear robust observers for state-of charge estimation of lithium-ion cells based on a reduced electrochemical model," *IEEE Transactions on Control Systems Technology*, vol. 23, no. 5, pp. 1935–1942, 2015.

[27] X. Hu, D. Cao, and E. Bo, "Condition monitoring in advanced battery management systems: moving horizon estimation using a reduced electrochemical model," *IEEE/ASME Transactions on Mechatronics*, vol. 23, no. 1, pp. 167–178, 2018.

[28] S. Dai, W. Yewen, X. Zhang et al., "Low-loss power equalization technique for Li-ion battery," *Battery Bimonthly*, vol. 48, no. 3, pp. 175–178, 2018.

[29] J. Li and J. Jiang, "Active capacitor voltage balancing methods based on dynamic model for five-level nested neutral point piloted converter," *IEEE Transactions on Power Electronics*, vol. 33, no. 8, pp. 6567–6581, 2018.

[30] D. Yang, N. Wu, L. Yin, and Z. Lu, "Natural frame control of single-phase cascaded H-bridge multilevel converter based on fictive-phases construction," *IEEE Transactions on Industrial Electronics*, vol. 65, no. 5, pp. 3848–3857, 2017.

[31] F. Mestrallet, L. Kerachev, J. C. Crebier, and A. Collet, "Multiphase interleaved converter for lithium battery active balancing," *IEEE Transactions on Power Electronics*, vol. 29, no. 6, pp. 2874–2881, 2014.

[32] V. Yuhimenko, G. Geula, G. Agranovich, M. Averbukh, and A. Kuperman, "Average modeling and performance analysis of voltage sensorless active supercapacitor balancer with peak current protection," *IEEE Transactions on Power Electronics*, vol. 32, no. 2, pp. 1570–1578, 2016.

[33] C. Liu, W. Liu, L. Wang, G. Hu, L. Ma, and B. Ren, "A new method of modeling and state of charge estimation of the battery," *Journal of Power Sources*, vol. 320, pp. 1–12, 2016.

[34] X. Liu, W. Li, and A. Zhou, "PNGV equivalent circuit model and SOC estimation algorithm for lithium battery pack adopted in AGV vehicle," *IEEE Access*, vol. 6, pp. 23639–23647, 2018.

[35] X. Hu, R. Xiong, and B. Egardt, "Model-based dynamic power assessment of lithium-ion batteries considering different operating conditions," *IEEE Transactions on Industrial Informatics*, vol. 10, no. 3, pp. 1948–1959, 2014.

7

A Double Update PWM Method to Improve Robustness for the Deadbeat Current Controller in Three-Phase Grid-Connected System

Ling Yang (iD),[1] Yandong Chen (iD),[1] An Luo,[1] Kunshan Huai,[2] Leming Zhou,[1] Xiaoping Zhou,[1] Wenhua Wu,[1] Wenjuan Tan,[1] and Zhiwei Xie[1]

[1]National Electric Power Conversion and Control Engineering Technology Research Center, Hunan University, Changsha 410082, China
[2]Guangzhou Power Supply Bureau, Guangzhou 510620, China

Correspondence should be addressed to Yandong Chen; yandong_chen@hnu.edu.cn

Academic Editor: Luigi Piegari

In the grid-connected inverter based on the deadbeat current control, the filter inductance variation and single update PWM affect the distortion of the grid current, stability, and dynamic of the system. For this, a double update PWM method for the deadbeat current controller in three-phase grid-connected system is proposed, which not only effectively decreases the grid current distortion and control delay, but also improves the system stability and dynamic response speed due to reducing the characteristic root equation order of the closed-loop transfer function. The influence of the filter inductance deviation coefficient on the system performance is analyzed. As a conclusion, the corresponding filter inductance deviation coefficient in the system critical stability increases with increase in the parasitic resistance of the filter inductance and line equivalent resistance and decreases with increase in the sampling frequency. Considering the system stability and dynamic response, the optimal range of the control parameters is acquired. Simulation and experimental results verify the effectiveness of the proposed method.

1. Introduction

With the increasingly serious energy crisis and environmental pollution, renewable energy distributed generation technology has been widely concerned and researched in [1–4]. The grid-connected inverter is the core of the distributed generation system in [5–8]. Its role is to convert the DC power generated by renewable energy to the AC power accepted by the grid. The deadbeat current control is based on the mathematical model of the grid-connected inverter and depends on the actual electrical parameters of the main circuit. Theoretically, no static-state error can be achieved in [9–12]. Because of its fast current transient response, accurate current tracking characteristic, and all digital control, the deadbeat current control has been applied more often.

However, reliance on the accurate electrical model and control delay are the main constraints in the deadbeat control.

On the one hand, the filter inductance value cannot be accurately detected. Even with the increasing of the grid current, the magnetic flux of the filter inductance tends to be saturated, resulting in the decrease of the filter inductance. This will lead to a certain deviation between the model filter inductance and actual filter inductance, affect the control accuracy of the deadbeat control, and cause the grid current distortion in [13, 14]. On the other hand, the inherent delay of the sampling and calculation limits the maximum duty cycle of the grid-connected inverter. For this, the delay caused by the proposed single update PWM in [15] may increase poles of the open-loop pulse transfer function, which affects the stability and dynamic response speed of the system. Therefore, how to improve the stability and dynamic response speed of the system and decrease the distortion of the grid current has become the research focus and goal of the grid-connected inverter.

In [16], a current prediction control method for the inductance online identification is proposed, which can accurately identify the inductance value in static and dynamic processes. An online estimation method of the filter inductance parameter is proposed in [17], which can prevent the current phase difference and system instability when the filter inductance parameter is not matched by the traditional prediction deadbeat current control method. However, the delay caused by the single update PWM has not been considered in the above methods.

The predictive current control technology is proposed to compensate the delay in [15, 18–20]. A supply current predictive controller that adopts the variable step-size adaptive algorithm is proposed in [15], which solves the problem that the maximum duty is limited by the delay caused by the sampling and calculation. Because the weighted coefficient is regulated by the error calculated in each sampling period, this algorithm shows well predictive accuracy. But a single voltage vector is only used in a control cycle, so it needs very high sampling frequency to get good performance [18]. In [19, 20], the linear prediction method is proposed, which can estimate the information of the next beat based on the control object model and the information of the current and past period. Although it eliminates the control delay, this method depends on the accuracy of the model, which may exist a certain estimation error.

At the same time, the current observer is used to predict the next beat current in [10, 21, 22], and the current reference is predicted in advance to compensate for the delay. In [10, 21], the current predictive algorithm based on repetitive control observer is proposed, which solves the problem of the current instability caused by the control delay and improves the current predictive precision. However, this method does not involve the dynamic analysis. In [22], an improved deadbeat current control scheme with a novel adaptive self-tuning load model is proposed. An improved deadbeat current controller with delay compensation is used to achieve high bandwidth current control characteristic, which compensates for the delay of the total system. However, a single prediction algorithm is adopted in the current reference variation, which may cause a large current overshoot or phase lag and deteriorate the system performance.

Digital PWM in double update mode is used to reduce the modulation delays and achieve performance. In [23], the effects of the measurement sampling and PWM updating methods on PI-based current control performance have been studied for a converter system. Reference [24] proposed proportional resonant controller implementation with double update mode digital PWM for single-phase grid-connected inverter. Digitally controlled grid-connected inverters with converter current control scheme and converter current plus grid current control scheme have been studied in [25]. A combination of two samples' time displacement and the line current PWM ripple was proposed to cancel that error and boost the performance of such drives in [26]. Reference [27] demonstrated the improved performance of a three-phase voltage source inverter when digital multi-sampled space vector modulation was used. The bandwidth expansion strategy was proposed to achieve the stator current double

sampling and PWM duty cycle double update in a carrier period in [28]. However, the deadbeat current control has not been considered in the above methods. A fast robust PWM method for photovoltaic grid-connected inverter is proposed in [29], which effectively solves the delay of the one-step-delay control and improves the system stability.

In this paper, the double update PWM method to improve robustness for the deadbeat current controller in three-phase grid-connected system is proposed. The paper is organized as follows. Section 2 presents the structure and control method for three-phase grid-connected system. The double update PWM method for the deadbeat current controller is proposed in Section 3. Finally, simulations and experiments are illustrated and discussed in Sections 4 and 5. Some conclusions are given in Section 6.

2. Control Method for Three-Phase Grid-Connected System

Structure of photovoltaic grid-connected system is shown in Figure 1, including the photovoltaic array, inverter circuit, and LC filter. C_{dc} is the DC side storage capacitance, which is used to stabilize the DC voltage U_{dc}. The power transistors Q_1-Q_6 constitute a three-phase full bridge inverter circuit that converts the DC voltage U_{dc} into the AC output voltage, which is the same as the amplitude and phase of the grid voltage. LC filter is formed by the inductance L and capacitance C. In the grid-connected mode, LC filter is equivalent to a single filter inductance. The resistance r is the sum of the parasitic resistance of the filter inductance L and line equivalent resistance. I_{dc} is the DC current. i_{invj} (j=a,b,c) is the inverter output current. i_{gj} is the grid current. Since the current flowing through the filter capacitance C is small, i_{invj} is approximately equal to i_{gj}.

Diagram of the double update PWM method for the deadbeat current controller is shown in Figure 2, including the double closed-loop control and double update PWM. The outer voltage loop adopts the PI control to stabilize U_{dc}. The inner current loop adopts the deadbeat control, which decreases the distortion of the grid current caused by the filter inductance variation. The double update PWM effectively solves the delay caused by the single update PWM and improves the stability and dynamic response speed of the system. U_{dcr} is the DC voltage command, u_{gj} is the grid voltage, ω_1 is the fundamental angular frequency of the grid, I_{gr} is the input amplitude command of the inner current loop, i_{gjr} is the grid current command, and D_j is the equivalent duty cycle.

3. The Double Update PWM Method for the Deadbeat Current Controller

3.1. The Single Update PWM Method. In digital control, the single update PWM is shown in Figure 3, where u_{ms} is the single update PWM wave. The sampling is carried out at the peak of the (k-1)th triangular carrier. The calculation is based on the sampling value. At the peak of the kth triangular carrier, the single update PWM wave $u_{ms}(k$-1) of the (k-1)th

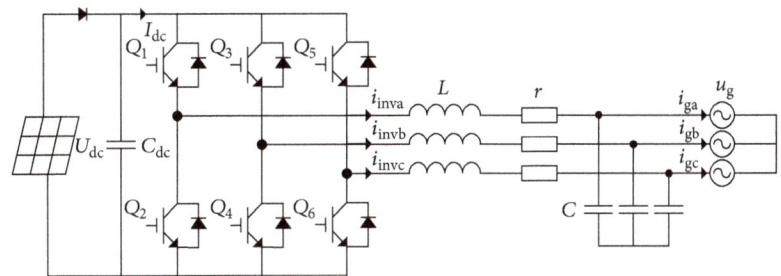

FIGURE 1: Structure of photovoltaic grid-connected system.

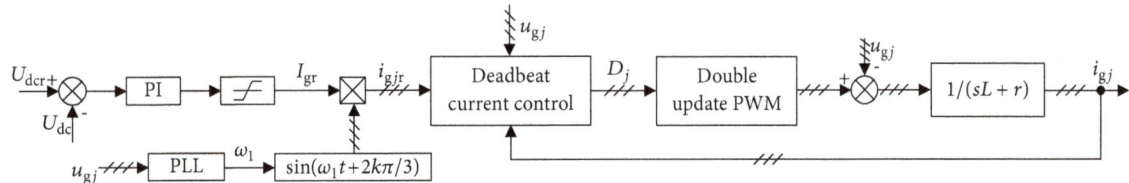

FIGURE 2: Diagram of the double update PWM method for photovoltaic grid-connected inverter based on the deadbeat control.

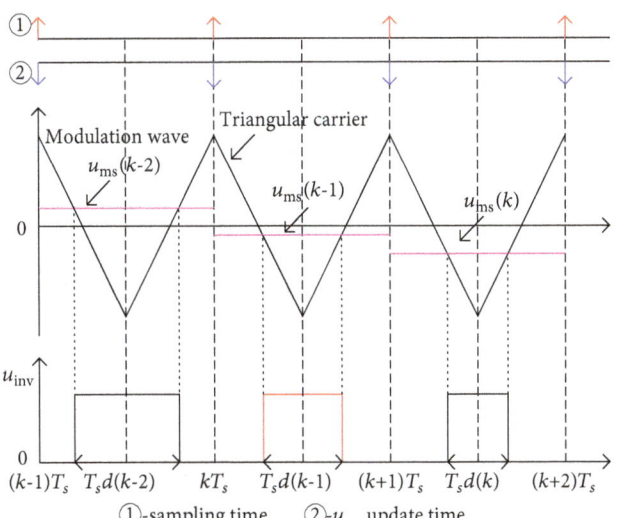

FIGURE 3: The single update PWM.

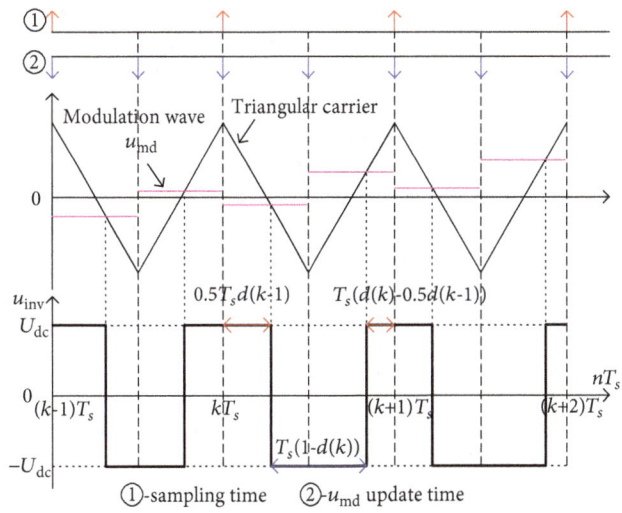

FIGURE 4: The double update PWM method.

carrier cycle is loaded. So the equivalent duty cycle $D(k)$ of the kth carrier cycle can be expressed as

$$D(k) = d(k-1) \tag{1}$$

where $d(k-1)$ is the duty cycle calculated by the sampling values at the peak of the $(k-1)$th triangular carrier.

Obviously, the loading time of the single update PWM wave is lagged behind a sampling period at the beginning of the sampling. So the one-step delay is caused by the single update PWM.

3.2. The Double Update PWM Method.

The double update PWM method is shown in Figure 4, where u_{md} is the double update PWM wave. The sampling is carried out at the peak of the triangular carrier. The loading is carried out at the

peak and valley of the triangular carrier. That is to say, one sampling and double loading are taken in each carrier cycle. The modulation process is as follows: the sampling is carried out at the peak of the kth triangular carrier. Meanwhile, the double update PWM wave corresponding to the $d(k-1)$ is loaded, so that the conduction time of the first half of the carrier cycle is $0.5T_s d(k-1)$. At the valley of the kth triangular carrier, the double update PWM wave corresponding to the difference calculated by subtracting the $d(k-1)$ from the $2d(k)$ is loaded, so that the conduction time of the second half of the carrier cycle is $T_s(d(k)-0.5d(k-1))$. So the equivalent duty cycle $D(k)$ of the kth carrier cycle can be expressed as

$$D(k) = \frac{0.5T_s d(k-1) + T_s(d(k) - 0.5d(k-1))}{T_s}$$

$$= d(k) \tag{2}$$

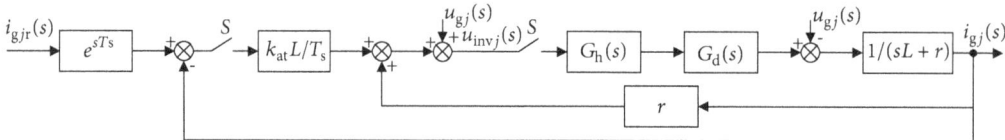

FIGURE 5: Diagram of the deadbeat current control for the grid-connected inverter.

TABLE 1: The delay comparison between the single update PWM and proposed method.

Serial number	$G_d(s)$	Method
Case 1	e^{-sT_s}	Single update PWM
Case 2	1	Proposed(double update PWM)

where $d(k)$ is the duty cycle calculated by the sampling values at the peak of the kth triangular carrier.

Therefore, the proposed double update PWM method eliminates the control delay. Meanwhile, the time margin of the sampling and calculation decreases from T_s to $T_s/2$.

Diagram of the deadbeat current control for the grid-connected inverter is shown in Figure 5, where $i_{gjr}(s)$ is the input quantity, $u_{gj}(s)$ is the disturbance input quantity, $i_{gj}(s)$ is the output quantity, T_s is the sampling cycle, and S is the synchronous sampling switch. With the increasing of the grid current, the magnetic flux of the filter inductance tends to be saturated, resulting in the decrease of the filter inductance. There is the deviation between the filter inductance L_1 and actual filter inductance L. k_{at} is the filter inductance deviation coefficient. $k_{at}=L_1/L$. $G_h(s)$ is the continuous domain transfer function of the zero order holder (ZOH).

$$G_h(s) = \frac{\left(1 - e^{-sT_s}\right)}{s} \tag{3}$$

$G_d(s)$ is the continuous domain transfer function of the single update PWM, which can be expressed as

$$G_d(s) = e^{-sT_s} \tag{4}$$

The delay comparison between the single update PWM and proposed method is shown in Table 1. Considering the parasitic resistance of the filter inductance and line equivalent resistance, the influence of the control delay on the stability and dynamic of the system is analyzed, and the appropriate control parameter range is given.

3.3. Control System Stability Analysis. In Figure 5, using the single update PWM, $G_d(s)=e^{-sT_s}$. $i_{gj}^*(s)$ is Laplace transform of $i_{gj}(s)$, which can be expressed as

$$i_{gj}^*(s) = \frac{k_{at}(L/T_s) \cdot G_h G_d G_L^*(s)}{1 + (k_{at}L/T_s - r) \cdot G_h G_d G_L^*(s)} \cdot e^{sT_s} \cdot i_{gjr}^*(s) \tag{5}$$

where $G_L(s)$ is the transfer function of the filter, $G_L(s)=sL+r$, $G_h G_d G_L^*(s)$ is Laplace transform of $G_h(s)G_d(s)G_L(s)$, and $i_{gjr}^*(s)$ is Laplace transform of the input quantity $i_{gjr}(s)$. By

substituting $z=e^{-sT_s}$ into (5), the closed-loop pulse transfer function in the z domain is obtained using the single update PWM.

$$\Phi(z) = \frac{k_{at}(L/T_s)\left(1 - e^{-T_s r/L}\right) \cdot z}{r\left(z - e^{-T_s r/L}\right)z + \left(k_{at}L/T_s - r\right)\left(1 - e^{-T_s r/L}\right)} \tag{6}$$

By substituting $z=(\omega+1)/(\omega-1)$ into (6), the characteristic root equation of the closed-loop transfer function of the system can be obtained as

$$A\omega^2 + B\omega + C = 0$$

$$A = k_{at}\left(\frac{L}{T_s}\right)\left(1 - e^{-T_s r/L}\right)$$

$$B = 2r - 2\left(k_{at}\frac{L}{T_s} - r\right)\left(1 - e^{-T_s r/L}\right) \tag{7}$$

$$C = r\left(1 + e^{-T_s r/L}\right)$$

$$+ \left(k_{at}\frac{L}{T_s} - r\right)\left(1 - e^{-T_s r/L}\right)$$

where ω is transform operator from the z domain to the ω domain.

Based on the Routh criterion [30], the range of the filter inductance deviation coefficient k_{at} can be expressed as

$$0 < k_{at} < \frac{r\left(2 - e^{-T_s r/L}\right)}{1 - e^{-T_s r/L}}\frac{T_s}{L} \tag{8}$$

In Figure 5, using the proposed method, $G_d(s)=1$. The closed-loop pulse transfer function in the z domain can be expressed as

$$\Phi(z) = \frac{k_{at}(L/T_s)\left(1 - e^{-T_s r/L}\right) \cdot z}{r\left(z - e^{-T_s r/L}\right) + \left(k_{at}L/T_s - r\right)\left(1 - e^{-T_s r/L}\right)} \tag{9}$$

In (6) and (9), compared to the single update PWM, the characteristic root equation order of the closed-loop transfer function of the system is reduced when the proposed method is used.

When $|z| <1$, the system is stable and the range of the filter inductance deviation coefficient k_{at} can be expressed as

$$0 < k_{at} < \frac{2r}{1 - e^{-T_s r/L}}\frac{T_s}{L} \tag{10}$$

The corresponding filter inductance deviation coefficient in the system critical stability $k_{at\_critical}$ can be expressed as

$$k_{at\_critical} = \frac{2r}{1 - e^{-T_s r/L}}\frac{T_s}{L} \tag{11}$$

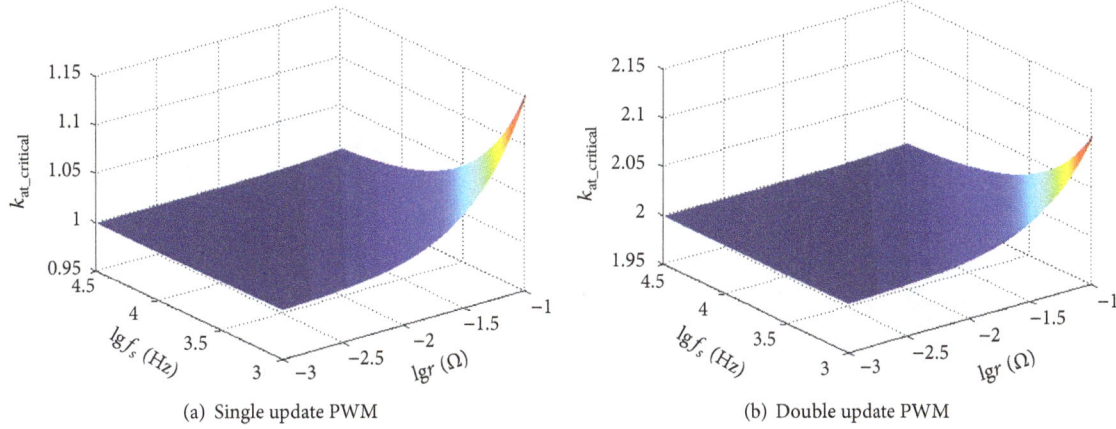

(a) Single update PWM

(b) Double update PWM

FIGURE 6: Relationship among the resistance r, sampling frequency f_s, and $k_{at\_critical}$.

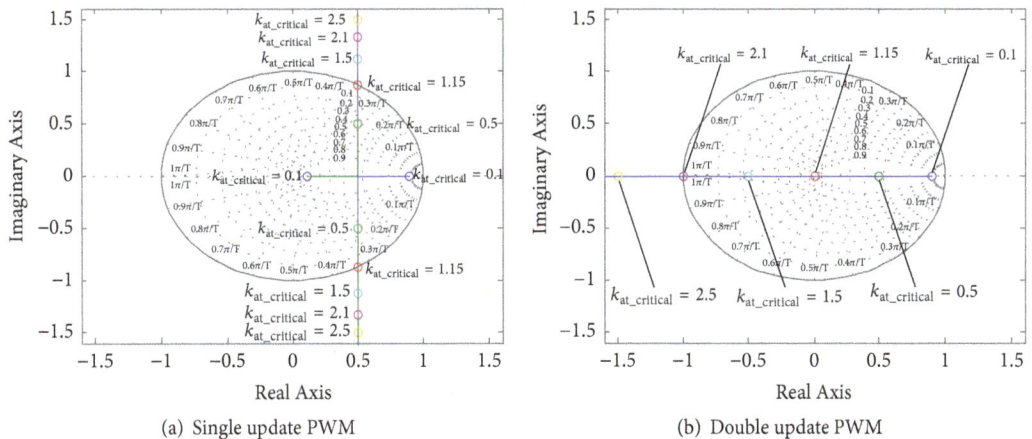

(a) Single update PWM

(b) Double update PWM

FIGURE 7: The contrast analysis of root locus with different $k_{at\_critical}$.

Relationship among the resistance r, sampling frequency f_s, and $k_{at\_critical}$ by the single update PWM and proposed method is shown in Figure 6. Figures 6(a) and 6(b) correspond to the single update PWM and proposed method, respectively. When f_s is constant, $k_{at\_critical}$ increases with increase in r. When r is constant, $k_{at\_critical}$ decreases with increase in f_s. When the proposed method is used, the range of $k_{at\_critical}$ is $0 < k_{at\_critical} \leq 2.1$, while the range of $k_{at\_critical}$ is $0 < k_{at\_critical} \leq 1.15$ by using the single update PWM. So the range of $k_{at\_critical}$ by using the proposed method is obviously larger than the range of $k_{at\_critical}$ by using the single update PWM.

When $k_{at\_critical}$=0.1, 0.5, 1.15, 1.5, 2.1, and 2.5, the root locus analysis with the single update PWM and proposed method is shown in Figure 7. Figures 7(a) and 7(b) correspond to the single update PWM and proposed method, respectively. In Figure 7(a), when $k_{at\_critical}$=0.1 and 0.5, the pole is inside the unit circle, and the system is in a stable state. When $k_{at\_critical}$=1.15, the pole is on the unit circle, and the system is in a critical stable state. When $k_{at\_critical}$=2.1 and 2.5, the pole is outside the unit circle, and the system is in an unstable state.

In Figure 7(b), when $k_{at\_critical}$=0.1, 0.5, 1.15, and 1.5, the pole is inside the unit circle, and the system is in a stable state. When $k_{at\_critical}$=2.1, the pole is on the unit circle, and the system is in a critical stable state. When $k_{at\_critical}$=2.5, the pole is outside the unit circle, and the system is in an unstable state. So the range of $k_{at\_critical}$ by using the proposed method is obviously larger than the range of $k_{at\_critical}$ by using the single update PWM.

3.4. Control System Dynamic Analysis. When the single update PWM is used, system response to unit step change with $k_{at\_critical}$ changing is shown in Figure 8, while the sampling frequency f_s=10kHz and resistance r=0.01Ω are constant. Figures 8(a) and 8(b) correspond to $0 < k_{at\_critical} \leq 1$ and $1 < k_{at\_critical} \leq 1.15$, respectively. When $0 < k_{at\_critical} \leq 1$, the dynamic response of the system is convergent. When $1 < k_{at\_critical} \leq 1.15$, the dynamic response of the system is divergent. Therefore, the range of $k_{at\_critical}$ is $0 < k_{at\_critical} \leq 1$.

In Figure 8(c), the curves *A, B, C, D, E* correspond to $k_{at\_critical}$=0.7, 0.5, 0.3, 0.2, 0.1, respectively. In curve *C*, the system has the best dynamic response and no overshoot when $k_{at\_critical}$ is equal to $k_{at\_critical0}$ ($k_{at\_critical}$=$k_{at\_critical0}$=0.3). In

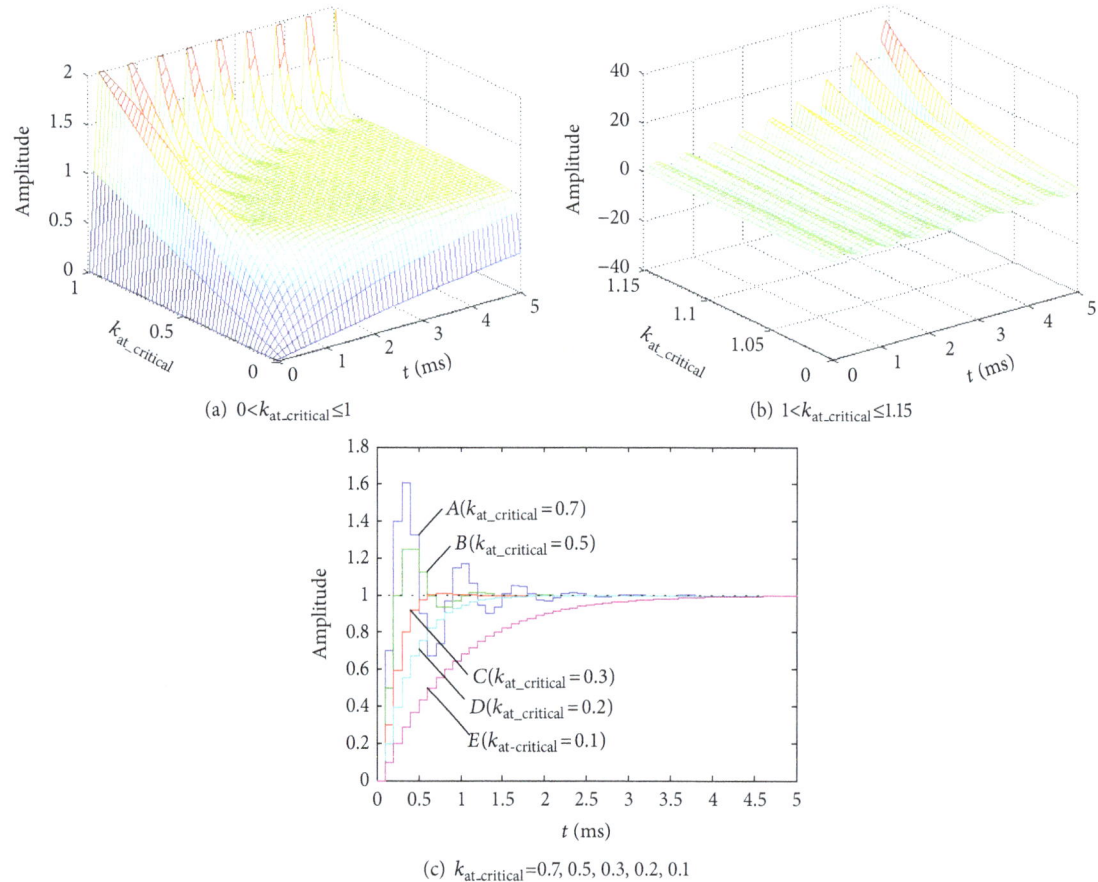

(a) $0<k_{at\_critical}\leq 1$

(b) $1<k_{at\_critical}\leq 1.15$

(c) $k_{at\_critical}=0.7, 0.5, 0.3, 0.2, 0.1$

FIGURE 8: Single update PWM system response to unit step change with the constant sampling frequency f_s=10kHz and resistance r=0.01 when $k_{at\_critical}$ changes.

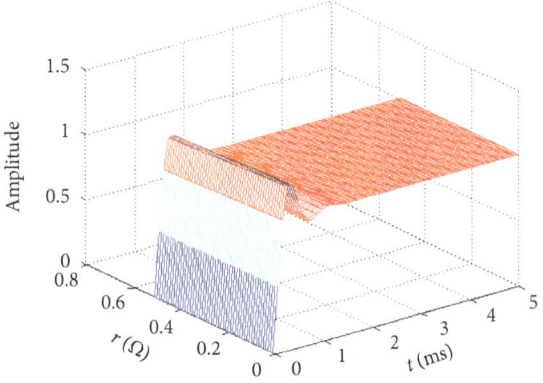

FIGURE 9: Single update PWM system response to unit step change with the constant sampling frequency f_s=10kHz and $k_{at\_critical}$=0.5 when the resistance r changes.

curves A and B, when $k_{at\_critical} > k_{at\_critical0}$, the closed-loop poles are conjugate complex poles located in the right half unit circle of the z plane, and the dynamic response of the system is the oscillation convergence pulse sequence. The smaller $k_{at\_critical}$ is, the closer poles are to the coordinate origin, the smaller the overshoot is, and the faster the

response speed is. In curves D and E, when $k_{at\_critical} < k_{at\_critical0}$, the closed-loop poles are located in the real axis of the right half unit circle of the z plane. The dynamic response of the system is unidirectional positive convergence pulse sequence and has no overshoot. The larger $k_{at\_critical}$ is, the closer poles are to the coordinate origin, and the faster the response speed is.

When the single update PWM is used, system response to unit step change with the resistance r changing is shown in Figure 9, while the sampling frequency f_s=10kHz and $k_{at\_critical}$=0.5 are constant. When the resistance r changes, the peak time and the adjustment time of the system do not change much.

When the proposed method is used, system response to unit step change with $k_{at\_critical}$ changing is shown in Figure 10, while the sampling frequency f_s=10kHz and resistance r=0.01Ω are constant. Figures 10(a) and 10(b) correspond to $0< k_{at\_critical} \leq 2$ and $2< k_{at\_critical} \leq 2.1$, respectively. When $0< k_{at\_critical} \leq 2$, the dynamic response of the system is convergent. When $2< k_{at\_critical} \leq 2.1$, the dynamic response of the system is divergent. Therefore, the range of $k_{at\_critical}$ is $0< k_{at\_critical} \leq 2$.

Contrast analysis of system response to unit step change between the single update PWM and proposed method is shown in Figure 11. Curves A and B correspond to

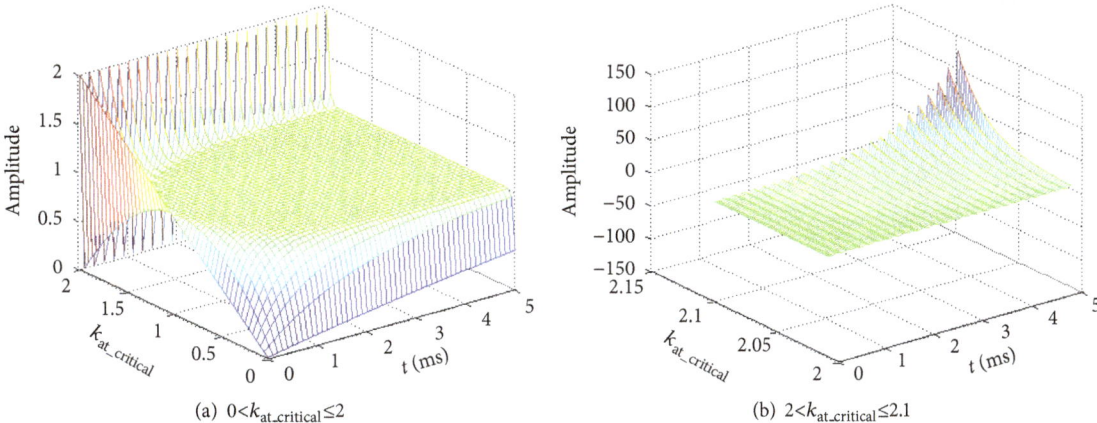

(a) $0<k_{at\_critical}\leq2$　　　　(b) $2<k_{at\_critical}\leq2.1$

FIGURE 10: Proposed control system response to unit step change with the constant sampling frequency f_s=10kHz and resistance r=0.01 when $k_{at\_critical}$ changes.

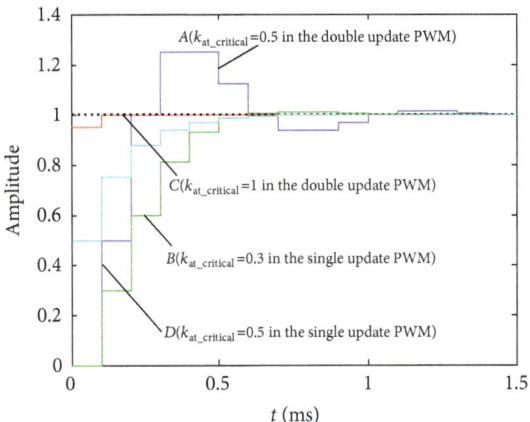

FIGURE 11: Contrast analysis of system response to unit step change between the single update PWM and proposed method.

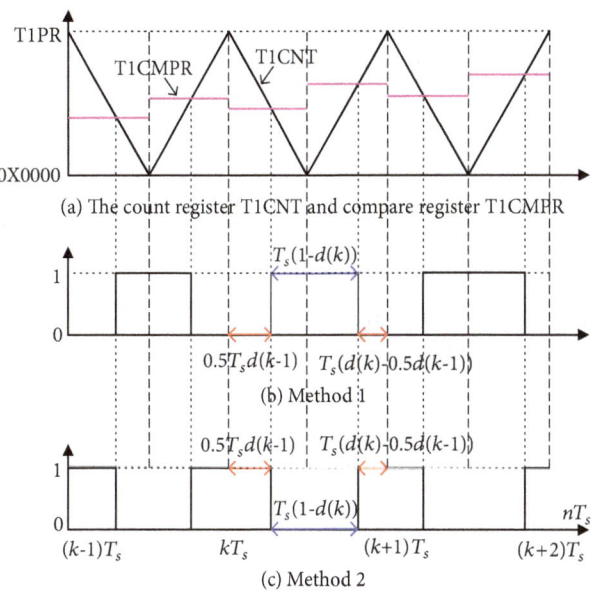

(a) The count register T1CNT and compare register T1CMPR

(b) Method 1

(c) Method 2

FIGURE 12: The implementation principle of the proposed modulation method based on the DSP2812 controller.

$k_{at\_critical}$=0.5 and 0.3 in the single update PWM, and curves C and D correspond to $k_{at\_critical}$=1 and 0.5 in the proposed method. Compared to curve D, the system has optimal response to unit step change, fast response speed, and no overshoot in curve C. Comparing curve C with curve B, the optimal response to unit step change of the system in curve C using the proposed method is superior. Comparing curve D with curve A, in curve A using the single update PWM, the unit step response of the system has overshoot, and the oscillation and adjustment time become longer, resulting in the poor dynamic of the system. In curve D using the proposed method, the unit step response of the system has no overshoot and the dynamic becomes better.

3.5. Project Implementation Design Method.

The implementation principle of the proposed modulation method based on the DSP2812 controller is shown in Figure 12. DSP adopts TI's TMS320F2812 chip, which is a 32-bit fixed point micro control unit (MCU) with a main frequency of up to 150MHz. And the selected switching frequency is up to 10kHz. The general-purpose timer T1 is set to operate in continuous up/down counting mode. When the value of the count register T1CNT is equal to the period register T1PR, the timer T1 has a cycle interrupt, which starts the AD sampling unit, and $(1-d(k-1))$T1PR is assigned to the compare register T1CMPR. When the value of the count register T1CNT is 0x0000, the timer T1 has an underflow interrupt. The time margin from the start of the sampling to the end of the calculation is $0.5T_s$. The duty cycle $d(k)$ is obtained, and $(1-2d(k)+d(k-1))$T1PR is assigned to the compare register T1CMPR. When the value of the count register T1CNT is equal to the compare register T1CMPR, the compare match event occurs, and the level of the pin T1PWM will jump.

Method 1. The compare output pin T1PWM of the timer is set to active high in the DSP2812 controller. The DSP pin output driver signal is shown in Figure 12(b). The drive signal is sent

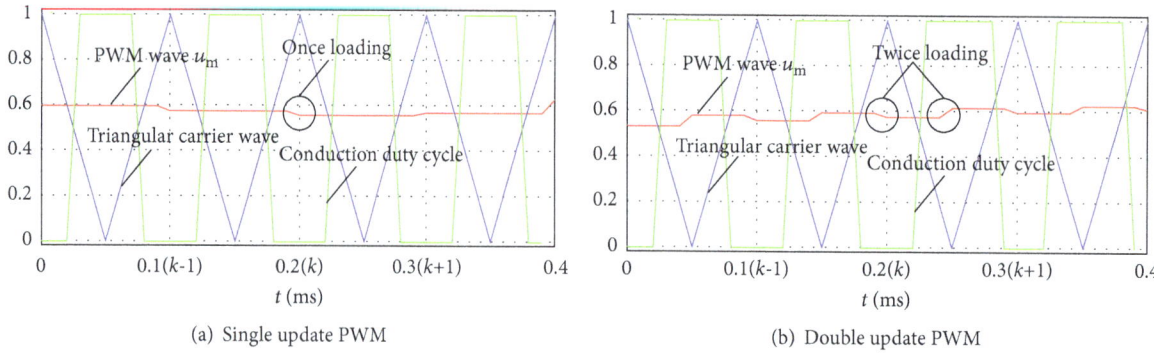

(a) Single update PWM   (b) Double update PWM

FIGURE 13: The simulation waveforms of the modulation wave and triangular carrier wave.

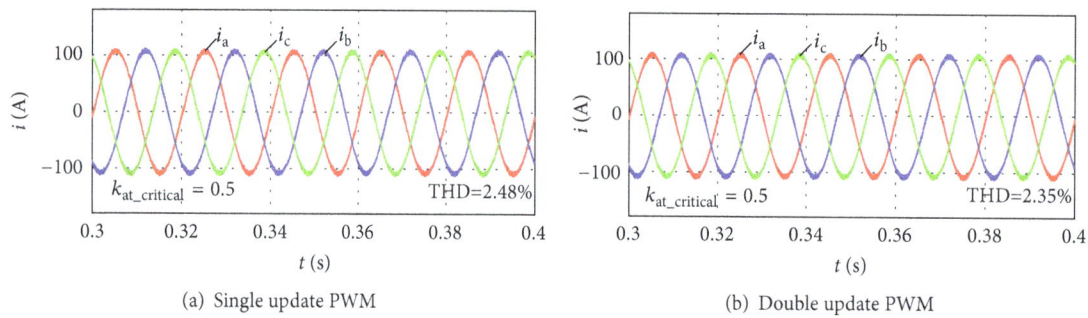

(a) Single update PWM   (b) Double update PWM

FIGURE 14: Steady state simulation waveforms of grid currents with $k_{at\_critical}$=0.5.

to the buffer for isolation and transformation. And then its level is reversed by the phase inverter. The final voltage signal is assigned to the corresponding IPM module.

Method 2. The compare output pin T1PWM of the timer is set to active low in the DSP2812 controller. The DSP pin output driver signal is shown in Figure 12(c). The drive signal is sent to the buffer for isolation and transformation. The final voltage signal is assigned to the corresponding IPM module.

4. Simulation Verification

In order to verify the effectiveness of proposed method, the simulation model of three-phase grid-connected inverter is built by using PSIM 9.0 based on Figure 1. The maximum output power of photovoltaic array is 50kW. System parameters are shown in Table 2.

In order to verify the implementation process of the proposed method, the simulation waveforms of the modulation wave and triangular carrier wave are shown in Figure 13. When the single update PWM and proposed method are used, the modulation processes are similar to the description of Figures 3 and 4. The correctness of the theoretical analysis is verified.

Steady state simulation waveforms of grid currents with $k_{at\_critical}$=0.5 and $k_{at\_critical}$=2 are shown in Figures 14 and 15, respectively. And the steady state simulation results of grid currents are shown in Table 3. In Figure 14(a), when the single update PWM is used with $k_{at\_critical}$=0.5, THD of the grid current is 2.48%. Meanwhile, in Figure 14(b), when the

TABLE 2: System parameters.

Parameters and units	Values
DC Voltage U_{dc} [V]	700
DC side storage capacitance C_{dc} [μF]	6000
Filter inductance L [mH]	1.0
Resistance r [Ω]	0.01
Outer voltage loop k_p, k_i	1.5, 0.3
Fundamental frequency f_1 [Hz]	50
Sampling frequency f_s [kHz]	10
Switching frequency f_{sw} [kHz]	10
Carrier frequency f_{tri} [kHz]	10

TABLE 3: The steady state simulation results of grid currents.

Method	$k_{at\_critical}$	THD
Single update PWM	0.5	2.48%
Proposed method	0.5	2.35%
Single update PWM	2	unstable
Proposed method	2	2.95%

proposed method is used with $k_{at\_critical}$=0.5, THD of the grid current is 2.35%. Under the two modulation methods, the system is in a stable state.

In Figure 15(a), when the single update PWM is used with $k_{at\_critical}$=2, the system is in an unstable state. However, in Figure 15(b), when the proposed method is used with $k_{at\_critical}$=2, THD of the grid current is 2.95%, and the system

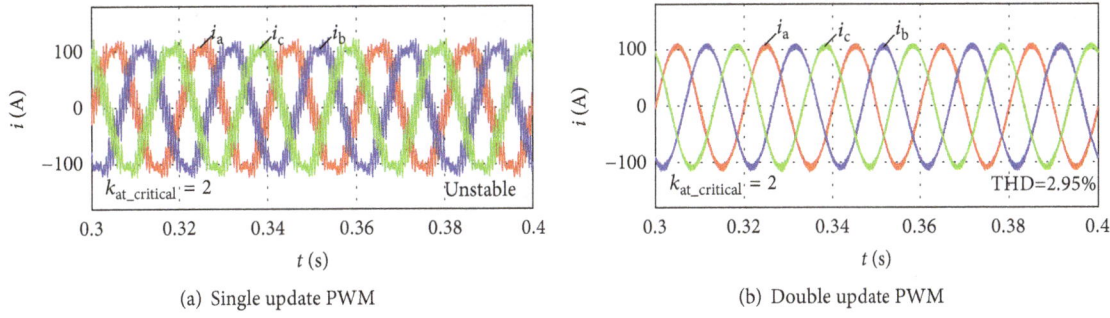

(a) Single update PWM   (b) Double update PWM

FIGURE 15: Steady state simulation waveforms of grid currents with $k_{at\_critical}$=2.

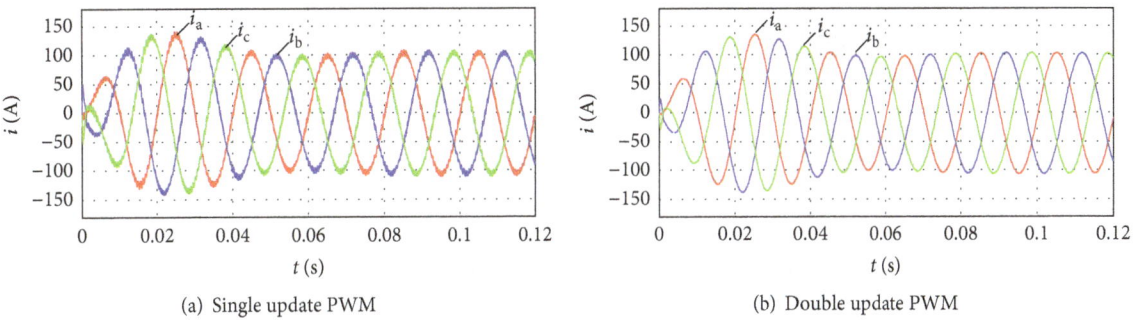

(a) Single update PWM   (b) Double update PWM

FIGURE 16: The dynamic simulation waveforms of grid currents.

TABLE 4: The dynamic simulation results of grid currents.

Method	Grid inrush current	THD
Single update PWM	50A	2.7%
Proposed method	25A	1.8%

is in a stable state. Therefore, if the value of $k_{at\_critical}$ exceeds its critical range with using the single update PWM, the instability will occur. But when this parameter is applied to the proposed method, the system is still in a stable state. So the range of $k_{at\_critical}$ by using the proposed method is obviously larger than the range of $k_{at\_critical}$ by using the single update PWM.

The dynamic simulation waveforms of grid currents using the single update PWM and proposed method are shown in Figure 16. And the dynamic simulation results of grid currents are shown in Table 4. In Figure 16(a), the grid inrush current is 50A and THD is 2.7% during the start-up process when the single update PWM is used. However, the grid inrush current can be decreased to 25A and THD can be reduced to 1.8% during the start-up process by using the proposed method, as shown in Figure 16(b).

5. Experimentation Verification

In order to verify the validity of simulation results, the experimental platform of three-phase grid-connected system is built shown in Figure 17, which includes the main circuit, the control board, and the LC filter circuit. DSP adopts TI's TMS320F2812 chip, and IGBT selects Infineon

FIGURE 17: Experimental platform of three-phase grid-connected system.

FF300R12ME4 module. The system parameters are shown as in Table 2.

By real-time debugging and setting breakpoints for the proposed method in CCStudio V3.3, the time spent on the sampling is 11.30μs, the time consumed by the outer voltage loop is 11.80μs, and the time required for the inner current loop is 7.60μs. The total time spent is 30.70μs. The selected switching frequency is up to 10kHz, and the switching period is 100.00μs. Therefore, the time consumed by the sampling

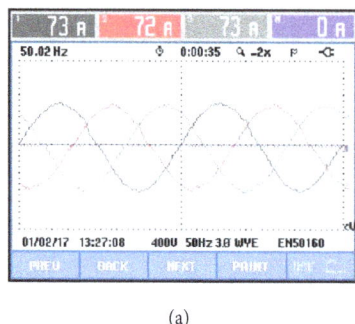 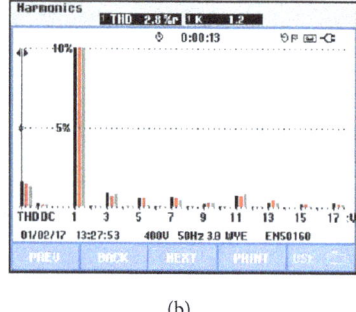

(a)   (b)

FIGURE 18: The experimental waveforms and THD of grid currents using the single update PWM ($k_{at\_critical}$=0.5).

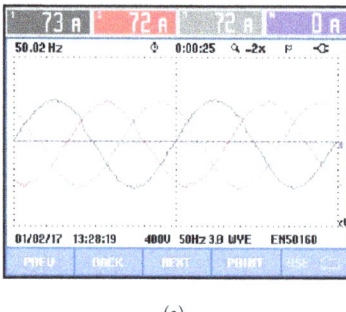 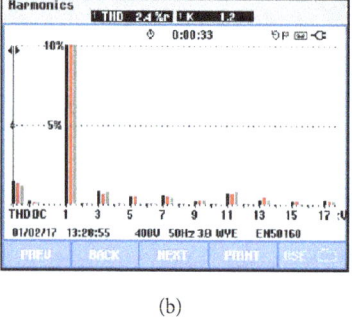

(a)   (b)

FIGURE 19: The experimental waveforms and THD of grid currents using the proposed method ($k_{at\_critical}$=0.5).

TABLE 5: The steady state experimental results of grid currents.

Method	$k_{at\_critical}$	THD
Single update PWM	0.5	2.8%
Proposed method	0.5	2.4%
Proposed method	1	2.1%
Proposed method	1.5	2.2%

TABLE 6: The dynamic experimental results of grid currents.

Method	$k_{at\_critical}$	Adjustment time	Overshoot
Single update PWM	0.5	75ms	yes
Single update PWM	0.3	55ms	yes
Proposed method	1	15ms	no
Proposed method	0.5	35ms	no

and calculation is only 30.70% of the switching period, which is less than 0.5 times the switching period. The sampling and calculation can be completed within 0.5 times the switching period.

The steady state experimental results of grid currents are shown in Table 5. And the experimental waveforms and THD of grid currents using the single update PWM and proposed method are shown in Figures 18 and 19, respectively. The theoretical root mean square (RMS) values of the grid currents are 50kW/(3×220V)≈75.76A. Considering some losses, the actual RMS values of grid currents are less than 75.76A. There is a measurement error among three current clamps of FLUKE, so three display values are not equal. Under the same $k_{at\_critical}$=0.5 and control parameters, THD of the grid current using the single update PWM is 2.8%, while THD of the grid current using the proposed method is only 2.4%.

The experimental waveforms and THD of grid currents using the proposed method with $k_{at\_critical}$=1, 1.5 are shown in Figures 20 and 21, respectively. In Figures 19, 20, and 21, THD of the grid current changes from 2.4% to 2.1%, and the next

is from 2.1% to 2.2%, showing a tendency to increase firstly and then decrease with the inductance deviation coefficient $k_{at\_critical}$ increasing. Therefore, THD of the grid current is the lowest at $k_{at\_critical}$=1.

Steady state experimental waveforms of grid currents with the single update PWM and proposed method are depicted in Figure 22. In Figure 22(a), when $k_{at\_critical}$ changes from 0.5 to 2 with the single update PWM, the system becomes unstable. However, when $k_{at\_critical}$ changes from 0.5 to 2 with the proposed method, the system can still be in a stable state, as shown in Figure 22(b).

Dynamic experimental waveforms between the single update PWM and proposed method are compared as shown in Figure 23. Figures 23(a) and 23(b) correspond to $k_{at\_critical}$=0.5 and 0.3 in the single update PWM. Figures 23(c) and 23(d) correspond to $k_{at\_critical}$=1 and 0.5 in the proposed method. And the dynamic experimental results of grid currents are as shown in Table 6. The adjustment time is 75ms, 55ms, 15ms, and 35ms, respectively. When the above two individual methods are adopted with $k_{at\_critical}$=0.5, in

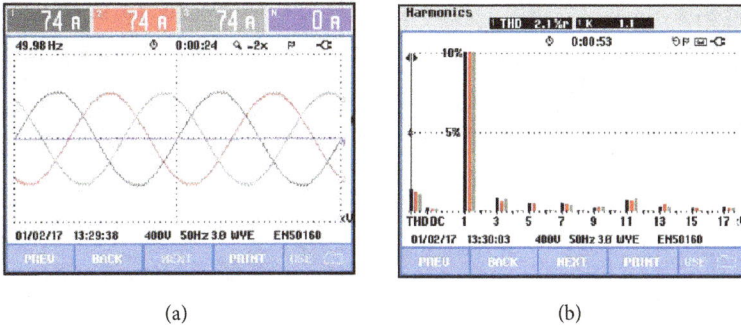

(a)   (b)

FIGURE 20: The experimental waveforms and THD of grid currents using the proposed method ($k_{at\_critical}$=1).

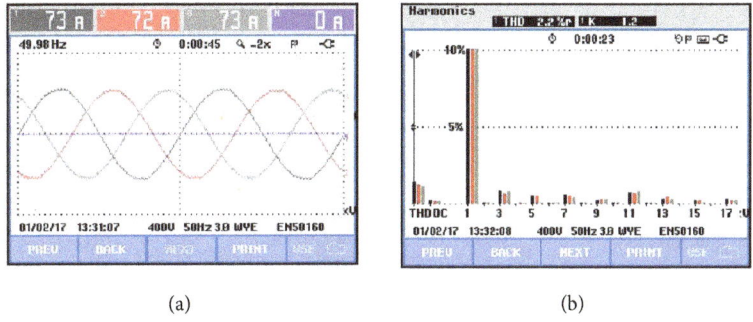

(a)   (b)

FIGURE 21: The experimental waveforms and THD of grid currents using the proposed method ($k_{at\_critical}$=1.5).

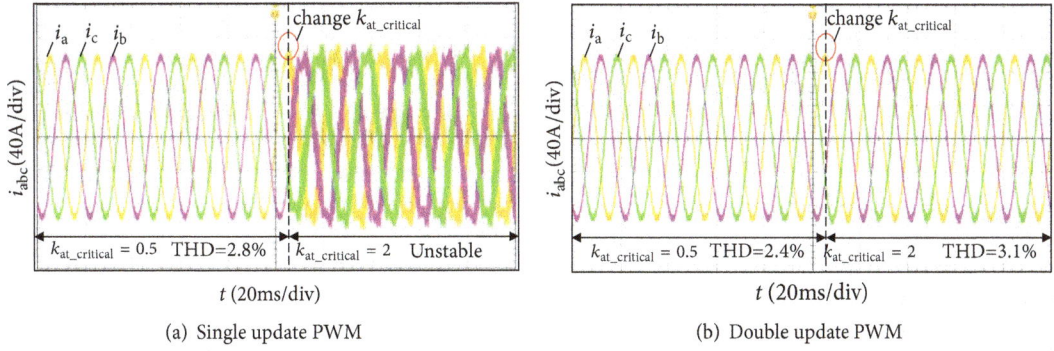

(a) Single update PWM   (b) Double update PWM

FIGURE 22: Steady state experimental waveforms of grid currents with the single update PWM and proposed method.

Figure 23(a), the system has overshoot, and the adjustment time becomes longer, resulting in the poor dynamic. However, in Figure 23(d), the system has no overshoot and the dynamic becomes better with using the proposed method. Comparing Figure 23(c) with Figure 23(d), the adjustment time of the former is only 15ms, while the latter is 35ms. Thus, the system has faster response speed when the proposed method is used with $k_{at\_critical}$=1. Comparing Figure 23(a) with Figure 23(b), the system has the better dynamic with $k_{at\_critical}$=0.3 rather than $k_{at\_critical}$=0.5 when the single update PWM is used. From Figures 23(c) and 23(b), it can be obtained that the optimal dynamic of the system is superior by using the proposed method. The experimental results prove the correctness of theoretical analysis for Figure 11.

6. Conclusion

In this paper, the proposed double update PWM method for the deadbeat current controller not only effectively decreases the grid current distortion and control delay, but also improves the system stability and dynamic response speed due to reducing the characteristic root equation order of the closed-loop transfer function. The influence of the filter inductance deviation coefficient on the system performance is analyzed. As a conclusion, the corresponding filter inductance deviation coefficient in the system critical stability increases with increase in the parasitic resistance of the filter inductance and line equivalent resistance and decreases with increase in the sampling frequency.

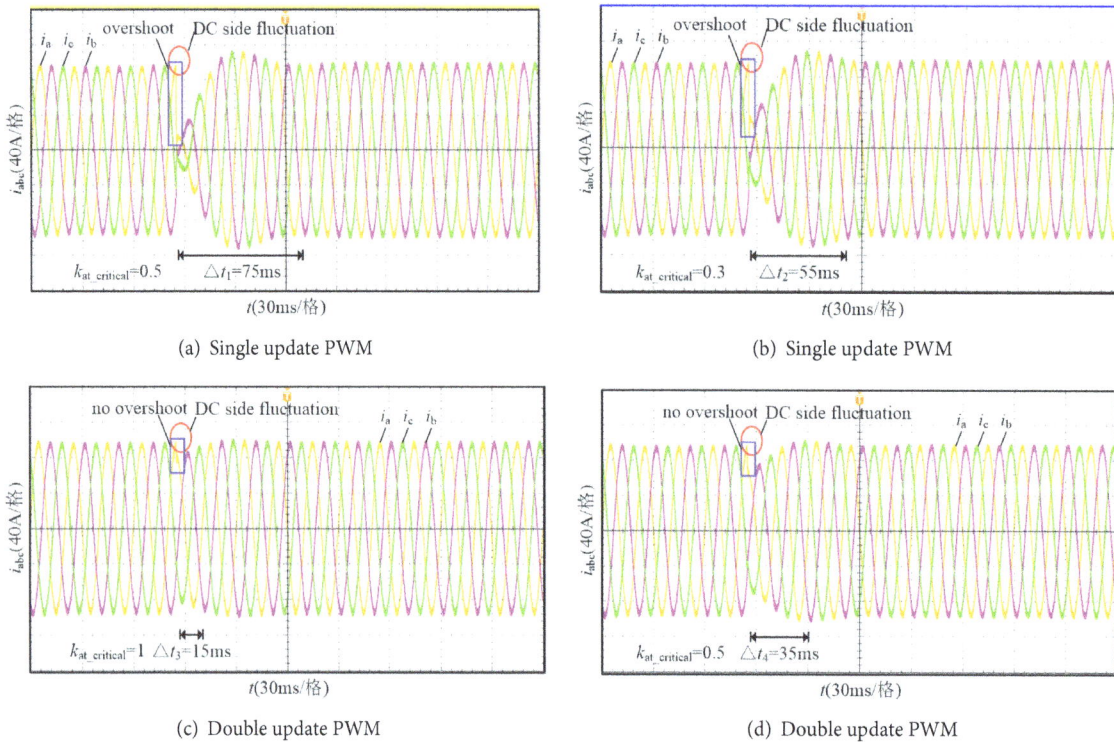

(a) Single update PWM

(b) Single update PWM

(c) Double update PWM

(d) Double update PWM

FIGURE 23: Contrast analysis of dynamic experimental waveforms between the single update PWM and proposed method.

Conflicts of Interest

The authors declare that they have no conflicts of interest.

Acknowledgments

This work was supported by the National Natural Science Foundation of China under Grant no. 51577056 and no. 51707061 and the Hunan Provincial Innovation Foundation for Postgraduate no. CX2017B104.

References

[1] M. Yazdanian and A. Mehrizi-Sani, "Distributed control techniques in microgrids," *IEEE Transactions on Smart Grid*, vol. 5, no. 6, pp. 2901–2909, 2014.

[2] Y. Li, Y. Zhou, F. Liu, Y. Cao, and C. Rehtanz, "Design and Implementation of Delay-Dependent Wide-Area Damping Control for Stability Enhancement of Power Systems," *IEEE Transactions on Smart Grid*, vol. 8, no. 4, pp. 1831–1842, 2017.

[3] L. Zhou, Y. Chen, A. Luo et al., "Robust two degrees-of-freedom single-current control strategy for LCL-type grid-connected DG system under grid-frequency fluctuation and grid-impedance variation," *IET Power Electronics*, vol. 9, no. 14, pp. 2682–2691, 2016.

[4] Z. Shuai, Y. Hu, Y. Peng, C. Tu, and Z. J. Shen, "Dynamic Stability Analysis of Synchronverter-Dominated Microgrid Based on Bifurcation Theory," *IEEE Transactions on Industrial Electronics*, vol. 64, no. 9, pp. 7467–7477, 2017.

[5] V. V. S. Pradeep Kumar and B. G. Fernandes, "A Fault-Tolerant Single-Phase Grid-Connected Inverter Topology with Enhanced Reliability for Solar PV Applications," *IEEE Journal of Emerging and Selected Topics in Power Electronics*, vol. 5, no. 3, pp. 1254–1262, 2017.

[6] A. Luo, Y. Chen, Z. Shuai, and C. Tu, "An improved reactive current detection and power control method for single-phase photovoltaic grid-connected dg system," *IEEE Transactions on Energy Conversion*, vol. 28, no. 4, pp. 823–831, 2013.

[7] C. Yoon, H. Bai, R. N. Beres, X. Wang, C. L. Bak, and F. Blaabjerg, "Harmonic stability assessment for multiparalleled, grid-connected inverters," *IEEE Transactions on Sustainable Energy*, vol. 7, no. 4, pp. 1388–1397, 2016.

[8] Y. Li, F. Liu, L. Luo, C. Rehtanz, and Y. Cao, "Enhancement of commutation reliability of an HVDC inverter by means of an inductive filtering method," *IEEE Transactions on Power Electronics*, vol. 28, no. 11, pp. 4917–4929, 2013.

[9] X. Zhang, B. Hou, and Y. Mei, "Deadbeat Predictive Current Control of Permanent-Magnet Synchronous Motors with Stator Current and Disturbance Observer," *IEEE Transactions on Power Electronics*, vol. 32, no. 5, pp. 3818–3834, 2017.

[10] W. Jiang, W. Ma, J. Wang, L. Wang, and Y. Gao, "Deadbeat Control Based on Current Predictive Calibration for Grid-Connected Converter under Unbalanced Grid Voltage," *IEEE Transactions on Industrial Electronics*, vol. 64, no. 7, pp. 5479–5491, 2017.

[11] M. Pichan, H. Rastegar, and M. Monfared, "Deadbeat Control of the Stand-Alone Four-Leg Inverter Considering the Effect of the Neutral Line Inductor," *IEEE Transactions on Industrial Electronics*, vol. 64, no. 4, pp. 2592–2601, 2017.

[12] F. Blaabjerg, R. Teodorescu, M. Liserre, and A. V. Timbus, "Overview of control and grid synchronization for distributed power generation systems," *IEEE Transactions on Industrial Electronics*, vol. 53, no. 5, pp. 1398–1409, 2006.

[13] J. C. Moreno, J. M. Espí Huerta, R. G. Gil, and S. A. Gonzalez, "A robust predictive current control for three-phase grid-connected inverters," *IEEE Transactions on Industrial Electronics*, vol. 56, no. 6, pp. 1993–2004, 2009.

[14] J. He, Y. W. Li, D. Xu, X. Liang, B. Liang, and C. Wang, "Deadbeat Weighted average current control with corrective feed-forward compensation for microgrid converters with nonstandard LCL filter," *IEEE Transactions on Power Electronics*, vol. 32, no. 4, pp. 2661–2674, 2017.

[15] B. Liu, Y. Zha, and T. Zhang, "D-Q frame predictive current control methods for inverter stage of solid state transformer," *IET Power Electronics*, vol. 10, no. 6, pp. 687–696, 2017.

[16] S. Liu, K. Wu, and Y. Tzou, "Control of a single-phase grid inverter with on-line inductance identification," in *Proceedings of the 2016 IEEE 25th International Symposium on Industrial Electronics (ISIE)*, pp. 454–459, Santa Clara, CA, USA, June 2016.

[17] C. Chen, L. Li, Q. Zhang et al., "Online inductor parameters identification by small-signal injection for sensorless predictive current controlled boost converter," *IEEE Transactions on Industrial Informatics*, vol. 13, no. 4, pp. 1554–1564, 2017.

[18] Z. Deng and W. Song, "Inductance sensitivity analysis of model predictive direct current control strategies for single-phase PWM converters," in *Proceedings of the 2015 IEEE 2nd International Future Energy Electronics Conference (IFEEC)*, pp. 1–6, Taipei, Taiwan, November 2015.

[19] S. Bibian and H. Jin, "Time delay compensation of digital control for DC switchmode power supplies using prediction techniques," *IEEE Transactions on Power Electronics*, vol. 15, no. 5, pp. 835–842, 2000.

[20] H. Kim, J. Han, Y. Lee, J. Song, and K. Lee, "Torque predictive control of permanent-magnet synchronous motor using duty ratio prediction," in *Proceedings of the 2013 IEEE 22nd International Symposium on Industrial Electronics (ISIE)*, pp. 1–5, Taipei, Taiwan, May 2013.

[21] C. Li, S. Shen, J. Zhang, M. Guan, and J. Lu, "Deadbeat Control Based on New State-observer for PWM Rectifier," in *Proceedings of the 2006 Chinese Control Conference*, pp. 1991–1995, Harbin, China, August 2006.

[22] J. M. Espí, J. Castelló, R. García-Gil, G. Garcerá, and E. Figueres, "An adaptive robust predictive current control for three-phase grid-connected inverters," *IEEE Transactions on Industrial Electronics*, vol. 58, no. 8, pp. 3537–3546, 2011.

[23] N. Hoffmann, F. W. Fuchs, and J. Dannehl, "Models and effects of different updating and sampling concepts to the control of grid-connected PWM convertersa study based on discrete time domain analysis," in *Proceedings of the European Conference on Power Electronics and Application*, p. 10, Appl., Birmingham, United kingdom, August-September, 2011.

[24] J. K. Pandit, B. Sakthisudhursun, and M. V. Aware, "PR controller implementation using double update mode digital PWM for grid connected inverter," in *Proceedings of the 2014 IEEE International Conference on Power Electronics, Drives and Energy Systems (PEDES)*, pp. 1–6, Mumbai, India, December 2014.

[25] X. Zhang, J. W. Spencer, and J. M. Guerrero, "Small-signal modeling of digitally controlled grid-connected inverters with LCL filters," *IEEE Transactions on Industrial Electronics*, vol. 60, no. 9, pp. 3752–3765, 2013.

[26] D. P. Marcetic and E. M. Adžić, "Improved three-phase current reconstruction for induction motor drives with DC-Link shunt," *IEEE Transactions on Industrial Electronics*, vol. 57, no. 7, pp. 2454–2462, 2010.

[27] G. Oriti and A. L. Julian, "Three-phase VSI with FPGA-based multisampled space vector modulation," *IEEE Transactions on Industry Applications*, vol. 47, no. 4, pp. 1813–1820, 2011.

[28] H. Wang, M. Yang, L. Niu et al., "Current-loop bandwidth expansion strategy for permanent magnet synchronous motor drives," in *Proceedings of the 2010 5th IEEE Conference on Industrial Electronics and Applications (ICIEA)*, pp. 1340–1345, Taichung, Taiwan, June 2010.

[29] L. Yang, A. Luo, Y. Chen et al., "A fast robust PWM method for photovoltaic grid-connected inverter," in *Proceedings of the IECON 2017 - 43rd Annual Conference of the IEEE Industrial Electronics Society*, pp. 4811–4816, Beijing, October 2017.

[30] E. J. Mastascusa, W. C. Rave, and B. M. Turner, "Polynomial Factorization using the Routh Criterion," *Proceedings of the IEEE*, vol. 59, no. 9, pp. 1358-1359, 1971.

8

Microgrids as Flexible and Network-Connected Grid Assets in Active Distribution Systems

Lei Shao, Xu Zhou, Ji Li⬤, Hongli Liu, and Xiaoqi Chen

Tianjin Key Laboratory for Control Theory & Application in Complicated Systems, Tianjin University of Technology, Tianjin 300384, China

Correspondence should be addressed to Ji Li; liji0606@163.com

Academic Editor: Mustafa I. Fadhel

The functionalities of microgrids are detailed and thereby expanded in active distribution systems. A versatile and coordinated operation among multiple microgrids is proposed to facilitate the integration of renewable energy sources (RES) in modern distribution grids. Particularly, to meet the requirements of high penetration level of renewables (e.g., photovoltaics and small hydro), more than one networked controlled microgrids are deployed simultaneously in different locations of a distribution system. Therefore, local RES can be aggregated in each microgrids. In order to satisfy the latest standards in terms of renewable energy integration (i.e., IEEE Std. 1547 Rev), an optimal operation strategy is designed to coordinate the operation of multiple microgrids. A simulation model is implemented in MATLAB to validate the proposed networked microgrids and the corresponding operation strategy.

1. Introduction

The existing planning and operation issues with respect to traditional fossil fuels have drawn increasing attention nowadays. These include the inevitable environmental pollution, increasing CO_2 emission, ever-decreasing fuel, and energy. In order to solve the present energy crisis, a paradigm shift in energy configuration should be initialized. Therefore, renewable energy sources (RES) have been gradually involved in the portfolio of energy deployment [1–3]. Compared to conventional energy sources, RES play a significant role in releasing the environmental stress in the whole energy cycle, including generation, transmission, distribution, and consumption. For example, large-scale wind farm can be used to combine with legacy thermal generation units in the power generation system, while high penetration level of different types of renewables is being used in the transmission and distribution grids. At the customer's side, residential renewables are also widely used to localize energy generation and consumption so that an active 'prosumer' is implemented.

Even though the increasing penetration level of RES has effectively solved the environmental issues in the past years and their advantages are obviously seen in modern grid operation, there are still issues that need further attention. Most importantly, it should be noticed that the RES does not have sufficient inertia, which is quite different from conventional synchronous generator based power systems and highly challenges the system stability [4–6]. To solve this potential stability issues, many approaches have been proposed and deployed in both academia and field applications. Energy storage systems (ESS) have been widely employed to compensate the intermittency in output power, which is induced by the low-inertia RES [7]. Hybrid energy sources, such as combined electrical and thermal generation systems, are being studied and gradually used in the actual power systems to leverage their complementary dynamic characteristics and mitigate the variations in their aggregated output power [8]. Meanwhile, it is worth mentioning that, to effectively integrate different types of sources and loads in a localized way, microgrids are being intensively studied over the past years [9–11]. A microgrid can be regarded as an integrated unit with both distributed sources and loads in it, and, in the meantime, with a point of common coupling (PCC), it can be regarded as a controllable unit and the output voltage, current, or power at the PCC can be controlled and flexibly adjusted. Given the tremendous advantages of microgrids compared to dispersed RES, microgrids have been used in many field test sites, including mission-critical applications (e.g., isolated power systems in airplanes), remote areas, or, most frequently, grid-connected systems with high requirements of power quality.

Note that the existing penetration level of RES in modern electric grids is high and the percentage of renewable energy in the total energy portfolio keeps increasing due to the ever-growing needs in electricity. Therefore, the current study of microgrids has been greatly advanced at the present. In the first stage, the research on microgrids mainly stays in individual one. In other words, single microgrids are the major research focus. However, with large-scale integration of RES, single microgrid integration cannot always meet the requirements. Hence, people are considering to further advance the concept of microgrids and proposing new technologies and expand the research and application horizon of microgrid deployment. Among the technologies, networked microgrid stays as one of the promising candidates [12]. Different from conventional single microgrids, networked microgrids enable multiple microgrids to coexist in the electric grids and coordinate with each other following a predesigned operation condition. Particularly, being geographically diversified, these networked microgrids can be distributed into different areas of a whole electric grids, so that the integration of renewables can be implemented in a simpler way. Effective control and coordination among multiple microgrids should be well established. Therefore, the penetration level of renewables in various areas of an electric grid can be balanced.

In this paper, the latest operation requirements of electric grids with high penetration level of renewable energy are reviewed along with the versatile functionalities of microgrids. Meanwhile, a problem formulation is established focusing on networked microgrids, especially for the coordinated operation among multiple microgrids. The interactions among multiple microgrids controllers are studied and modeled in the above problem formulation. A simulation test based on MATLAB is built up to validate the proposed coordinated control strategy.

The remainder of the paper is summarized below. Section 2 formulates the problem and defines the networked microgrids under study. Section 3 establishes the optimization formulation of networked microgrids to achieve the optimal operation in active distribution grids. The detailed constraints and optimal objectives are listed to implement the comprehensive formulation. Simulation models of 28-bus and 123-bus test systems rebuilt up in MATLAB in Section 4 to verify the proposed algorithm. Finally, Section 5 draws the conclusion of the whole paper.

2. Problem Formulation and Definition of Networked Microgrids

As in the precedent discussion, microgrids can be used as versatile units in distribution systems to facilitate the integration of RES and enhance the performance of distribution operation. Conventional RES is commonly integrated into distribution grids in a dispersed way, which is hard to control

TABLE 1: Variables and parameters.

$p_{dp,i}$	Dispatchable power #i
$q_{dp,i}$	Dispatchable reactive power #i
$p_{1,ex}$	Injected power into the microgrid
$p_{2,ex}$	Output power from the microgrid
$p_{d,i}$	Load active power #i
$q_{d,i}$	Load reactive power #i
$p_{ndp,i}$	PV or other non-dispatchable output power #i
c_g	Cost: dispatchable load
c_e	Cost: power exchange
c_d	Cost: demand side
V_i	Voltage #i
V_ε	Maximum voltage deviation
P_i	Active power flow #i
Q_i	Reactive power flow #i
p_i^{max}	Maximum output active power
q_i^{max}	Maximum output reactive power
r_i	Line resistance
x_i	Line inductance

and manage and may induce unexpected operational issues. To better utilize their advantages, networked microgrids can be deployed to achieve the optimal and coordinated control, as depicted in Figure 1. The problem formulation of networked microgrids is established based on using multiple interconnected microgrids to achieve aggregation of local DERs. Therefore, these DERs can be better operated and controlled without violating any operation constraints, e.g., potential voltage violations and active and reactive power violations. For a clear narrative, the related system parameters and variables are shown in Table 1.

3. Model Development of Networked Microgrids

The objective of the optimization problem is established to minimize the cost for exchanging power locally and between the network-connected microgrids. As shown below in (1), the first two terms represent the cost of local sources and loads, and the third to sixth terms show the cost induced by power exchange between networked microgrids. Note that the last two terms in (1) are only used when solving the optimization algorithm, which can be regarded as penalty items to enhance the convergence rate of the optimization problem.

$$\min \quad \sum_{i \in N} c_g p_{dp,i} - \sum_{i \in N} c_d p_{d,i}$$
$$+ \left(c_e p_{1,ex} + \sum_m c_e p_{1,ex}^m - c_e p_{2,ex} - \sum_m c_e p_{2,ex}^m \right) \tag{1}$$

The power balance is the most critical constraint that needs to be considered in the optimization problem. As shown in (2), given the flexibility of the output power of the sources and the power consumption of the loads, the

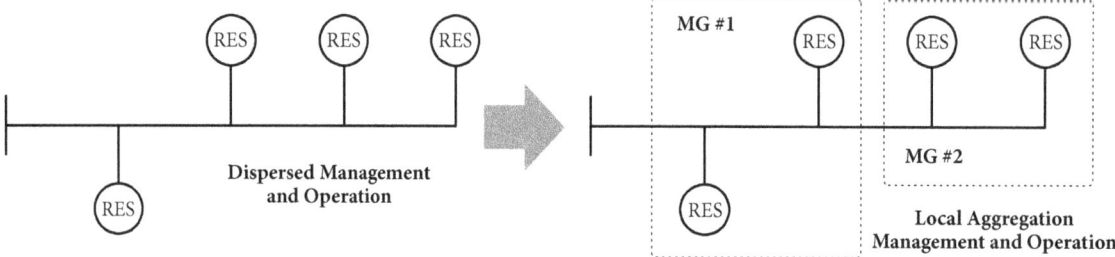

FIGURE 1: Conventional RES integration and the configuration with networked microgrids.

total power generation should be no less than the total load consumption. Note that the power sources include the local power generation units (e.g., the dispatchable sources) and the power injection from the other microgrids.

$$\sum_{i\in N} p_{dp,i} + \sum_{i\in N} p_{ndp,i} + \sum_m p_{1,ex}^m + p_{1,ex}$$

$$\geq \sum_{i\in N} p_{d,i} + \sum_m p_{2,ex}^m + p_{2,ex} \tag{2}$$

The distribution model should be established in the formulation so that the variations and limitations induced by the model itself can be considered. These constraints are formulated as shown in (3)-(5). Note that, in order to maintain a convex optimization problem, the distribution model is approximated and linearized using the format shown in (3)-(5), which follows the typical linearization in DistFlow.

$$P_{i+1} = P_i - p_{d,i+1} + p_{ndp,i+1} + p_{dp,i+1} \tag{3}$$

$$Q_{i+1} = Q_i - q_{d,i+1} + q_{dp,i+1} \tag{4}$$

$$V_{i+1} = V_i - \frac{(r_i P_i + x_i Q_i)}{V_1} \tag{5}$$

Besides the constraints above, it is also necessary to consider the operation limitations of each source, i.e., the voltage and power generation limits, as detailed in (6)-(8). Note that to maximize the power generation of nondispatchable sources, such as solar generation units and wind turbines, the output power of nondispatchable units is not limited. In other words, the output power of the nondispatchable unit follows its maximum power point and is not controlled.

$$1 - V_\varepsilon \leq V_i \leq 1 + V_\varepsilon \tag{6}$$

$$0 \leq p_{dp,i} \leq p_i^{\max} \tag{7}$$

$$0 \leq q_{dp,i} \leq q_i^{\max} \tag{8}$$

In the meantime, a penalty term is involved to facilitate the convergence of the formulated optimization problem. To limit these penalty terms, the power exchange determined in each microgrid controller is monitored and their differences are controlled within the acceptable limits.

Therefore, by considering the above formula together, the optimization formulation can be established as below and the

corresponding constraints and objectives can be summarized in Figure 2.

$$\min \quad \sum_{i\in N} c_g p_{dp,i} - \sum_{i\in N} c_d p_{d,i}$$

$$+ \left(c_e p_{1,ex} + \sum_m c_e p_{1,ex}^m - c_e p_{2,ex} - \sum_m c_e p_{2,ex}^m \right)$$

$$s.t. \quad \sum_{i\in N} p_{dp,i} + \sum_{i\in N} p_{ndp,i} + \sum_m p_{1,ex}^m + p_{1,ex}$$

$$\geq \sum_{i\in N} p_{d,i} + \sum_m p_{2,ex}^m + p_{2,ex}$$

$$P_{i+1} = P_i - p_{d,i+1} + p_{ndp,i+1} + p_{dp,i+1} \tag{9}$$

$$Q_{i+1} = Q_i - q_{d,i+1} + q_{dp,i+1}$$

$$V_{i+1} = V_i - \frac{(r_i P_i + x_i Q_i)}{V_1}$$

$$1 - V_\varepsilon \leq V_i \leq 1 + V_\varepsilon$$

$$0 \leq p_{dp,i} \leq p_i^{\max}$$

$$0 \leq q_{dp,i} \leq q_i^{\max}$$

4. Simulation Verification and Validation

4.1. Small-Scale Test System: 28-Bus Test Feeder. In order to verify the proposed optimization formulation and minimize the operation cost using networked microgrids, a numerical model is established in MATLAB. The configuration of the testing system is shown in Figure 3.

As depicted in Figure 3, the testing feeder is a 28-bus system. In addition to the conventional passive loads and power cables between neighboring buses, six DGs are incorporated in the system to further study their impacts on the system and the effectiveness of using MGs to aggregate the local sources and loads. Particularly, there are two MGs in the system. MG #1 locates near the feeder head, while MG #2 locates near the feeder end. Besides these two MGs, two additional DGs are also dispersedly connected to the main feeder. The system parameter is shown in Table 2.

By running the overall comprehensive optimization problem in (9), the results can be summarized in Table 3, where the output power of each dispatchable sources are calculated.

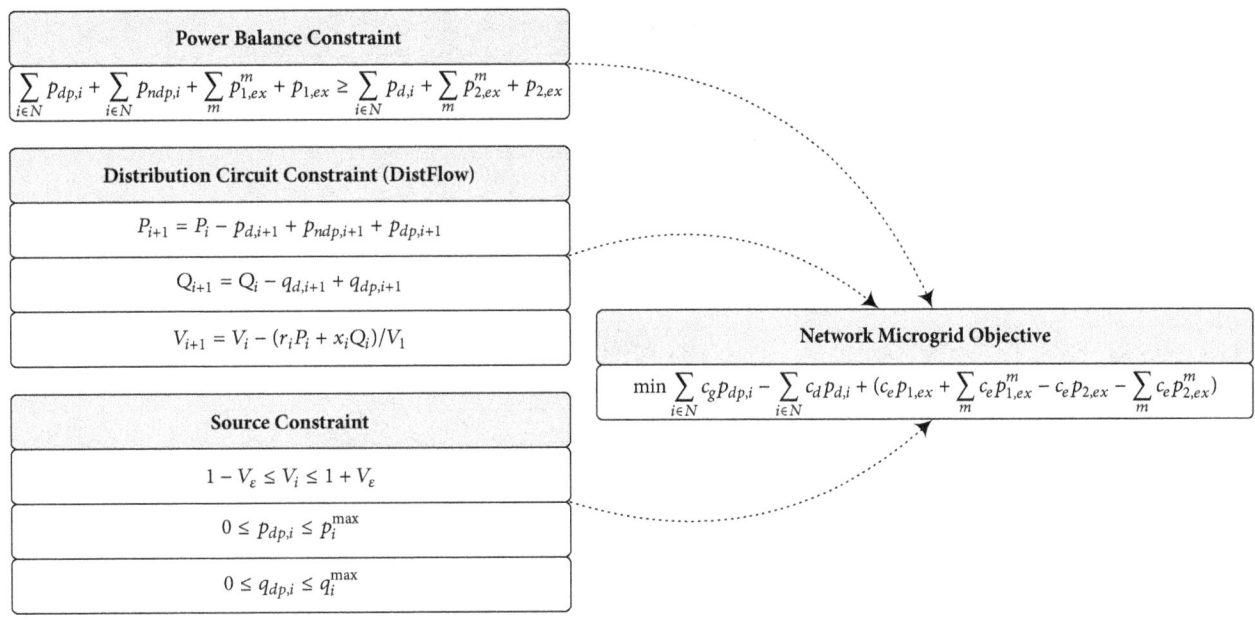

FIGURE 2: Optimization model of networked microgrids.

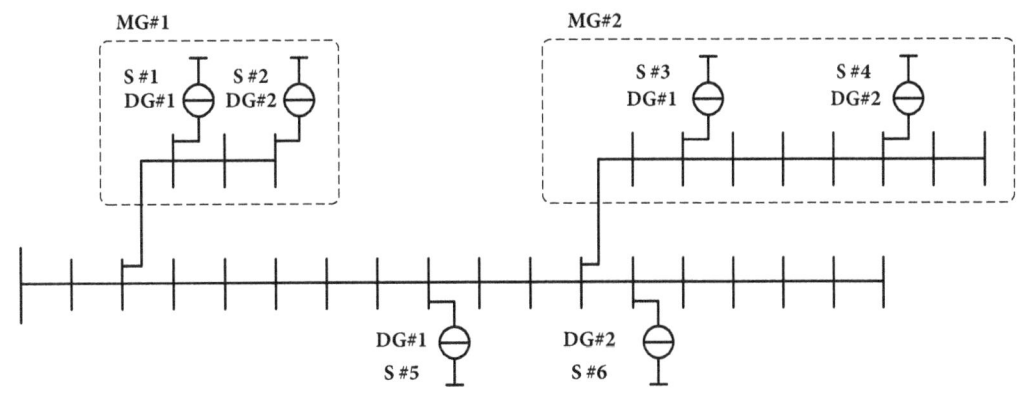

FIGURE 3: Configuration of the testing system.

TABLE 2: System parameters.

Parameter	Value	Unit
Base power	1	MW
Base voltage	110	kV
Line resistance	0.014	p.u.
Line inductance	0.026	p.u.
Active load in MG #1	2	p.u.
Active load in MG #2	1.8	p.u.
Active load outside the two MGs	2	p.u.
Cost: dispatchable load	0.12	$/kW
Cost: power exchange	0.2	$/kW
Cost: demand side	0.4	$/kW

Here, sources #1 and #2 belong to microgrid #1, while sources #3 and #4 belong to microgrid #2. It is seen that the output power converge in the first three steps.

Meanwhile, maximum and minimum of the output power and voltage amplitude in each microgrid are depicted in Figures 4(a) and 4(b). Per unit values are used here. It can be seen that the output power and voltage do not violate their upper and lower boundaries during the interaction of the optimization algorithm.

It should be noticed that, by using the proposed approach, the bus voltage throughout the test system can be well regulated within the acceptable range, and the output power of DGs can finally converge without triggering any stability issues. Conventional approaches still lack of consideration of microgrids and DGs when regulating voltage and power in distribution grids, which will lead to voltage violation and the sequential misfunctional behavior of protective devices. By using the proposed method, the DGs and microgrids can be well controlled and managed to contribute to voltage regulation in distribution grids. As a further numerical study, the proposed system configuration and the conventional system without microgrids (i.e., with all the microgrids

TABLE 3: Power generation of each source (p.u.).

Interaction	1	2	3	4	5
Source #1 (MG #1)	0.837	0.801	0.792	0.792	0.792
Source #2 (MG #1)	0.96	0.936	0.882	0.882	0.882
Source #3 (MG #2)	0.882	0.864	0.774	0.774	0.774
Source #4 (MG #2)	0.948	0.858	0.846	0.846	0.846
Source #5	0.891	0.864	0.846	0.846	0.846
Source #6	0.846	0.81	0.792	0.792	0.792

TABLE 4: Voltage profile comparison between conventional approach with only dispersed DGs and the developed approach with networked microgrids (p.u.).

	Source #1	Source #2	Source #3	Source #4
w/ Networked Microgrids	**1.0200**	**1.0220**	**0.9820**	**0.9800**
w/o Networked Microgrids	0.9989	1.0023	0.9960	0.9978

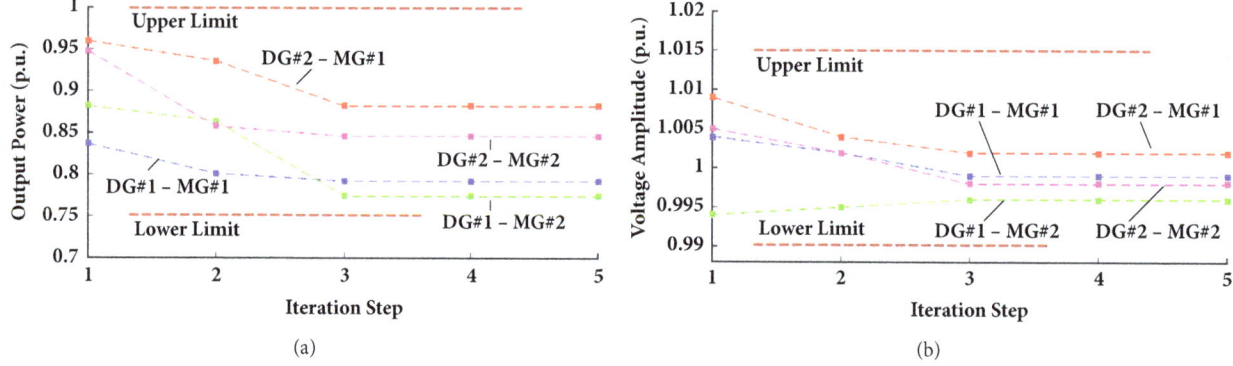

FIGURE 4: Maximum and minimum of power and voltage during interaction. (a) Output power. (b) Voltage amplitude.

in Figure 3 removed and only keeping the DGs in the system), the results of the comparative study are shown in Table 4. It can be seen that the final voltage values violate the upper or lower boundary at DG #1 and #2 since the dispatchable sources cannot be successfully managed, and more importantly, they cannot be coordinated using networked microgrids to maintain the acceptable voltage level. Therefore, the power injected from some DGs violates the upper voltage limit (1.015 p.u.) by injecting higher power into the point of connection, while some other DGs violate the lower voltage limit (0.9900 p.u.) by injecting insufficient power into the point of connection. The verification and validation of the proposed method are further demonstrated in the above numerical results.

4.2. Large-Scale Test System: 123-Bus Test Feeder. After the testing in the small-scale 28-bus system, a test in larger-scale test feeder, i.e., 123-bus system, is conducted to further examine the effectiveness of the proposed algorithm. The configuration of the 123-bus system is shown in Figure 5, and the capacity of the dispatchable DGs and structured microgrids are detailed in Table 5.

With the above interconnected microgrids considered, by solving the optimization problem as shown in the above sections, similar to the results in Table 3, the power generation

of each DG is finally converged, as shown in Table 6, and the voltage of the characteristic DG bus is detailed in Table 7. It can be seen that the generated power and voltage magnitude all satisfy the predefined operation constraints, i.e., 0.75~ 1 p.u. for power generation and 0.99~1.015 p.u. for voltage magnitude.

5. Conclusion

In this paper, the concept of microgrids has been expanded in a networked connected configuration. The latest operation standards and the functionalities of microgrids are reviewed in detail. Meanwhile, a comprehensive optimization problem is established to derive the optimal operation condition of networked microgrids. A test model is established in MATLAB and numerical study is conducted to determine the optimal operation conditions of networked microgrids. The operation of distribution grids is challenged by microgrids and DGs with increasing penetration level, which will lead to unexpected issues, e.g., voltage violations. In the proposed work, the optimal operation between local microgrids and the central control system in distribution grids is implemented, which is effective in coordinating multiple systems in the control hierarchy in distribution grids. Future work includes the coordination among multiple microgrids in various

TABLE 5: Dispatchable DGs and microgrids in the 123-bus testing system.

DG Capacity and Location					
DG	Location	Initial Capacity (p.u.)	DG	Location	Initial Capacity (p.u.)
DG #1	Bus 1	0.90	DG #10	Bus 21	0.98
DG #2	Bus 12	0.69	DG #11	Bus 18	0.74
DG #3	Bus 6	0.86	DG #12	Bus 27	0.98
DG #4	Bus 92	0.95	DG #13	Bus 36	0.97
DG #5	Bus 89	1.04	DG #14	Bus 41	1.06
DG #6	Bus 86	1.03	DG #15	Bus 46	0.94
DG #7	Bus 108	1.06	DG #16	Bus 67	0.76
DG #8	Bus 197	0.96	DG #17	Bus 72	0.86
DG #9	Bus 103	0.65	DG #18	Bus 76	0.96
Microgrids and Location					
Microgrid	Location		Microgrid	Location	
MG #1	Bus 1-17, Bus 34		MG #4	Bus 35-51, Bus 135, Bus 151	
MG #2	Bus 86-84, Bus 195		MG #5	Bus 67-83	
MG #3	Bus 18-33, Bus 250, Bus 251		MG #6	Bus 101-114	

TABLE 6: Power generation of each DG (p.u.), 123-bus system.

DG	Output Power (p.u.)	DG	Output Power (p.u.)
DG #1	0.92	DG #10	0.99
DG #2	0.86	DG #11	0.90
DG #3	0.90	DG #12	0.98
DG #4	0.96	DG #13	0.97
DG #5	0.98	DG #14	0.98
DG #6	0.98	DG #15	0.96
DG #7	0.99	DG #16	0.88
DG #8	0.94	DG #17	0.82
DG #9	0.86	DG #18	0.90

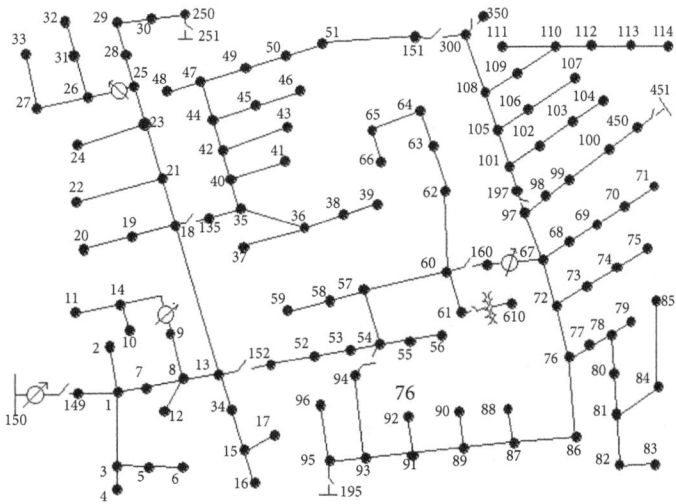

FIGURE 5: 123-bus testing system.

TABLE 7: Voltage magnitude of each DG (p.u.), 123-bus system.

DG	Voltage Mag. (p.u.)	DG	Voltage Mag. (p.u.)
DG #1	1.012	DG #10	1.013
DG #2	0.996	DG #11	0.992
DG #3	0.998	DG #12	0.998
DG #4	1.003	DG #13	0.993
DG #5	1.004	DG #14	1.012
DG #6	1.004	DG #15	0.994
DG #7	1.013	DG #16	0.992
DG #8	1.006	DG #17	0.992
DG #9	0.994	DG #18	0.999

grid conditions (balanced and unbalanced fault conditions, etc.) and topologies (i.e., lightly meshed grids). Regulatory constraints should be also considered when further updating the proposed method.

Conflicts of Interest

The authors declare that there are no conflicts of interest regarding the publication of this paper.

References

[1] D. Baimel, J. Belikov, J. M. Guerrero, and Y. Levron, "Dynamic modeling of networks, microgrids, and renewable sources in the dq0 reference frame: a survey," *IEEE Access*, vol. 5, pp. 21323–21335, 2017.

[2] M. A. Azzouz, H. E. Farag, and E. F. El-Saadany, "Real-time fuzzy voltage regulation for distribution networks incorporating high penetration of renewable sources," *IEEE Systems Journal*, vol. 11, no. 3, pp. 1702–1711, 2017.

[3] M. S. Hasan, Y. Kouki, T. Ledoux, and J.-L. Pazat, "Exploiting renewable sources: When green SLA becomes a possible reality in cloud computing," *IEEE Transactions on Cloud Computing*, vol. 5, no. 2, pp. 249–262, 2017.

[4] S. Kazemlou, S. Mehraeen, H. Saberi, and S. Jagannathan, "Stability of the small-scale interconnected DC grids via output-feedback control," *IEEE Journal of Emerging and Selected Topics in Power Electronics*, vol. 5, no. 3, pp. 960–970, 2017.

[5] L.-Y. Lu and C.-C. Chu, "Consensus-based secondary frequency and voltage droop control of virtual synchronous generators for isolated ac micro-grids," *IEEE Journal on Emerging and Selected Topics in Circuits and Systems*, vol. 5, no. 3, pp. 443–455, 2015.

[6] L. Ding, Z. Ma, P. Wall, and V. Terzija, "Graph spectra based controlled islanding for low inertia power systems," *IEEE Transactions on Power Delivery*, vol. 32, no. 1, pp. 302–309, 2017.

[7] Y. Lu, Q. Wu, Q. Wang, D. Liu, and L. Xiao, "Analysis of a novel zero-voltage-switching bidirectional DC/DC converter for energy storage system," *IEEE Transactions on Power Electronics*, vol. 33, no. 4, pp. 3169–3179, 2018.

[8] A. Rong and P. B. Luh, "A dynamic regrouping based dynamic programming approach for unit commitment of the transmission-constrained multi-site combined heat and power system," *IEEE Transactions on Power Systems*, vol. 33, no. 1, pp. 714–722, 2018.

[9] M. Farrokhabadi, B. V. Solanki, C. A. Canizares et al., "Energy storage in microgrids: compensating for generation and demand fluctuations while providing ancillary services," *IEEE Power & Energy Magazine*, vol. 15, no. 5, pp. 81–91, 2017.

[10] G. Strbac, N. Hatziargyriou, J. P. Lopes, C. Moreira, A. Dimeas, and D. Papadaskalopoulos, "Microgrids: enhancing the resilience of the european megagrid," *IEEE Power & Energy Magazine*, vol. 13, no. 3, pp. 35–43, 2015.

[11] K. Ravindra, B. Kannan, and N. Ramappa, "Microgrids: a value-based paradigm: the need for the redefinition of microgrids," *IEEE Electrification Magazine*, vol. 2, no. 1, pp. 20–29, 2014.

[12] Z. Wang, B. Chen, J. Wang, and J. Kim, "Decentralized energy management system for networked microgrids in grid-connected and islanded modes," *IEEE Transactions on Smart Grid*, vol. 7, no. 2, pp. 1097–1105, 2016.

9

Harmonic Susceptibility Study of DC Collection Network based on Frequency Scan and Discrete Time-Domain Modelling Approach

Carlos Enrique Imbaquingo, Eduard Sarrà, Nicola Isernia, Alberto Tonellotto, Yu-Hsing Chen, Catalin Gabriel Dincan ⓘ, Philip Kjær, Claus Leth Bak, and Xiongfei Wang

Department of Energy Technology, Aalborg University, Aalborg, Denmark

Correspondence should be addressed to Catalin Gabriel Dincan; cgd@et.aau.dk

Guest Editor: Luigi P. Di Noia

The equivalent model of offshore DC power collection network for the harmonic susceptibility study is proposed based on the discrete time-domain modelling technique and frequency scan approach in the frequency domain. The proposed methodology for modelling a power converter and a DC collection system in the frequency domain can satisfy harmonic studies of any configuration of wind farm network and thereby find suitable design of power components and array network. The methodology is intended to allow studies on any configuration of the wind power collection, regardless of choice of converter topology, array cable configuration, and control design. To facilitate harmonic susceptibility study, modelling DC collection network includes creating the harmonic model of the DC turbine converter and modelling the array network. The current harmonics within the DC collection network are obtained in the frequency domain to identify the resonance frequency of the array network and potential voltage amplification issues, where the harmonic model of the turbine converter is verified by the comparison of the converter switching model in the PLECS™ circuit simulation tool and laboratory test bench, and show a good agreement.

1. Introduction

Offshore wind power is becoming an important energy resource in Europe. However, electrical power losses are always a concern in the operation of wind farm with long-range power transmission. To reduce losses, researchers look to replace AC with DC in the entire path from the wind turbine, through power collection and transmission to shore. Medium-voltage DC (MVDC) collection of wind power is an attractive solution to reduce overall losses and installation cost [1, 2]. The article [2] addresses cost-efficient solution to form the multiterminals of DC grid with the diode rectifier unit (DRU). However, the solution of DRU can produce a considerable harmonic current within the array network if the 6-pulse DRU topology is used.

Figure 1 illustrates a DC collection network of radial type of offshore wind farm with the HVDC transmission. This general configuration in Figure 1 supports any choice of MVDC collector voltage level as well as converter topology

and the number of wind power plant (WPP) clusters. For example, the turbine and offshore substation can employ any DC/DC converter topology with multiple degrees of freedom: uni- vs bidirectional power flow; modular (parallel/series connection) vs monolithic; topology (single active bridge, dual active bridge, and more [3–5]). Figures 1 and 2 illustrate the details of wind farm configuration including wind turbine DC/DC converters, offshore substation converters, and onshore substation converters. The offshore wind farm is designed to generated and deliver 700 MW power from offshore to onshore [1].

The series resonant converter (SRC#) is selected as the turbine DC/DC converter based on the converter efficiency, high-voltage transfer ratio, scale of converter, and galvanic isolation property. Considering the existing converter topology and its power rating in high-power applications, the modular multilevel converter (MMC) with the galvanic isolation and unidirectional power flow control is selected to serve as the offshore substation converter to deliver the

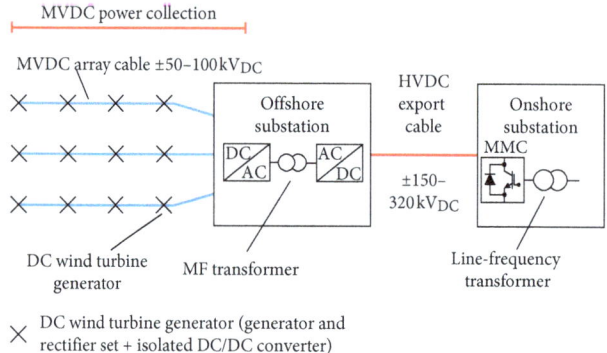

FIGURE 1: General configuration of the wind power plant with MVDC power collection.

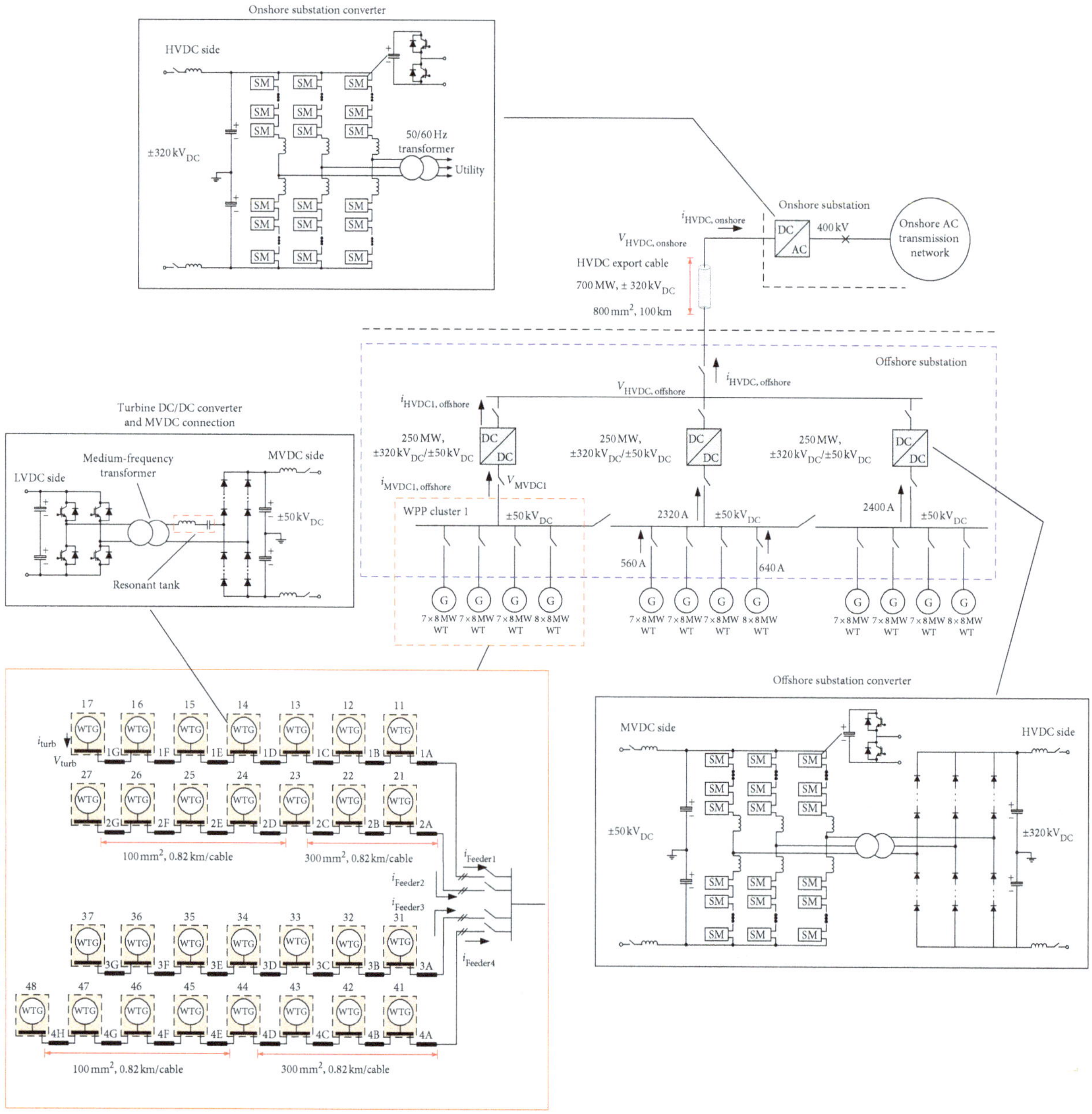

FIGURE 2: Single line diagram of example 700 MW offshore DC wind farm [1].

offshore wind power to the onshore grid via the HVDC transmission line. The onshore substation captures the power from the offshore wind farm by the onshore MMC with a grid-connected 50/60 Hz transformer. The onshore substation is designed to match with onshore AC grid voltage (utility) and deliver the offshore power to the onshore AC transmission network and alleviate the influence from the grid fault.

In this study, a series resonant converter topology (named SRC#) shown in Figure 3 is selected as the candidate for DC power conversion via $\pm 50\,kV_{DC}$ MVDC array network. With the series resonant converter, the DC turbine converter can take advantages of high efficiency, high-voltage transformation ratio, and galvanic fault isolation for different ratings of turbine generator [6–9]. A series resonant converter SRC# with the resonant tank on the secondary side and governed by the phase shift and a frequency-depended power flow control technique is here considered [10, 11].

To synthesize the DC-connected wind turbine converter and the MVDC gird, the wind turbine converter has to meet the following requirements:

(1) Control of the LVDC bus
(2) High-voltage transformation from the LVDC to MVDC
(3) Galvanic isolation
(4) Roust and compact design

Large amount of possible topology of converter topologies for a DC turbine is investigated. However, all of existing converter designs are immature technology, thus selecting an optimal topology is not a straightforward solution for converter design [12]. The turbine converter is designed to deliver the captured wind energy produced by the generator to the MVDC gird and then control the LVDC bus. Regarding power rating, number of components, and the possible solution of the turbine converter topologies in the early DC wind farm design, the hard-switched full-bridge power converter is selected as preferred topology [4, 13–17]. Therefore, the proposed series resonant converter # (SRC#) combined with the medium-frequency transformer (MF transformer) which serves as the DC turbine converter is selected to boot the low-voltage DC (LVDC) to high-voltage DC (MVDC-side of SRC#) with a high-voltage transferring ratio and provides a galvanic isolation to make sure that the line is immune from line faults [4, 18, 19].

The concept of SRC is widely used in the traction applications, which is usually operated at constant frequency and subresonant mode to archive a roust and compact design target in high-voltage specifications [20–24]. One of famous modulation techniques (or topologies) is the half-cycle discontinuous-conduction-mode series resonant converter (HC-DCM-SRC). Under the traction applications, the DC/DC converter is tied between two DC voltages with a fixed voltage transfer ratio and the open-loop control. Relying on the pulse removal technique, the bulky transformer in the classic SRC can be voided in the proposed SRC#, and thus the compact convert design can

be archived [25]. More details of operation principle of SRC# is addressed Section 2.

To archive the study of harmonic susceptibility in offshore wind farm with the selected turbine converter, the analysis of array network in wind farm includes modelling the selected turbine converter (SRC#), array cable, and substation converter, where the substation converter in this study is considered as an ideal voltage source to produce DC component of MVDC and harmonic components. The objective of the study is to investigate how the harmonic affects the turbine converter and array networks. In the power collection system, the wind turbines and substation are connected together by medium-voltage DC cables. However, there are no standards or guidelines for such DC array network design, and thus some studies are still made by simply selecting commercially available medium-voltage AC submarine cables without giving any additional consideration in the cable's properties such as insulation [26].

Modelling of the array cables is an essential task for evaluating results of cable sizing by computer simulation. The complexity of modelling is dependent on the phenomenon, which will be conducted in the study case. For example, the travelling waves (voltage and current) with a finite propagation speed cause reflections. This phenomenon has to be represented on the cable model when study cases are intended to conduct the overvoltage or switching transient issues. Typically, three types of cable models, Pi Model, Bergeron model, and frequency-dependent models, which are available in the electromagnetic transient simulation program PSCAD™ (EMTDC) for studying AC and DC power collection systems are used [27–30]. To adequately represent the cable parameters for the harmonic susceptibility studies over different frequency ranges, the frequency-dependent array cable models are generated from geometric cable data based on the method [31] and recommendations proposed by the articles [32] and verified by the generated cable model in PSCAD™.

Although these DC alternatives promise lower electrical losses and lower bill-of-material, it is still an immature technology. As the many existing AC wind power plants, the harmonic pollution is an important issue since wind power collection systems serve by large amount of turbine converters [33–36]. Depending on the configuration of the wind power plant, every power collection system has its own inherent resonance behavior. The network, which is rich in the harmonic current, can increase the power losses and the stress of power components and cause unpredicted equipment trip. The resonance behavior inside the collection network is affected by properties of the array cable, filter designs, converter topologies, and PWM schemes. Therefore, the objective of this paper is to develop a methodology to identify potential harmonic resonant problems in the DC collection network. For example, the phenomenon of harmonic amplification relates to grid contingencies, variation of load flow, cable length, and uncertainties of network [33, 37]. The article [38] focuses on the stability issues of DC grid related to the travel waves on the transmission line and harmonic content of converters. The paper identifies the travelling wave in the long submarine array cable

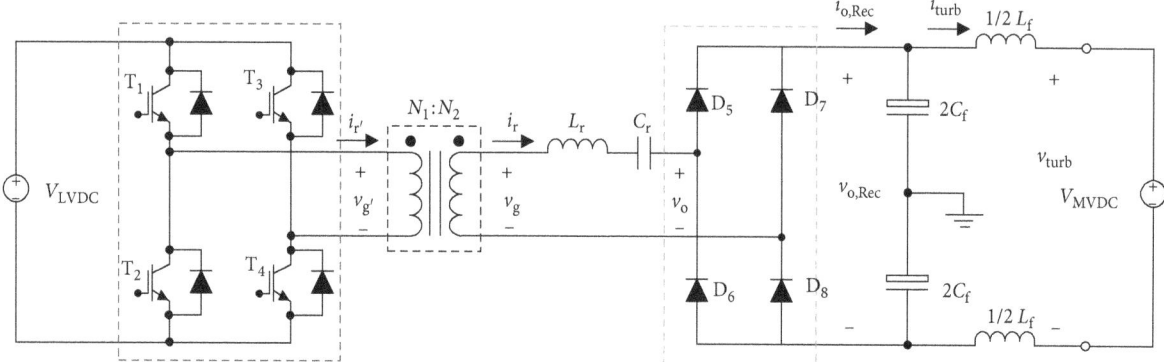

FIGURE 3: Circuit topology of the DC wind turbine converter (SRC#).

and resonant problem between the cable length and the switching action of the substation converter. According to the practical cases, the DC wind farm topologies for the case study mainly focus on the offshore wind farm with a short DC power transmission path.

Reviewing the common techniques for wind farm analysis, the frequency scan techniques are usually used for identifying harmonic voltage distortion of specific harmonic component in wind farms [34, 39, 40]. In this case, the wind turbines are represented for a specific frequency with an admittance equivalent circuit. The array cable model is selected according to the cable length and interesting frequency ranges of harmonic analysis. To adequately represent the cable parameters in the study, both equivalent PI section and frequency-dependent phase model are *generally* used to consider the frequency dependence of the cable parameters [28, 31, 41]. Additionally, the wind farm configuration could be changed via the operation of disconnectors or circuit breakers due to maintenance or forced outages of fault. Therefore, the harmonic study of wind farm must include all agreed switching configuration [39].

2. Review of Operation Principle of Series Resonant Converter

The circuit topology of series resonant converter # (SRC#) with the medium-frequency transformer is depicted in Figure 3, and the topology comprises a full-bridge inverter, a medium-frequency transformer, resonant LC tank, and medium-voltage output diode rectifier. The switch pairs T_1 (leading leg) and T_2 and T_3 (lagging leg) and T_4 operate at a 50% duty cycle as in Figure 4(e). Commutation of switches on the leading leg (T_1, and T_2) is shifted by an angle of δ, which is respect to the conduction of the switches on the lagging leg (T_3, and T_4). The duration of phase shift equals to half cycle of the $L_r C_r$ tank resonant period, resulting in a quasisquare excitation voltage as seen in Figure 4(a), and then, the voltage passes through the medium-frequency transformer and excites $L_r C_r$ resonant tank where the resonant tank current i_r is given in Figure 4(b). Via the diode rectifier and the output filter, the current (or power) is delivered to the medium-voltage network, V_{MVDC}. Up to

this point, there is no difference in the operation compared to a constant frequency and phase-shift control of classical SRC, which is usually operated in the *super resonant mode*, to achieve ZVS at turn on.

Considering the switching losses with IGTB applications, the subresonant mode is preferred in SRC# which allows *ZCS* or a *low current at turnoff* (Figure 5(a)); regardless of switching frequency, a full-resonant current pulse is delivered to the load [12]. The control law of SRC# is also allowed *ZVS at turnon* and *ZCS at turnoff* for the diode rectifier, as shown in Figure 5(b). It should be noted that the output power has a positive relation with the switching frequency, which means that the output power depends how much energy (current pulse) transfers to the output stage. SRC# operated in a low frequency means lower output power, while high-frequency operation will deliver a higher output power.

3. Linearized Model and Closed-Loop Control of Series Resonant Converter

The objective of the study is to understand the harmonics distribution of offshore DC wind farm and how the DC wind turbines are affected by harmonics from the MVDC gird. Figure 6 summarizes the derivation of the plant model of the DC wind turbine based on the discrete time-domain modelling approach (Steps 1–8) [42], the transfer function of the output LC filter (Step 9), and SRC#'s control design (Steps 10–12). The details of the derivation (or equations) are address in Appendices A, B, and C which can help reader to reach the harmonic model of the DC wind turbine and then conduct harmonic susceptibility study of the DC wind farm. The following discussion including the flow chart only shows most important conclusions (or equations) of the derivation due to the limitation of page length.

The ideal of the harmonic model of the DC wind turbine converter is developed based on the harmonic model of the AC wind turbine and linearized model of the series resonant converter (SRC#) with discrete time-domain modelling approach [42, 43]. A linearized state-space model of SRC# in subresonant mode is reported in (1) and in (A.37) in Appendix:

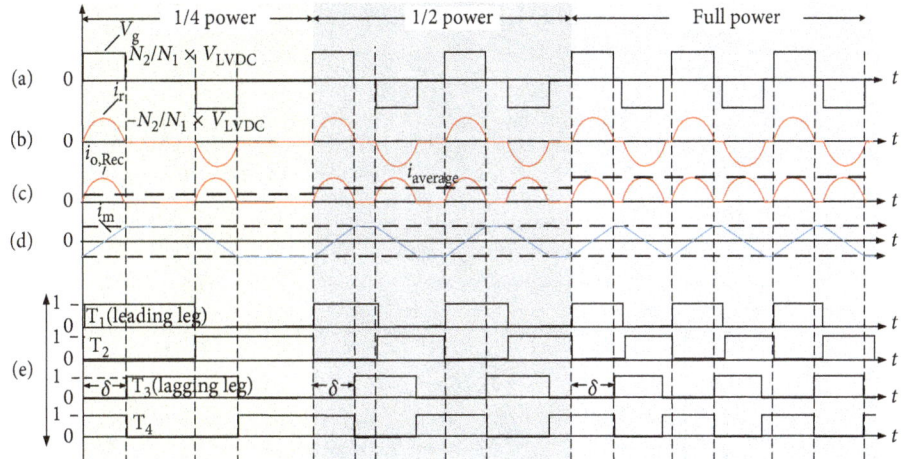

FIGURE 4: Operation waveform of SRC#: (a) inverter output voltage V_g; (b) resonant tank current i_r; (c) rectifier output current $i_{o,Rec}$ and output averaged current $i_{average}$; (d) transformer magnetizing current i_m; (e) switching patterns (T$_1$, T$_2$, T$_3$, and T$_4$) [12].

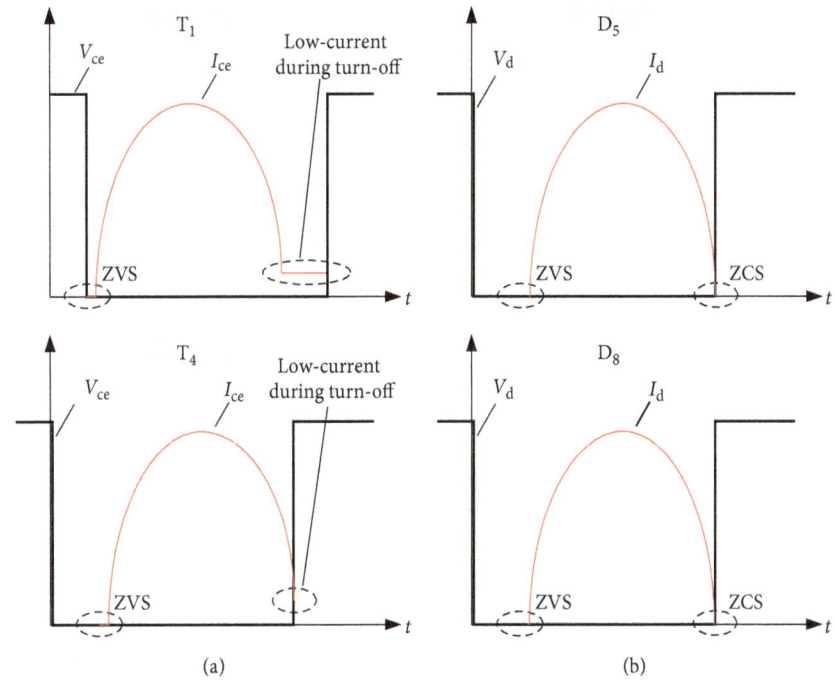

FIGURE 5: Current waveform during switching of SRC# [12]. (a) Example of ZCS or a small current at turnoff: switching waveforms T$_1$ and T$_4$ IGBTs. (b) Example of ZCS at turn-off for the rectifier diodes: switching waveforms D$_5$ and D$_8$ diodes.

$$\begin{bmatrix} \dot{\tilde{x}}_1 \\ \dot{\tilde{x}}_2 \end{bmatrix} = [A]\begin{bmatrix} \tilde{x}_1 \\ \tilde{x}_2 \end{bmatrix} + [B]\begin{bmatrix} \tilde{f}_s \\ \tilde{V}_g \\ \tilde{V}_o \end{bmatrix},$$

$$\tilde{I}_o = [C]\begin{bmatrix} \tilde{x}_1 \\ \tilde{x}_2 \end{bmatrix} + [D]\begin{bmatrix} \tilde{f}_s \\ \tilde{V}_g \\ \tilde{V}_o \end{bmatrix}, \qquad (1)$$

where $\tilde{x}_1 = \tilde{I}_r$ and $\tilde{x}_2 = \tilde{v}_{cr}$.

It is enough to recall here that the duty cycle is fixed to 50%, and one event is defined as a whole time interval in which the IGBT T$_1$ is pulsed or not, lasting half of the switching period [10]. The state variables are meaningful of the tank current and resonant capacitor voltage at the beginning of one event, while the rectified output current is averaged over one event. The average output current is regulated by the proper action of the switching frequency. Transfer functions between converter output current and input variables are

$$\tilde{I}_{o,Rec}(s) = \begin{bmatrix} g_1(s) & g_2(s) & g_3(s) \end{bmatrix}\begin{bmatrix} \tilde{f}_s \\ \tilde{V}_g \\ \tilde{V}_o \end{bmatrix}. \qquad (2)$$

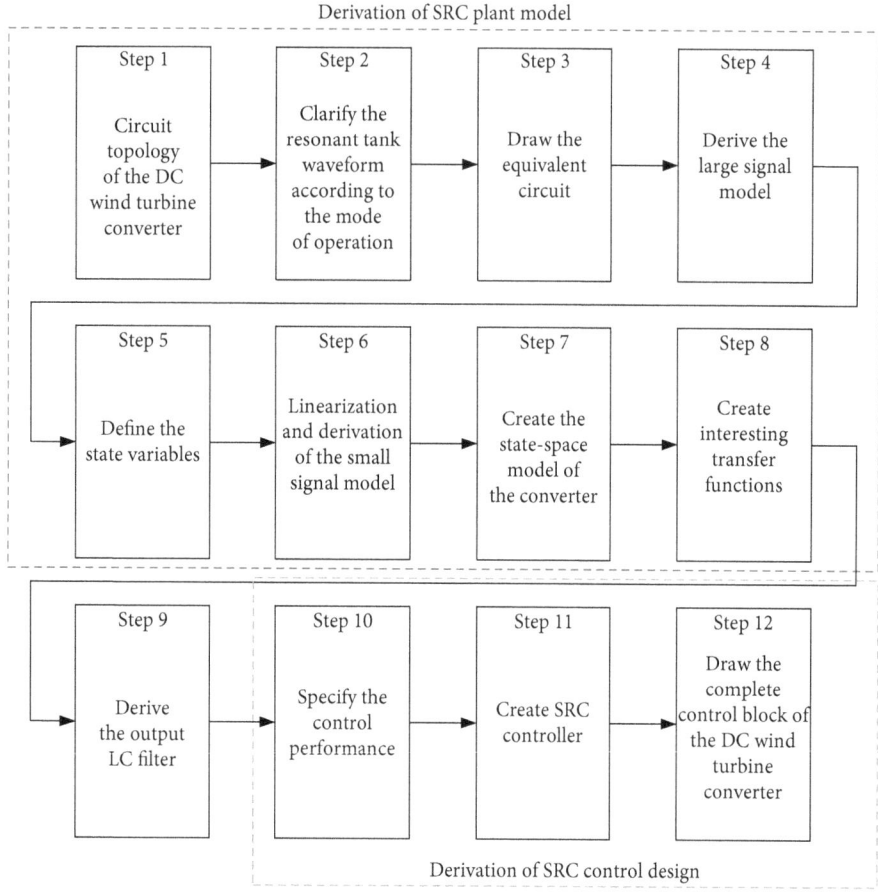

FIGURE 6: Flow chart of mathematical derivation of the converter plant model and the digital controller.

The transfer function between output current and input variables (i.e., $g_1(s)$, $g_2(s)$, and $g_3(s)$) can be obtained via

$$\left[\, g_1(s) \;\; g_2(s) \;\; g_3(s) \,\right] = C\,(sI - A)^{-1}B + D, \qquad (3)$$

where

$$g_1(s) = \left.\frac{\widetilde{I}_{o,Rec}(s)}{\widetilde{f}_s(s)}\right|_{\widetilde{V}_o(s),\widetilde{V}_g(s)=0},$$

$$g_2(s) = \left.\frac{\widetilde{I}_{o,Rec}(s)}{\widetilde{V}_g(s)}\right|_{\widetilde{f}_s(s),\widetilde{V}_o(s)=0}, \qquad (4)$$

$$g_3(s) = \left.\frac{\widetilde{I}_{o,Rec}(s)}{\widetilde{V}_o(s)}\right|_{\widetilde{f}_s(s),\widetilde{V}_g(s)=0}.$$

Those transfer functions describe how the averaged output current of the converter, before the output LC filter, is influenced by input disturbances. Regarding the voltage harmonics in the array network (MVDC network), the transfer function $g_3(s)$ can be used to evaluate the effect of the voltage harmonics on the converter output current. Detailed derivation of the linearized state-space model for SRC# from (1) to (3) and control design including the expression of elements in [A], [B], [C], and [D] matrixes are revealed in Appendix.

The output LC filter is designed with the natural frequency of 100 Hz (i.e., $f_f = 1/(2\pi\sqrt{(L_fC_f)}) \approx 100$ Hz) in order to have constant power delivered to the MVDC grid and attenuate high frequency harmonics, around twice the switching frequency. Based on the plant model of SRC#, the transfer function of the compensator (or controller) $g_c(s)$ in the control block shown in Figure 7 for a specific operating point (OP) is set as

$$g_c = \frac{K}{s \cdot \left(1 - \left(s/\omega_p\right)\right)}. \qquad (5)$$

The transfer function g_c in (5) is designed based on predetermined specifications, such as the ability to reach its reference output current with least ringing, overshoot, and fastest dynamic during perturbations in input variables. The compensator is composed by an integrator to minimize the steady-state error, a pole around the resonant frequency of the filter to lower the resonant magnitude peak influence (ω_p), and finally a gain K, necessary to set a proper crossover frequency, which in this case needs to be lower than the resonant frequency of the filter to avoid instability. Control deign of SRC# is revealed in Appendix, and the parameters of the controller are described in Section 6 that operates in the working conditions specified in Table 1 are shown in Table 2.

Figure 8 demonstrates the transfer function $g_c(s)$ at the operation point ($\omega_p = -250$ rad/s, $K = 50.8$ dB, and $P_{out} = 12.7$ MW). The result shows that the gain margin G_M is

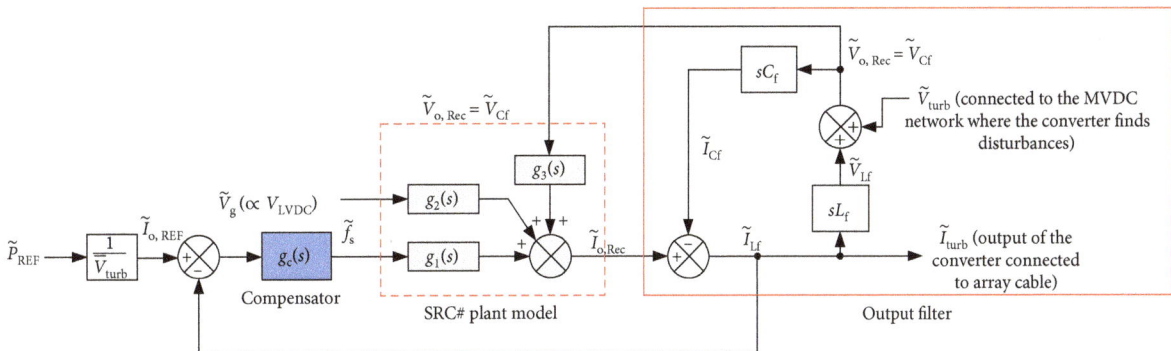

FIGURE 7: Block diagram of the the DC wind turbine converter (SRC#) in the closed-loop control with the output filter.

TABLE 1: Parameters of the plant model of SRC#.

Turns ratio of transformer ($N_1 : N_2$)	$1 : 25$
Resonant inductor, L_r	78.1 (mH)
Resonant capacitor, C_r	0.25 (μF)
Output filter inductor, L_f	250 (mH)
Output filter capacitor, C_f	10 (μF)

TABLE 2: Specifications of operating point and control parameters.

Low-voltage DC, V_{LVDC}	4.04 (kV$_{DC}$)
Medium-voltage DC, V_{MVDC}	100.0 (kV$_{DC}$)
Switching frequency, f_s	800 (Hz)
Rated output power, P_{out}	8.2 (MW)
Crossover frequency	15.91 (Hz)
Gain, K	58.6 (dB)
ω_p	−400 (rad/s)

infinite (>0 dB) and the phase margin Φ_M is around 78.7 deg. at 49.8 rad/s which represents that the controller is stable. The design criterion has been implemented to select the parameters for the different operation points (different output powers of the DC wind turbine), which have enough gain margin and the phase margin to deal with the parameter uncertainty and variation.

4. Modelling of DC Wind Turbine Converter for Harmonic Susceptibility Study

To identify the harmonic model of the DC wind turbine converter for the harmonic susceptibility study, the voltage harmonics within the MVDC network are considered as a perturbation on the steady-state operation point. According to the control block diagram of the SRC# given in Figure 7, the harmonic components (disturbance component $\tilde{I}_{turb}$) on the output current are dominated by the disturbance on the power reference signal P_{REF}, the disturbance in the low-voltage DC network V_g, and the disturbance in the output terminal voltage of diode rectifier V_{turb}.

Without the disturbance component in the power reference signal P_{REF} and in the low voltage DC network V_g ($\tilde{P}_{REF} = 0$ and $\tilde{V}_g = 0$), therefore, the harmonic model of WTG (wind turbine generator) considering the output LC filter (L_f and C_f) can be represented by an admittance equivalent Y_{eq} in (6) as given in Figure 9, where the harmonic admittance Y_{eq} is affected by the circuit parameters of

DC wind turbine converter topology (plant model), the control design, and the output LC filter:

$$Y_{eq} = \frac{-\tilde{I}_{turb}}{\tilde{V}_{turb}} = \frac{sC_f - g_3(s)}{1 + s^2 L_f C_f + g_c(s)g_1(s) - sL_f g_3(s)}, \quad (6)$$

where the admittance of the converter without considering the output filter can be represented by the following equation:

$$Y_c = \frac{-\tilde{I}_{o,Rec}}{\tilde{V}_{o,Rec}} = \frac{-g_3(s)}{1 + g_c(s) \cdot g_1(s)}. \quad (7)$$

Equation (6) describes satisfactorily the variation of the converter current delivered to the grid in response to grid voltage disturbances where the transfer functions g_1 and g_3 in (6) and (7) (the details of expression is in (3)) describe how the output current $\tilde{I}_{o,Rec}$ is influenced by the disturbances in the control input signal $\tilde{f}_s$ and the output voltage $\tilde{V}_o$ ($\propto$ voltage V_{MVDC}), respectively. In practical applications, the output LC filter contains a tiny amount of the parasitic resistances which can alleviate the peak value of admittance magnitude Y_{eq} at the resonances frequency, as done in the following sections for the study case described in Section 6.

Figure 10(a) shows the variation of admittance magnitude $|Y_{eq}|$ of the converter with/without the output LC filter. It also represents the magnitude $|Y_c|$ of the converter without the output LC filter to observe the filter's influence on the equivalent admittance of the converter. Admittance magnitude of the converter with the output filter is high as it approaches the natural frequency of the output LC filter ($f_f \approx$ 100 Hz), while the magnitude of admittance of the converter itself (without output filter) is low in the whole spectrum and reaches a constant value at low frequencies. Regarding the phase angle of the admittance equivalent, the behavior of the converter with output filter changes from inductive, at low frequencies, to capacitive, at the natural frequency of the filter as shown in Figure 10(b).

5. Harmonic Susceptibility Study of DC Collection Network

In this section, the harmonic susceptibility study of DC collection network is revealed based on the harmonic model of the turbine converter developed previously and compared

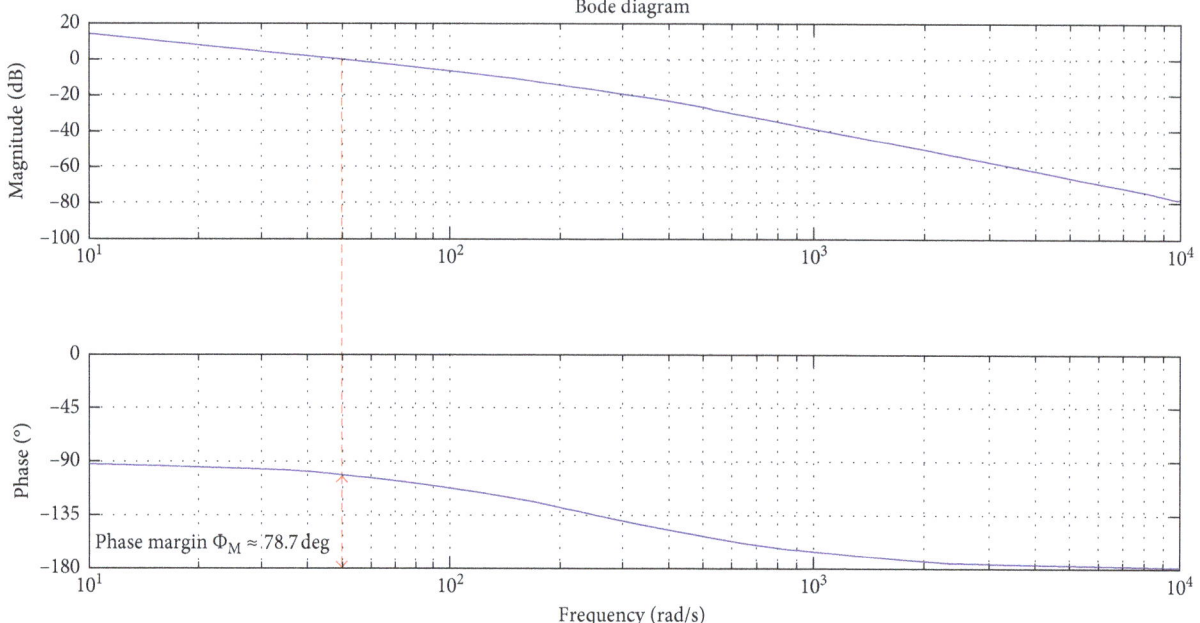

FIGURE 8: Bode plot of $g_c(s)$.

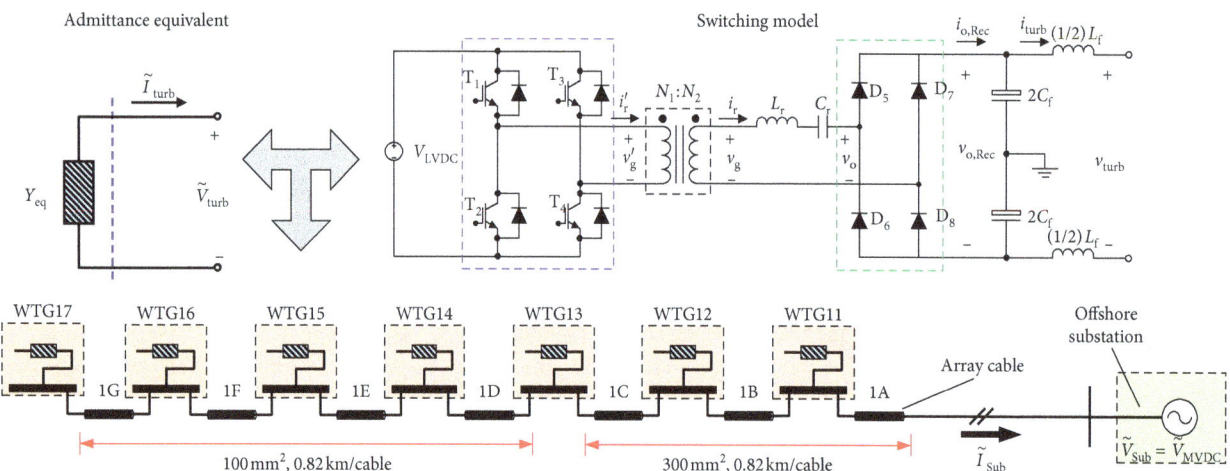

FIGURE 9: Study case of DC collection network for one single cluster of wind turbines with the proposed harmonic model (admittance equivalent) of series resonant converter SRC# (DC wind turbine generator, WTG) with the output filter included (L_f and C_f) for the harmonic susceptibility study.

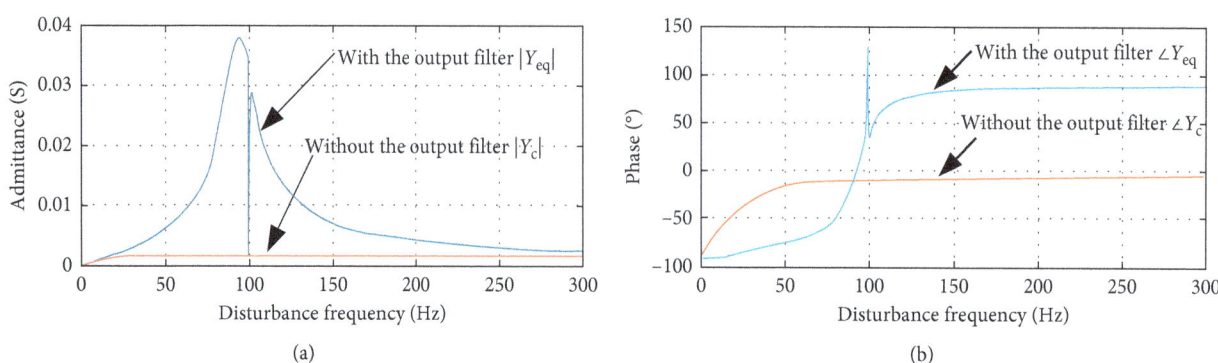

FIGURE 10: Characteristic of equivalent admittance (harmonic model) of the converter for the study case of Tables 1 and 2. (a) Magnitude of admittance. (b) Phase angle of admittance.

to a switching model implemented in PLECS™. According to Figure 9, a radial-type cluster of seven wind turbines of an offshore wind power plant (OWPP) is given. It will represent the study case for the comparison of the harmonic model and switching model. The parameters for the turbine converter and operating point of the plant model with controller are given in Section 6. Since different configurations of collection network and wind turbine converter topologies could generate different harmonic spectra, background harmonics of the DC grid (correction/array network) from the offshore substation DC/DC converter and DC wind turbine generator (DC WTG) can cause undesirable resonant phenomena and raise stability issues. One of the major tasks of the harmonic susceptibility study is to identify resonance frequencies of network and investigate potential voltage amplification issues.

5.1. Cable Model. Array cables in the following study cases are modelled as a pi-equivalent model with longitudinal admittance parallel branches in order to consider the frequency dependency of distributed cable parameters and represent correctly the behavior in the frequency ranging from 0 to 10 kHz [31, 44]. In the study, the electromagnetic transient program *PSCAD* is usefully employed to extract the longitudinal admittance and the shunt capacitance from the geometric data of array cable [41]. The values of resistance, inductance, and capacitance to use at each frequency are calculated by the software individually.

According to the cable model in *PSCAD*, the capacitance is not varying within the range of interesting frequency. Instead, as the frequency rises, the resistance keeps increasing, due to the skin effect, and the inductance decreases its value due to the eddy currents in the outer conductors. For the considered study case, eight *RL* series branches are used to approximate the longitudinal impedance.

The *PSCAD* generates the admittance date (i.e., matrix $[Z_{PSCAD}]$ and $[Y_{PSCAD}]$) from the geometrical data of the marine cable (in frequency-dependent (phase) model) in certain operational frequency to form "pi-equivalent cable model" [45]. An equivalent pi-equivalent model is built based on the cable model, which can fit the admittance date (matrix $[Z_{PSCAD}]$ and $[Y_{PSCAD}]$) from the geometrical data of the submarine cable (implemented in frequency-dependent (phase) model in *PSCAD*) to make sure that the adequate cable model is implemented in the array cable network. Actually, the frequency-dependent phase model in *PSCAD* is also regarded as a most numerically accurate cable model thus far. The formulation of models has been proved that the bandwidths of the model can be up to 1 MHz, which can satisfy most of transient studies and harmonics studies [41, 45].

5.2. Frequency Scan Approach. The WPP cluster consists of seven turbines with seven MVDC cable segments, where the offshore substation converter is represented by an ideal harmonic voltage source. According to the frequency scan approach for network analysis addressed in Figure 11, the harmonic voltage source (disturbance component) in the output DC voltage of the offshore substation converter is activated and injects harmonic voltage component into the MVDC grid with 0.5% (=0.005 p.u) of voltage magnitude ($|V_{MVDC,h,n}|$ = 500 V) and the frequency of injected harmonic voltage raises until 300 Hz, by steps of 20 Hz. The output voltage and output current traces of each wind turbine model in each frequency step are extracted for *FFT* analysis after the steady-state condition is reached. This approach allows an arbitrary choice of frequency and magnitude of the disturbance in the offshore substation. Therefore, it does not intend to represent only a specific range of harmonics in the DC collection network. Eventually, models of the DC collection system (or network) could be established for different frequencies of the injected disturbances. As an example, for the harmonic model, the wind turbine converter can be described by a particular *RL* or *RC* series branch for disturbance frequency.

Based on the harmonic model of DC collection network in Figure 9, Figure 11 gives a flow chart of frequency scan approach for the DC collection system in the harmonic susceptibility study. The process includes the identification of the harmonic model of the DC wind turbine for certain frequency of the injected disturbance (harmonic) and the establishment of the harmonic model of DC collection network. The DC collection network in Figure 9 is set as an example to explain the sequence of frequency scan illustrated in Figure 11. *Step 1*: the harmonic voltage source in the offshore substation ($\tilde{V}_{MVDC}$) is connected to the network and injects a 20 Hz harmonic disturbance into the MVDC network with a specific voltage magnitude $|V_{MVDC}|$. The harmonic model of the DC wind turbine generator (WTG) under 20 Hz perturbation (harmonic frequency) in MVDC network is identified by connecting to the MVDC network at $t = t_1$. *Step 2*: the voltage and current within network are extracted for *FFT* analysis after the voltage and current have reached the steady-state condition. *Step 3*: at $t = t_2$, the frequency for which the harmonic model of the DC wind turbine is evaluated and is increased by 20 Hz, as the frequency of the injected disturbance. The algorithm turns back to *Step 2*. Eventually, the frequency sweep continues until 300 Hz (at $t = t_{n-1}$), which depends on the control bandwidth of the DC wind turbine converter and interesting frequency range in the harmonic susceptibility study.

6. Verification

Based on the DC collection network in Figure 9 and the converter control design addressed in (5), Tables 1 and 2 give the parameters used in the study of the DC collection grid, where the geometric parameters of cables are shown in Figure 12 which is implemented in the created equivalent PI section according to the frequency-dependent (phase) model in *PSCAD*. For the turbine harmonic model, all the turbines are considered operating in the same operating point, described in Table 2. As far for the switching model, wind turbine converters are operated with a switching frequency of 800 Hz that guarantees around an output power of 8.2 MW when the output voltage of the converter is 100 kV. The LVDC source (input voltage) of each converter

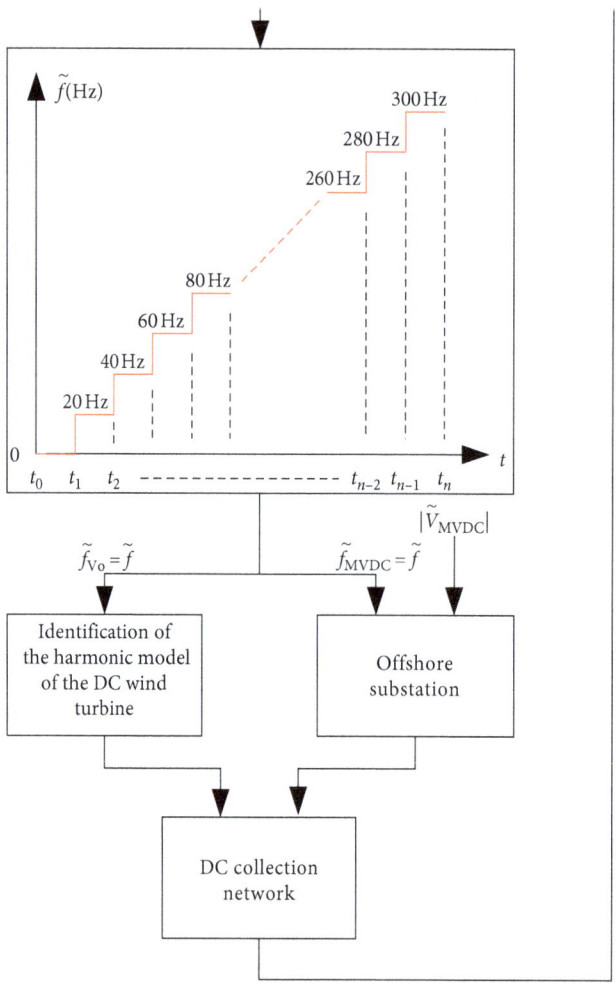

FIGURE 11: Frequency scan approach in the harmonic susceptibility study.

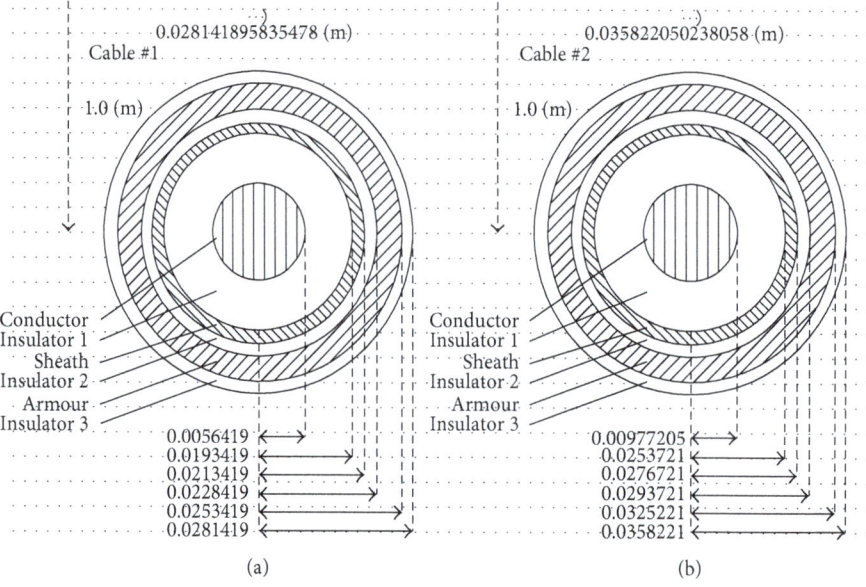

FIGURE 12: MVDC submarine array cables spacing and dimensions (XLPE, pairs of single core conductor cross section 100 mm^2 (a) and 300 mm^2 (b)) [41].

is 4.04 kV, while the voltage drops across cables are responsible for different voltages at the output terminal of the wind turbines. The controller of each turbine regulates the switching frequency in order to control the output power to follow its reference. In particular, the frequencies are higher as wind turbines move further away from the offshore substation, to compensate the lower voltage difference on the two sides of the converter. According to the frequency scan approach illustrated in Figure 11, additionally, a perturbation with a specific sequence of frequency in the range of 20 to 300 Hz is applied to the corresponding harmonic model of turbine converter and the results are given in Figure 13 for a single wind turbine.

Figure 13 gives a verification of the turbine harmonic model (equivalent admittance) by comparing the harmonic spectrum of output current with the results generated by the corresponding switching model of the turbine converter in PLECS™. The current spectrum shows that the derived harmonic model and converter switching model are very close up to 300 Hz. The magnitude of output current spectrum is reduced as the frequency of the harmonic component is far from the natural frequency of the output LC filter, where the mismatch between the harmonic model and its switching model in PLECS™ becomes higher. Furthermore, the turbine output current trace shows that the harmonic component of output current is amplified, while the frequency of MVDC harmonic voltage ($V_{\text{MVDC,h,n}}$) approaches to the natural frequency of output LC filter of the turbine converter.

The results of network (array network) simulation in Figure 14 between both models (harmonic model and the corresponding switching model) show a good agreement. However, some mismatch coming from the harmonic model developed can be seen in all turbines in cluster. Furthermore, as observed along the cluster, last wind turbines (shown in Figures 14(f) and 14(g)) have second- and third-order current harmonics due to nonlinear behavior considered by the turbine converter switching model. This is because the proposed harmonic model cannot predict these higher order harmonics since the modelling approach for the turbine converter in MVDC network are linear. Figure 15 gives a comparison of the harmonic spectrum of output current for wind turbines at different locations to study the error of the harmonic model. While in the closest WTG to the offshore substation, a sinusoidal wave is observed (WTG11); in the last ones, sinusoidal waves have a considerable amount of harmonic components at multiples of the injected disturbance frequency (WTG17). Furthermore, the amplitudes of the fundamental current harmonics decrease along the cluster, approaching the farthest wind turbine from the offshore substation. This phenomenon is related to the rising amplitude of higher order harmonics as depicted in Figure 15. Additionally, it is worth mentioning that, in the switching model, there are current harmonics around 1600 Hz, twice of the switching frequency of converter, as the filter at the output of each converter cannot cancel completely these current harmonics.

The harmonic current spectrum at natural frequency of the output LC filter ($\approx$100 Hz) was not shown in the grid simulation in Figure 14 because it was in the equivalent turbine model (admittance equivalent), since it was "ideal" current source (linear model), the parameters were really high and were not matching the switching model at al. Probably the filter damping is too small which results in very high current (approx. 500 A) at resonance of the output LC filter for a single turbine in grid simulation (equivalent model). However, the results of grid simulation still shows that the admittance equivalent follows results from the switching model for a single turbine in all frequencies; at natural frequency of the turbine output LC filter, the admittance equivalent is clearly not well represented.

In the verification of modelling of the array cable with equivalent PI-cable section based on the method proposed by Beerten et al. [31], the values of resistance and inductance of each branch are calculated in order to fit the behavior of the longitudinal admittance in the frequency domain calculated by *PSCAD's cable model* (frequency-dependent (phase) model). Figure 16 shows a good matching of the fitted values of the admittance by the eight parallel branches ($N = 8$) and the values calculated by software (*PSCAD's cable*). The error on the magnitude of the admittance between two models is less than 1.0%, within the range of 0.001 Hz–10000 Hz. The error in phase is always less than 1.7°, in the range of interesting frequency.

Since the calculation of the admittance date from the frequency-dependent (phase) model in *PSCAD* is not straightforward and is extremely complicate, there are some articles which are useful to conduct the calculation of admittance date from the geometry date of the cable model [28, 30, 45].

7. Laboratory Test

A laboratory test is performed to prove the validity of the proposed harmonic model of the turbine converter. A schematic of the setup of the DC wind turbine converter for SRC# is depicted in Figure 17. The down-scale prototype circuit with 216 V_{DC} (V_{LVDC}) to 400 V_{DC} (V_{MVDC}) DC wind turbine converter is built to verify the harmonic distribution of the turbine converter. The wind turbine converters, whatever AC or DC turbines, are the boost structures in order to increase the efficiency of power transmission. Therefore, the down-scale prototype circuit is designed to match with this character (boost the low-voltage DC, V_{LVDC}, from the generator) to match with the DC grid voltage (medium-voltage DC, V_{MVDC}) from the offshore substation. Although only a single turbine converter is implemented in the test bench, the harmonic signatures of the wind turbine converter (in the test bench) in the interesting frequency range have been confirmed which match with the simulation model as in the following results.

A programmable AC source is used to emulate the harmonic voltage source ($V_{\text{MVDC,h}}$) with sinusoidal waves at different frequencies on the output terminal of the converter. A module of resistors R_{load} and a diode D_{Aux} are placed on the output side of the converter to simulate the character of unidirectional power flow of MVDC network. Table 3

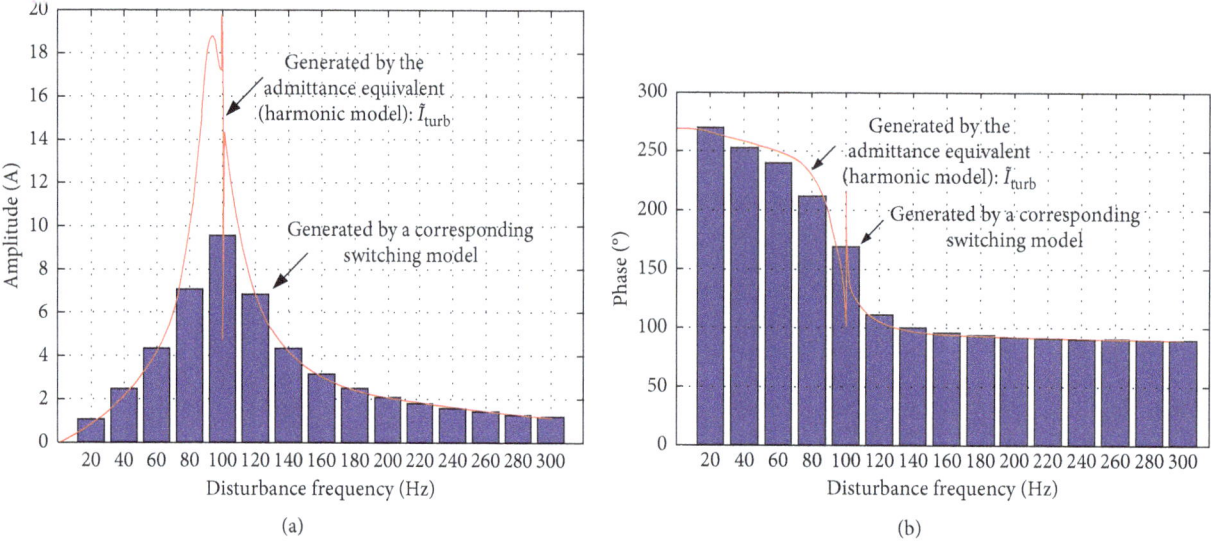

FIGURE 13: Spectrum of turbine output current with a 0.005 p.u harmonic disturbance on the output voltage terminal of the single wind turbine converter ($V_{\text{MVDC}} = \pm 50$ kV). (a) Magnitude of current harmonic. (b) Phase angle of current harmonic.

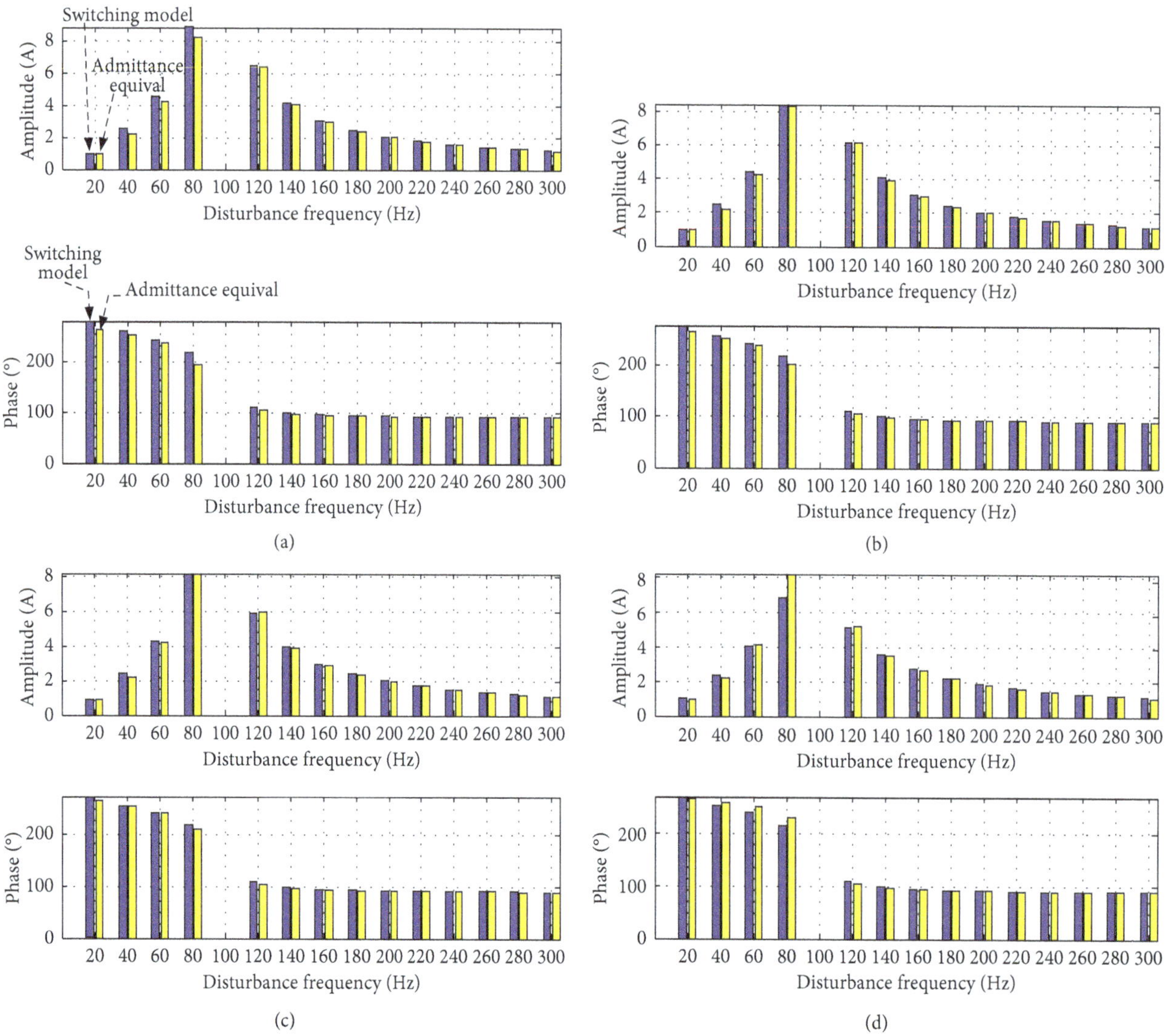

FIGURE 14: Continued.

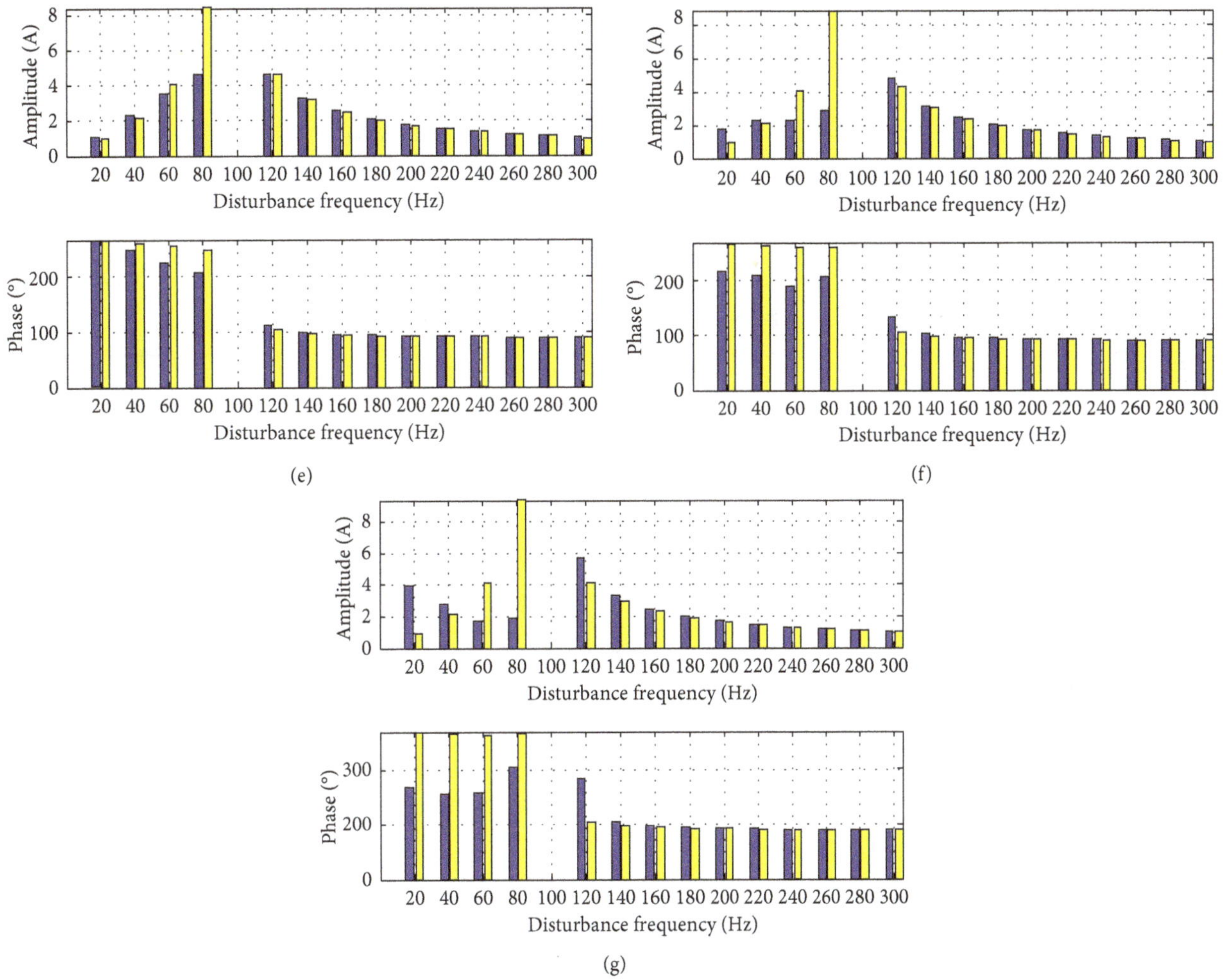

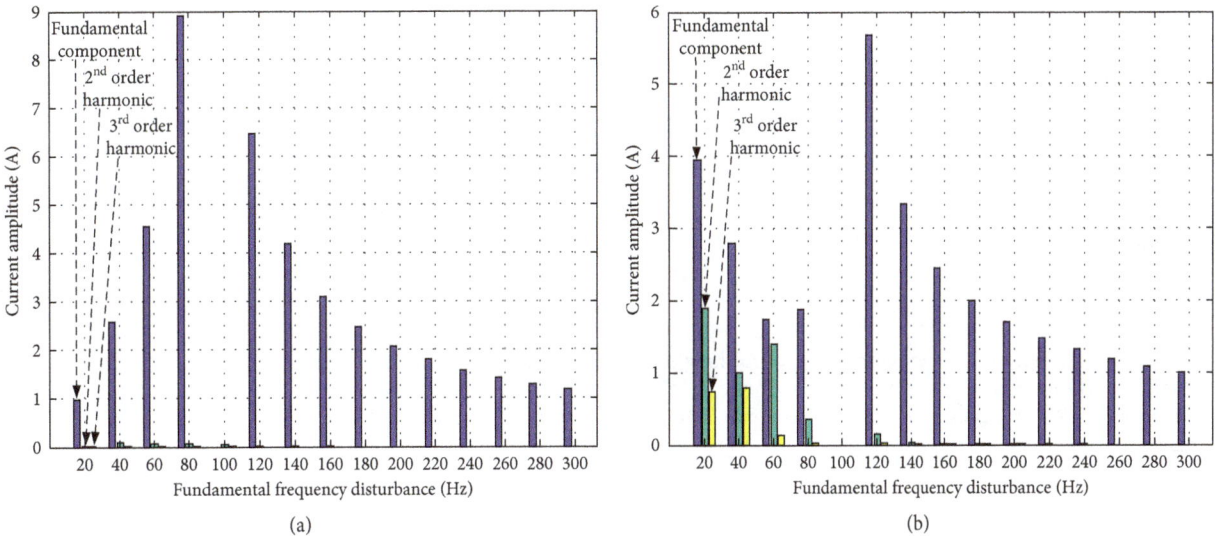

FIGURE 14: Output current spectrum of individual turbines in the operating point described in Tables 1 and 2, being WTG11 the closest to the offshore substation and WTG17 the furthest. Harmonic current amplitude and phase angle are represented for a switching model of the DC wind turbine converter ($|I_{turb,sw,WTGxx}| \angle \theta_{turb,sw,WTGxx}$) and for the admittance equivalent model (harmonic model) ($|I_{turb,eq,WTGxx}| \angle \theta_{turb,eq,WTGxx}$). (a) WTG11. (b) WTG12. (c) WTG13. (d) WTG14. (e) WTG15. (f) WTG16. (g) WTG17.

FIGURE 15: A comparison of output current magnitudes for the switching model in PLECS™ for the WTG11 (first WTG) (a) and WTG17 (seventh WTG) (b) (fundamental component: injected harmonic component).

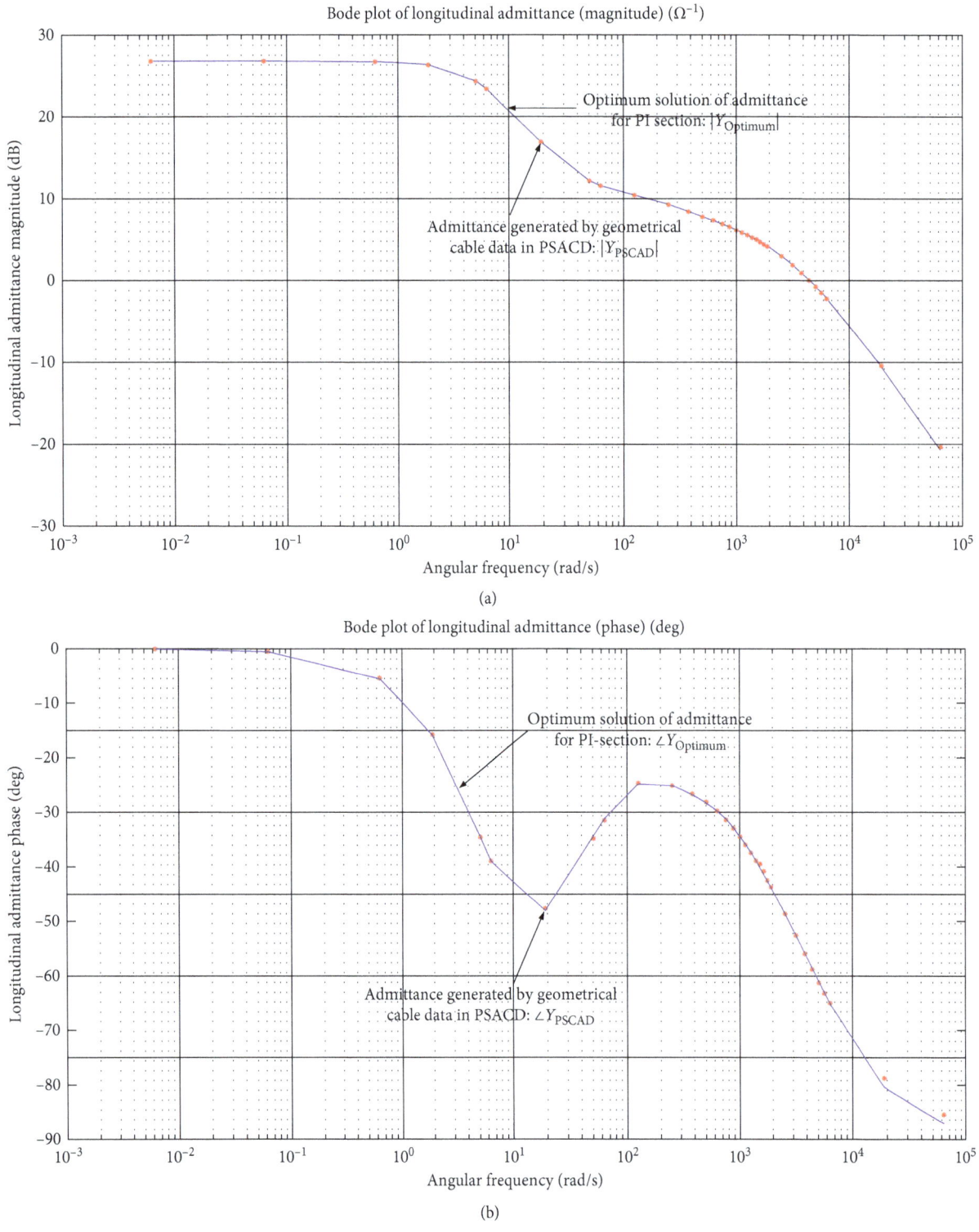

FIGURE 16: Results of the optimum vector fitting for the longitudinal admittance of the submarine array cable with the parallel of *eight* RL series branches. The longitudinal admittance values found in *PSCAD* are highlighted in red star (points), while the admittance of the parallel of the *eight* series RL branches is depicted in blue curve (line). (a) Magnitude of calculated admittance. (b) Phase angle of calculated admittance.

describes the power rating and circuit parameters of the laboratory test bench of *SRC#* where the average of *output voltage* is around 400 V_{DC} ($V_{MVDC} = V_{turb}$, DC component of output voltage of the turbine) and the *output current* (DC component of output current of the turbine) is around 1.4 A_{DC}.

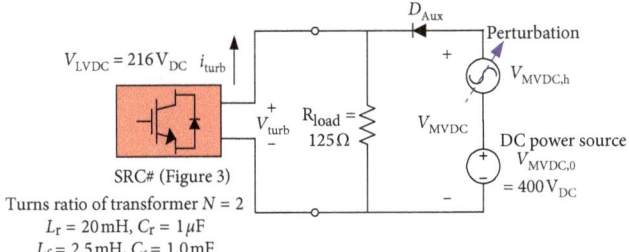

$V_{LVDC} = 216\,V_{DC}$ i_{turb}

SRC# (Figure 3)

Turns ratio of transformer $N = 2$
$L_r = 20\,\text{mH}$, $C_r = 1\,\mu\text{F}$
$L_f = 2.5\,\text{mH}$, $C_f = 1.0\,\text{mF}$

FIGURE 17: Schematic of the laboratory setup for experiment.

TABLE 3: Specifications of the laboratory test bench.

Low-voltage DC source, V_{LVDC}	216 (V_{DC})
DC component of medium-voltage source, $V_{MVDC,0}$	400 (V_{DC})
Transformer winding voltage ratio, $N_1 : N_2$	1 : 2
Output power, P_{out}	550 (W)
Resonant inductor, L_r	20.0 (mH)
Resonant capacitor, C_r	1.0 (μF)
Output filter inductor, L_f	2.5 (mH)
Output filter capacitor, C_f	1.0 (mF)
Resistive load, R_{load}	125 (Ω)
Duty cycle	50%

A harmonic model and switching model in PLECS™ are also developed with the parameters of the laboratory setup for the comparison. The operating switching of SRC# is set to 800 Hz with 50% duty cycle and an output power of 550 W. Input voltage (V_{LVDC}) is set to 216 V, which is 11 V slightly higher than the input voltage (205 V) in the switching model developed in PLECS™ to take into account voltage drops in the practical setup. The DC component of converter's output voltage ($V_{MVDC,0}$) is set to 400 V. A harmonic voltage source ($V_{MVDC,h}$) of $3.5\,V_{RMS}$, with frequency varying in the range of 20 to 300 Hz, is used for harmonic susceptibility tests. Harmonics current in the output are recorded and the results from the laboratory experiment, PLECS™ simulations, and the admittance equivalent developed for the laboratory setup are compared in Figure 18. The results show that there is a good agreement with the proposed DC turbine harmonic model. However, there is a mismatch in the amplitudes (Figure 18(a)) when the injected frequencies approach the natural frequency of the output LC filter. As given in Section 4, the admittance of the DC turbine harmonic model is heavily influenced by the output LC filter. The behavior of the converter changes from inductive to capacitive at the natural frequency of the filter as in Figure 18(b).

It should be noted that there is no standard for selecting the medium-voltage DC (MVDC) grid. The MVDC of $400\,V_{DC}$ is chosen as output voltage of the down-scale prototype circuit of SRC# from the available test equipment in the laboratory for simulating the MVDC grid (controllable DC power source). In practical applications, the selection of MVDC is based on the electromagnetic transient (EMT) simulation of the grid and existing offshore wind farm technologies such as power converters, protection devices, available offshore platform, and array cables [1].

8. Conclusion

A model-based methodology in conducting the harmonic susceptibility study is revealed based on the discrete time-domain modelling technique and the frequency scan approach in the frequency domain. The developed harmonic model in the paper can be used to perform a harmonic susceptibility study of power collection network in an early stage of a wind farm project because of its simplicity and low computational cost compared to a switching model. In the later stage, the proposed method is recommended to consider the switching model of the converters in order to be closer to reality. The process of harmonic susceptibility study contains identification of the harmonic model of the turbine converter and the establishment of array network in the frequency domain. The validation of the harmonic model of the turbine converter is given in Sections 6 and 7 via computer simulation implemented in PLECS™ and the laboratory tests, respectively. Additionally, the distribution of current harmonics within DC collection network is investigated including the identification of possible resonance frequency within array network. Moreover, analytical data from simulation and the laboratory tests give indications to set a proper design and natural frequency for the output LC filter of the DC turbine converter and a matched switching frequency for the converter placed in the offshore substation.

Appendix

According to the flow chart of mathematical derivation in Figure 6 in Section 3, firstly, the circuit topology of the turbine converter and mode of operations are decided as shown in Figures 3 and 19, and then the equivalent circuit based on the switching sequence of transistors is created as shown in Figure 20. In Figure 20, the large-signal model of the converter is generated (Step 4) and then the interesting state variables (Step 5) are defined to create the small-signal equations. Steps 6 and 7 establish the space model of the plant. Finally, the converter plant mode is established based on the *discrete time-domain modelling* approach proposed in (Step 8) [42, 46]. Step 9 gives a set of interesting transfer function of the output LC filter. Eventually, Steps 10 to 12 focus on the control design for SRC#.

A. Derivation of the Plant Model

Steps 1–3. Decide the circuit topology of DC turbine converter, resonant tank waveform, and equivalent circuit.

Steps 1–3 describe how to obtain the large-signal model and the corresponding equivalent circuit in subresonant CCM in Figure 20 from the circuit topology of SRC#, which is operated in subresonant CCM as in Figures 3 and 20, respectively. The waveform in Figure 19 is divided by different time zones (different switching sequences) based on the *discrete time-domain modelling* approach. The figures are eventually used to generate the large-signal model of SRC#.

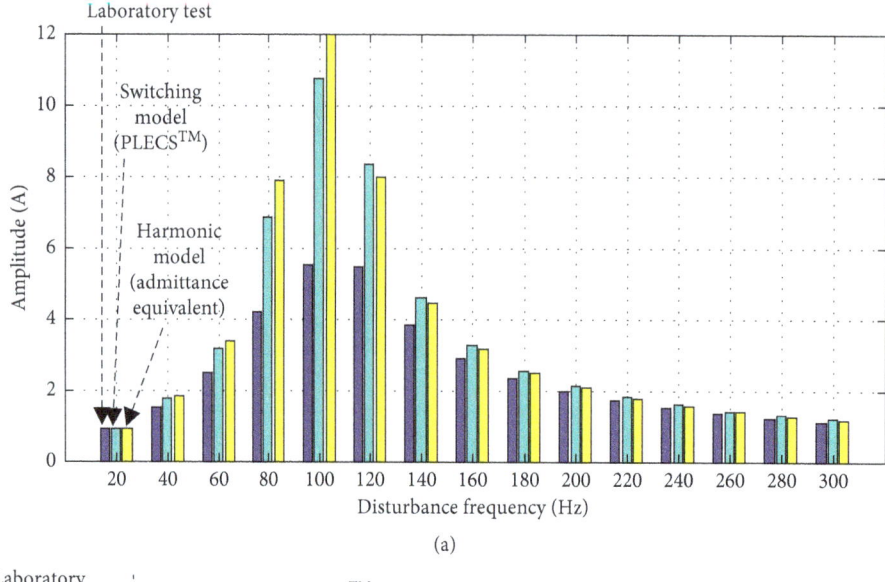

(a)

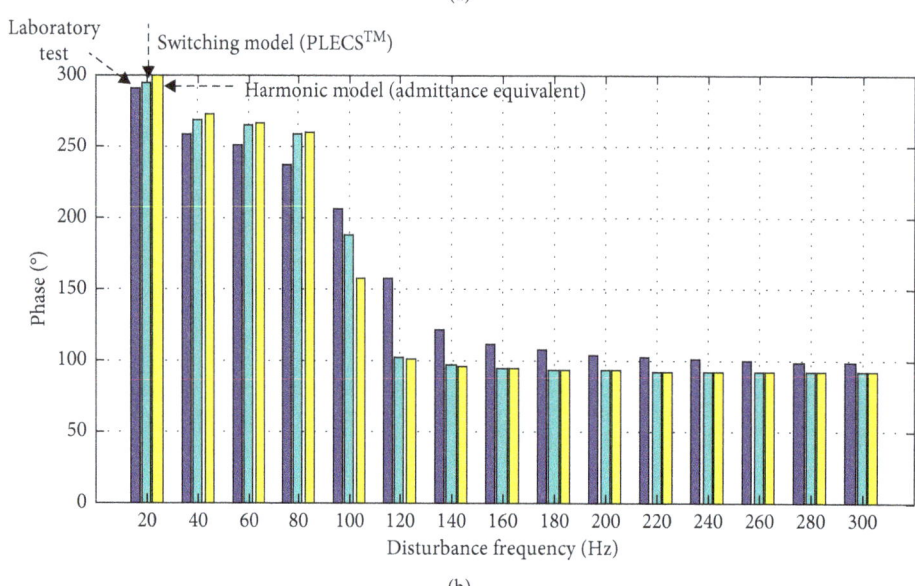

(b)

FIGURE 18: Harmonic spectrum of output currents for the scaled-down experiment, with 5.0 V perturbation in amplitude in the range of 20 to 300 Hz: laboratory test result, result generated by the switching model in PLECS™ simulation and result generated by the proposed harmonic model. (a) Magnitude of current harmonics. (b) Phase angle of current harmonics.

Step 4. Large-signal model.

In the large-signal model, the final value of interesting state variables in each switching interval is represented by the initial values of states. According to *discrete time-domain modelling* approach, the procedure is only valid when the variations in MVDC grid voltage $v_o(t)$ (output terminal voltage) and LVDC voltage $v_g(t)$ (input terminal voltage) in the event (switching) are relatively small than their initial and final values [42]. According to Figures 19 and 20, Equations (A.1), (A.3), (A.4), (A.6)–(A.10), (A.12), and (A.13) show the details of derivation of the large-signal model of resonant inductor current $i_r(t)$, capacitor voltage $v_{Cr}(t)$, and their end values at k_{th} event in terms of initial values of k_{th} event.

For $t_{0(k)} \leq t \leq t_{1(k)}$ (T$_1$, T$_4$ ON),

$$v_g = L_r \frac{di_r}{dt} + v_{Cr} + v_o,$$

$$i_r = C_r \frac{dv_{Cr}}{dt}, \tag{A.1}$$

where

$$v_g = V_{g,0(k)},$$

$$v_o = V_{o,0(k)}. \tag{A.2}$$

The resonant inductor current $i_r(t)$ and resonant capacitor voltage $v_{Cr}(t)$ can be obtained by (A.1):

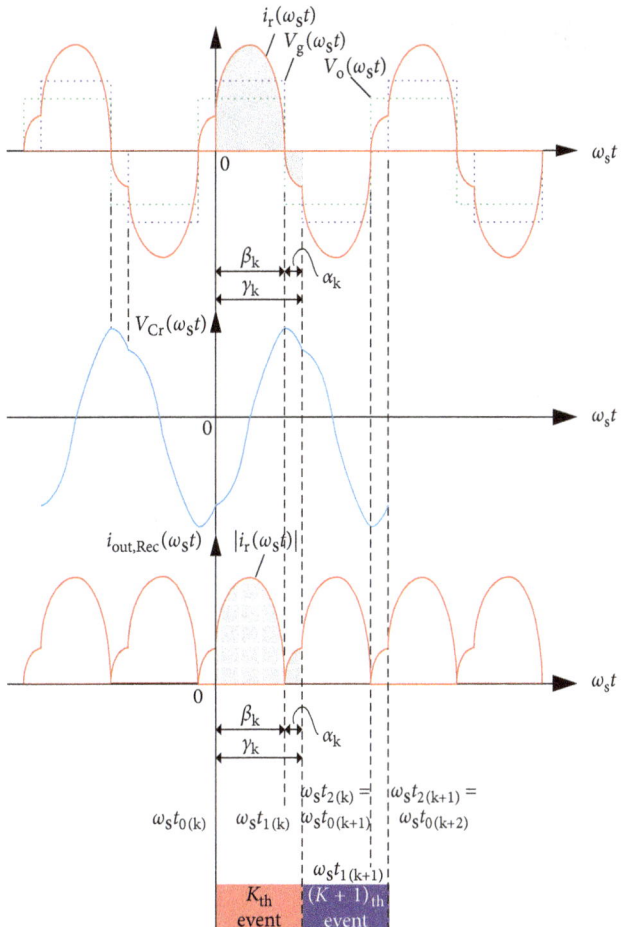

FIGURE 19: Resonant inductor current and resonant capacitor voltage waveforms of SRC# in subresonant CCM.

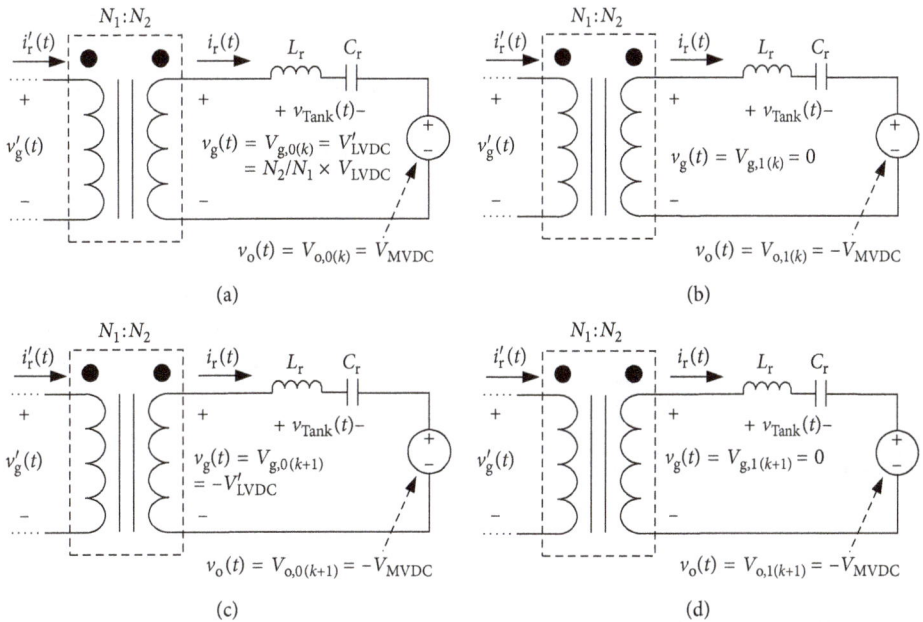

FIGURE 20: Equivalent circuit of SRC# for large-signal analysis of conduction intervals in subresonant CCM. (a) $t_{0(k)} \leq t \leq t_{1(k)}$ (T_1, T_4 ON). (b) $t_{1(k)} \leq t \leq t_{2(k)}$ (D_1, T_3 ON). (c) $t_{0(k+1)} \leq t \leq t_{1(k+1)}$ (T_2, T_3 ON). (d) $t_{1(k+1)} \leq t \leq t_{2(k+1)}$ (D_2, T_4 ON).

$$i_r = \frac{1}{Z_r}\left(V_{g,0(k)} - V_{o,0(k)} - V_{Cr,0(k)}\right)\sin\left(\omega_r t\right)$$

$$+ I_{r,0(k)}\cos\left(\omega_r t\right), \tag{A.3}$$

$$v_{Cr} = V_{g,0(k)} - \left(V_{g,0(k)} - V_{o,0(k)} - V_{Cr,0(k)}\right)\cos\left(\omega_r t\right)$$
$$+ I_{r,0(k)}Z_r \sin\left(\omega_r t\right) - V_{o,0(k)}, \tag{A.4}$$

where

$$Z_r = \sqrt{\frac{L_r}{C_r}}, \tag{A.5}$$

$$\omega_r = \frac{1}{\sqrt{L_r C_r}}.$$

At time $t = t_{1(k)}$, the tank current i_r reaches the point of zero crossing, commutating the IGBT switches T_1 and T_4 turnoff and turning on body-diode D_1, and IGBT switch T_3. Therefore, the resonant inductor current $i_r(t)$ can be expressed by

$$I_{r,1(k)} = i_r\left(t_{1(k)}\right) = \frac{1}{Z_r}\left(V_{g,0(k)} - V_{o,0(k)} - V_{Cr,0(k)}\right)\sin\left(\omega_r t_{1(k)}\right)$$

$$+ I_{r,0(k)}\cos\left(\omega_r t_{1(k)}\right) = 0, \tag{A.6}$$

where

$$\omega_r t_{1(k)} = \frac{\omega_r}{\omega_s}\beta_K = \omega_{rs}\cdot\beta_K \quad \text{for } t_{0(k)} = 0, \tag{A.7}$$

$$\tan\left(\omega_{rs}\beta_K\right) = \frac{-I_{r,0(k)}Z_r}{\left(V_{g,0(k)} - V_{o,0(k)} - V_{Cr,0(k)}\right)},$$

$$0 < \left(\omega_{rs}\beta_K\right) \le \pi, \quad t_{1(k)} = \beta_K/\omega_s,$$

$$V_{Cr,1(k)} = v_{Cr}\left(t_{1(k)}\right) = V_{g,0(k)} - \left(V_{g,0(k)} - V_{o,0(k)} - V_{Cr,0(k)}\right)$$
$$\cdot \cos\left(\omega_{rs}\beta_K\right) + I_{r,0(k)}Z_r\cdot\sin\left(\omega_{rs}\beta_K\right) - V_{o,0(k)}. \tag{A.8}$$

For $t_{1(k)} \le t \le t_{2(k)}$ (D_1, T_3 ON),

$$i_r(t') = \frac{-1}{Z_r}\left(V_{o,1(k)} + V_{Cr,1(k)}\right)\sin\left(\omega_r t'\right), \tag{A.9}$$

where

$$t' = t - t_{1(k)},$$

$$v_{Cr}(t') = \left(V_{o,1(k)} + V_{Cr,1(k)}\right)\cos\left(\omega_r t'\right) - V_{o,1(k)}, \tag{A.10}$$

where

$$I_{r,1(k)} = 0,$$
$$V_{g,1(k)} = 0, \tag{A.11}$$
$$V_{o,1(k)} = -V_{MVDC} = -V_{o,0(k)}.$$

Eventually, the end values of inductor current $i_r(t)$ and capacitor voltage $v_{Cr}(t)$ (at time $t = t_{2(k)}$) can be written as

$$I_{r,2(k)} = \left[-\sin\left(\omega_{rs}\beta_K\right)\cdot\sin\left(\omega_{rs}\alpha_K\right)\right]\cdot I_{r,0(k)}$$

$$+ \left[-\frac{1}{Z_r}\cdot\cos\left(\omega_{rs}\beta_K\right)\cdot\sin\left(\omega_{rs}\alpha_K\right)\right]\cdot V_{Cr,0(k)}$$

$$+ \left[\frac{2}{Z_r}\cdot\sin\left(\omega_{rs}\alpha_K\right) + \frac{-1}{Z_r}\cdot\cos\left(\omega_{rs}\beta_K\right)\cdot\sin\left(\omega_{rs}\alpha_K\right)\right]$$

$$\cdot V_{o,0(k)} + \left[\frac{-1}{Z_r}\cdot\sin\left(\omega_{rs}\alpha_K\right) + \frac{1}{Z_r}\cdot\cos\left(\omega_{rs}\beta_K\right)\right]$$

$$\cdot \sin\left(\omega_{rs}\alpha_K\right)\right]\cdot V_{g,0(k)}, \tag{A.12}$$

$$V_{Cr,2(k)} = \left[Z_r\cdot\sin\left(\omega_{rs}\beta_K\right)\cdot\cos\left(\omega_{rs}\alpha_K\right)\right]\cdot I_{r,0(k)}$$

$$+ \left[\cos\left(\omega_{rs}\beta_K\right)\cdot\cos\left(\omega_{rs}\alpha_K\right)\right]\cdot V_{Cr,0(k)} + \left[-2\right.$$

$$\cdot \cos\left(\omega_{rs}\alpha_K\right) + \cos\left(\omega_{rs}\beta_K\right)\cdot\cos\left(\omega_{rs}\alpha_K\right) + 1\right]$$

$$\cdot V_{o,0(k)} + \left[\cos\left(\omega_{rs}\alpha_K\right) - \cos\left(\omega_{rs}\beta_K\right)\right.$$

$$\cdot \cos\left(\omega_{rs}\alpha_K\right)\right]\cdot V_{g,0(k)}, \tag{A.13}$$

where

$$\omega_s\cdot\left(t_{2(k)} - t_{1(k)}\right) = \alpha_K. \tag{A.14}$$

The same process of derivation as given in (A.1), (A.3), (A.4), (A.6)–(A.10), (A.12), and (A.13) can be applied for obtaining the large-signal expression of resonant inductor current $i_r(t)$ and resonant capacitor voltage $v_{Cr}(t)$ in $(k + 1)$th event ($t_{0(k+1)} \le t \le t_{1(k+1)}$ and $t_{1(k+1)} \le t \le t_{2(k+1)}$).

A.1. Steady-State Solution of Large-Signal Model. The following equation expresses the conditions for obtaining steady-state solution (at certain operating points of CCM) of the discrete state equation of the large-signal model:

$$I_{r,2(k)} = -I_{r,0(k)},$$
$$V_{Cr,2(k)} = -V_{Cr,0(k)}. \tag{A.15}$$

By substituting (A.15) into (A.12) and (A.13), the steady-state solution of $I_{r,0(k)}$ and $V_{Cr,0(k)}$ can be written in term of $V_{o,0(k)}$, $V_{g,0(k)}$, β_k, and α_k:

$$\bar{I}_r = I_{r,0(k)} = f\left(V_{o,0(k)}, V_{g,0(k)}, \beta_K, \alpha_k\right),$$
$$\bar{V}_{Cr} = V_{Cr,0(k)} = f\left(V_{o,0(k)}, V_{g,0(k)}, \beta_K, \alpha_k\right), \tag{A.16}$$

where the overbar used in equations is to indicate the steady-state value of the interesting state variables. For simplifying the derivation of the plant model, the output LC filter of SRC# (i.e., L_f and C_f) is neglected due to a very slow dynamics in voltage and current comparing with the dynamics in the resonant inductor current $i_r(t)$ and resonant capacitor voltage

$v_{Cr}(t)$. Therefore, only the DC component of the output current diode rectifier $i_{out,Rec}$ is considered as an output variable which is expressed by i_o. Eventually, the output current equation delivered by the SRC# during the K_{th} event is express as

$$
i_o = \frac{1}{\gamma_k} \int_0^{\beta_k} i_{out,Rec}(\theta_s)\, d\theta_s + \frac{1}{\gamma_k} \int_{\beta_k}^{\gamma_k} i_{out,Rec}(\theta_s)\, d\theta_s
$$

$$
= \frac{1}{\gamma_k} \cdot \left\{ \frac{1}{\omega_{rs}} \sin(\omega_{rs}\beta_k) + \frac{1}{\omega_{rs}} \sin(\omega_{rs}\beta_K) \cdot [1 - \cos(\omega_{rs}\alpha_k)] \right\}
$$

$$
\cdot I_{r,0(k)} + \frac{1}{\gamma_k} \cdot \left\{ -\frac{1}{\omega_{rs}} \frac{1}{Z_r} [1 - \cos(\omega_{rs}\beta_k)] + \frac{1}{\omega_{rs}} \frac{1}{Z_r} \cos \right.
$$

$$
\left. (\omega_{rs}\beta_K) \cdot [1 - \cos(\omega_{rs}\alpha_k)] \right\} \cdot V_{Cr,0(k)} + \frac{1}{\gamma_k}
$$

$$
\cdot \left\{ -\frac{1}{\omega_{rs}} \frac{1}{Z_r} [1 - \cos(\omega_{rs}\beta_k)] + \frac{1}{\omega_{rs}} \frac{1}{Z_r} (\cos(\omega_{rs}\beta_K) - 2) \right.
$$

$$
\left. \cdot [1 - \cos(\omega_{rs}\alpha_k)] \right\} \cdot V_{o,0(k)} + \frac{1}{\gamma_k}
$$

$$
\cdot \left\{ \frac{1}{\omega_{rs}} \frac{1}{Z_r} [1 - \cos(\omega_{rs}\beta_k)] + \frac{1}{\omega_{rs}} \frac{1}{Z_r} (1 - \cos(\omega_{rs}\beta_K)) \right.
$$

$$
\left. \cdot [1 - \cos(\omega_{rs}\alpha_k)] \right\} \cdot V_{g,0(k)},
$$

$$
\tag{A.17}
$$

where

$$
\begin{aligned}
\theta_s &= \omega_s t, \\
\alpha_k &= \gamma_k - \beta_k.
\end{aligned}
\tag{A.18}
$$

It should be noted that the initial value of inductor current is represented by $I_{r,0(k)}$, $V_{Cr,0(k)}$ is the initial value of capacitor voltage, the initial value of rectifier output voltage is written as $V_{o,0(k)}$, and $V_{g,0(k)}$ is then the initial value of input voltage of SRC#. Characteristic impedance Z_r is defined by the parameter of resonant tanks ($= \sqrt{L_r C_r}$), α_k is the transistor and diode conduction angle during the switching interval event k, and the switching frequency of the converter is represented by θ_s ($=\omega_s \cdot t$). Finally, the steady-state solution of the discrete-state equation for the output variable i_o can be obtained by substituting the steady-state condition into (A.17) as

$$
\bar{I}_o = i_o \big|_{(\beta_k = \bar{\beta}, \alpha_k = \bar{\alpha}, \gamma_k = \bar{\gamma}, I_{r,0(k)} = \bar{I}_r, V_{Cr,0(k)} = \bar{V}_{cr}, V_{o,0(k)} = \bar{V}_o, V_{g,0(k)} = \bar{V}_g)}.
$$

$$
\tag{A.19}
$$

Step 5. Define state variable.

Control design technique based on the linear control theory cannot directly be applied for, mainly because of the high nonlinearity in the discrete large-signal state equations in (A.12), (A.13), and (A.17). Therefore, the linearization of the above large-signal equations is necessary. Equation (A.20) shows the definitions of interesting state variables in both k_{th} switching event ($t_{0(k)} \le t \le t_{2(k)}$) and $(k + 1)_{th}$ switching event ($t_{2(k)} \le t \le t_{2(k+1)}$) which is used in the

linearization process. Eventually, the following equations of approximation of derivative are applied for converting the discrete large-signal model into the continuous time:

$$
\begin{aligned}
x_{1(k)} &= I_{r,0(k)}, \\
x_{2(k)} &= V_{Cr,0(k)}, \\
I_{r,2(k)} &= -x_{1(k+1)}, \\
V_{Cr,2(k)} &= -x_{2(k+1)},
\end{aligned}
\tag{A.20}
$$

$$
\dot{x}_i(t_k) = \frac{x_{i,(k+1)} - x_{i,(k)}}{t_{0(k+1)} - t_{0(k)}} = \frac{\omega_s}{\gamma_k} \left(x_{i,(k+1)} - x_{i,(k)} \right),
\tag{A.21}
$$

where

$$
\gamma_k = \omega_s \big(t_{2(k)} - t_{0(k)} \big) = \omega_s \big(t_{0(k+1)} - t_{0(k)} \big).
\tag{A.22}
$$

Instead of using the state variables in (A.12) and (A.13), the state variables in (A.12) and (A.13) are replaced by the defined state given in (A.20). With the approximation of derivation in (A.21), the nonlinear state-space model is given by

$$
\dot{x}_{1(k)} = \frac{\omega_s}{\gamma_k} \cdot [\sin(\omega_{rs}\beta_K) \cdot \sin(\omega_{rs}\alpha_K) - 1] \cdot x_{1(k)}
$$

$$
+ \frac{\omega_s}{\gamma_k} \cdot \left[\frac{1}{Z_r} \cdot \cos(\omega_{rs}\beta_K) \cdot \sin(\omega_{rs}\alpha_K) \right] \cdot x_{2(k)} + \frac{\omega_s}{\gamma_k}
$$

$$
\cdot \left[\frac{-2}{Z_r} \cdot \sin(\omega_{rs}\alpha_K) + \frac{1}{Z_r} \cdot \cos(\omega_{rs}\beta_K) \cdot \sin(\omega_{rs}\alpha_K) \right]
$$

$$
\cdot V_{o,0(k)} + \frac{\omega_s}{\gamma_k} \cdot \left[\frac{1}{Z_r} \cdot \sin(\omega_{rs}\alpha_K) + \frac{-1}{Z_r} \cdot \cos(\omega_{rs}\beta_K) \right.
$$

$$
\left. \cdot \sin(\omega_{rs}\alpha_K) \right] \cdot V_{g,0(k)},
$$

$$
= f_1 \{ x_{1(k)}, x_{2(k)}, V_{o,0(k)}, V_{g,0(k)}, \alpha_k \}
$$

$$
= f_1 \{ x_1, x_2, v_o, v_g, \alpha \}
$$

$$
= \frac{\omega_s}{\gamma_k} \cdot f_1^* \{ x_1, x_2, v_o, v_g, \alpha \},
$$

$$
\dot{x}_{2(k)} = \frac{\omega_s}{\gamma_k} \cdot [-Z_r \cdot \sin(\omega_{rs}\beta_K) \cdot \cos(\omega_{rs}\alpha_K)] \cdot x_{1(k)}
$$

$$
+ \frac{\omega_s}{\gamma_k} \cdot [-\cos(\omega_{rs}\beta_K) \cdot \cos(\omega_{rs}\alpha_K) - 1] \cdot x_{2(k)} + \frac{\omega_s}{\gamma_k}
$$

$$
\cdot [2 \cdot \cos(\omega_{rs}\alpha_K) - \cos(\omega_{rs}\beta_K) \cdot \cos(\omega_{rs}\alpha_K) - 1]
$$

$$
\cdot V_{o,0(k)} + \frac{\omega_s}{\gamma_k} \cdot \left[-\cos(\omega_{rs}\alpha_K) + \cos(\omega_{rs}\beta_K) \right.
$$

$$
\left. \cdot \cos(\omega_{rs}\alpha_K) \right] \cdot V_{g,0(k)},
$$

$$
= f_2 \{ x_{1(k)}, x_{2(k)}, V_{o,0(k)}, V_{g,0(k)}, \alpha_k \}
$$

$$
= f_2 \{ x_1, x_2, v_o, v_g, \alpha \}
$$

$$
= \frac{\omega_s}{\gamma_k} \cdot f_2^* \{ x_1, x_2, v_o, v_g, \alpha \},
$$

$$
\tag{A.23}
$$

where the output equation is

$$i_o = f_{out}\{x_{1(k)}, x_{2(k)}, V_{o,0(k)}, V_{g,0(k)}, \alpha_k\},$$

$$= f_{out}\{x_1, x_2, v_o, v_g, \alpha\} \tag{A.24}$$

$$= \frac{1}{\gamma_k} \cdot f_{out}^*\{x_1, x_2, v_o, v_g, \alpha\}.$$

Step 6. Linearization and small-signal model.

Inject a small perturbation in all the interesting state variables in Step 5 during the steady state (near the certain operating point, OP). The nonlinear state equations can be rewritten with *Taylor series expansion* in terms of the operating point (OP) and the perturbations:

(i) Taylor series expansion of resonant inductor current:

$$[\dot{\overline{x}}_1 + \dot{\tilde{x}}_1] = f_1\{\overline{x}_1 + \tilde{x}_1, \overline{x}_2 + \tilde{x}_2, \overline{V}_o + \tilde{V}_o, \overline{V}_g$$

$$+ \tilde{V}_g, \overline{\alpha} + \tilde{\alpha}\},$$

$$= \frac{\omega_s}{\gamma_k} \cdot f_1^*\{\overline{x}_1 + \tilde{x}_1, \overline{x}_2 + \tilde{x}_2, \overline{V}_o + \tilde{V}_o, \overline{V}_g$$

$$+ \tilde{V}_g, \overline{\alpha} + \tilde{\alpha}\}, \tag{A.25}$$

where

$$x_1 = \overline{x}_1 + \tilde{x}_1,$$
$$x_2 = \overline{x}_2 + \tilde{x}_2,$$
$$v_g = \overline{V}_g + \tilde{V}_g, \tag{A.26}$$
$$v_o = \overline{V}_o + \tilde{V}_o,$$
$$\alpha = \overline{\alpha} + \tilde{\alpha},$$

then

$$[\dot{\overline{x}}_1 + \dot{\tilde{x}}_1] = f_1\{\overline{x}_1, \overline{x}_2, \overline{V}_o, \overline{V}_g, \overline{\alpha}\} + \left.\frac{\partial f_1}{\partial x_1}\right|_{OP} \tilde{x}_1$$

$$+ \left.\frac{\partial f_1}{\partial x_2}\right|_{OP} \tilde{x}_2 + \left.\frac{\partial f_1}{\partial v_g}\right|_{OP} \tilde{V}_g + \left.\frac{\partial f_1}{\partial v_o}\right|_{OP} \tilde{V}_o$$

$$+ \left.\frac{\partial f_1}{\partial \alpha}\right|_{OP} \tilde{\alpha} + \frac{1}{2!} \left.\frac{\partial^2 f}{\partial x_1^2}\right|_{OP} \tilde{x}_1^2 + \cdots, \tag{A.27}$$

where the subscript OP in equations indicate that the equation is evaluated at that steady-state point:

$$OP = \overline{I}_r, \overline{V}_{Cr}, \overline{V}_o, \overline{V}_g, \overline{\alpha}, \tag{A.28}$$

$$f_1\{\overline{x}_1, \overline{x}_2, \overline{V}_o, \overline{V}_g, \overline{\alpha}\} = \dot{\overline{I}}_r = 0.$$

(ii) Taylor series expansion of resonant capacitor voltage:

$$[\dot{\overline{x}}_2 + \dot{\tilde{x}}_2] = f_2\{\overline{x}_1 + \tilde{x}_1, \overline{x}_2 + \tilde{x}_2, \overline{V}_o + \tilde{V}_o, \overline{V}_g$$

$$+ \tilde{V}_g, \overline{\alpha} + \tilde{\alpha}\},$$

$$= \frac{\omega_s}{\gamma_k} \cdot f_2^*\{\overline{x}_1 + \tilde{x}_1, \overline{x}_2 + \tilde{x}_2, \overline{V}_o + \tilde{V}_o, \overline{V}_g$$

$$+ \tilde{V}_g, \overline{\alpha} + \tilde{\alpha}\}, \tag{A.29}$$

where

$$x_1 = \overline{x}_1 + \tilde{x}_1,$$
$$x_2 = \overline{x}_2 + \tilde{x}_2,$$
$$v_g = \overline{V}_g + \tilde{V}_g, \tag{A.30}$$
$$v_o = \overline{V}_o + \tilde{V}_s,$$
$$\alpha = \overline{\alpha} + \tilde{\alpha},$$

then

$$[\dot{\overline{x}}_2 + \dot{\tilde{x}}_2] = f_2\{\overline{x}_1, \overline{x}_2, \overline{V}_o, \overline{V}_g, \overline{\alpha}\} + \left.\frac{\partial f_2}{\partial x_1}\right|_{OP} \tilde{x}_1$$

$$+ \left.\frac{\partial f_2}{\partial x_2}\right|_{OP} \tilde{x}_2 + \left.\frac{\partial f_2}{\partial v_g}\right|_{OP} \tilde{V}_g + \left.\frac{\partial f_2}{\partial v_o}\right|_{OP} \tilde{V}_o$$

$$+ \left.\frac{\partial f_2}{\partial \alpha}\right|_{OP} \tilde{\alpha} + \frac{1}{2!} \left.\frac{\partial^2 f_2}{\partial x_1^2}\right|_{OP} \tilde{x}_1^2 + \cdots, \tag{A.31}$$

where

$$f_2\{\overline{x}_1, \overline{x}_2, \overline{V}_o, \overline{V}_g, \overline{\alpha}\} = \dot{\overline{V}}_{Cr} = 0. \tag{A.32}$$

(iii) Taylor series expansion of the output current equation:

$$\overline{I}_o + \tilde{I}_o = f_{out}\{\overline{x}_1 + \tilde{x}_1, \overline{x}_2 + \tilde{x}_2, \overline{V}_o + \tilde{V}_o, \overline{V}_g + \tilde{V}_g, \overline{\alpha} + \tilde{\alpha}\},$$

$$= \frac{1}{(\overline{\gamma} + \tilde{\gamma})} \cdot f_{out}^*\{\overline{x}_1 + \tilde{x}_1, \overline{x}_2 + \tilde{x}_2, \overline{V}_o + \tilde{V}_o, \overline{V}_g$$

$$+ \tilde{V}_g, \overline{\alpha} + \tilde{\alpha}\}, \tag{A.33}$$

where

$$x_1 = \overline{x}_1 + \tilde{x}_1,$$
$$x_2 = \overline{x}_2 + \tilde{x}_2,$$
$$v_g = \overline{V}_g + \tilde{V}_g, \tag{A.34}$$
$$v_o = \overline{V}_o + \tilde{V}_o,$$
$$\alpha = \overline{\alpha} + \tilde{\alpha},$$
$$\tilde{\gamma} = \tilde{\alpha} + \tilde{\beta},$$

then

$$\bar{I}_\mathrm{o} + \tilde{I}_\mathrm{o} = f_\mathrm{out}\{\bar{x}_1, \bar{x}_2, \bar{V}_\mathrm{o}, \bar{V}_\mathrm{g}, \bar{\alpha}\} + \left.\frac{\partial f_\mathrm{out}}{\partial x_1}\right|_\mathrm{OP} \tilde{x}_1$$

$$+ \left.\frac{\partial f_\mathrm{out}}{\partial x_2}\right|_\mathrm{OP} \tilde{x}_2 + \left.\frac{\partial f_\mathrm{out}}{\partial v_\mathrm{g}}\right|_\mathrm{OP} \tilde{V}_\mathrm{g} + \left.\frac{\partial f_\mathrm{out}}{\partial v_\mathrm{o}}\right|_\mathrm{OP} \tilde{V}_\mathrm{o}$$

$$+ \left.\frac{\partial f_\mathrm{out}}{\partial \alpha}\right|_\mathrm{OP} \tilde{\alpha} + \frac{1}{2!}\left.\frac{\partial^2 f_\mathrm{out}}{\partial x_1^2}\right|_\mathrm{OP} \tilde{x}_1^2 + \cdots,$$

$$= \frac{1}{\bar{\gamma}} \cdot \left\{ 1 - \left(\frac{\tilde{\gamma}}{\bar{\gamma}}\right) + \left(\frac{\tilde{\gamma}}{\bar{\gamma}}\right)^2 - \cdots + \cdots \right\}$$

$$\cdot \left\{ f^*_\mathrm{out}\{\bar{x}_1, \bar{x}_2, \bar{V}_\mathrm{o}, \bar{V}_\mathrm{g}, \bar{\alpha}\} + \left.\frac{\partial f^*_\mathrm{out}}{\partial x_1}\right|_\mathrm{OP} \tilde{x}_1 \left.\frac{\partial f^*_\mathrm{out}}{\partial x_2}\right|_\mathrm{OP} \right.$$

$$\cdot \tilde{x}_2 + \left.\frac{\partial f^*_\mathrm{out}}{\partial v_\mathrm{g}}\right|_\mathrm{OP} \tilde{V}_\mathrm{g} + \left.\frac{\partial f^*_\mathrm{out}}{\partial v_\mathrm{o}}\right|_\mathrm{OP} \tilde{V}_\mathrm{o} + \left.\frac{\partial f^*_\mathrm{out}}{\partial \alpha}\right|_\mathrm{OP} \tilde{\alpha}$$

$$\left. + \frac{1}{2!}\left.\frac{\partial^2 f^*_\mathrm{out}}{\partial x_1^2}\right|_\mathrm{OP} \tilde{x}_1^2 + \cdots \right\},$$

$$(A.35)$$

where

$$\bar{I}_\mathrm{o} = f_\mathrm{out}\{\bar{x}_1, \bar{x}_2, \bar{V}_\mathrm{o}, \bar{V}_\mathrm{g}, \bar{\alpha}\},$$

$$\tilde{I}_\mathrm{o} = \frac{1}{\bar{\gamma}} \cdot \left\{ \left[\left.\frac{\partial f^*_\mathrm{out}}{\partial x_1}\right|_\mathrm{OP} - \bar{I}_\mathrm{o} \cdot \left.\frac{\partial \beta}{\partial x_1}\right|_\mathrm{OP} \right] \cdot \tilde{x}_1 \right.$$

$$+ \left[\left.\frac{\partial f^*_\mathrm{out}}{\partial x_2}\right|_\mathrm{OP} - \bar{I}_\mathrm{o} \cdot \left.\frac{\partial \beta}{\partial x_2}\right|_\mathrm{OP} \right] \cdot \tilde{x}_2$$

$$+ \left[\left.\frac{\partial f^*_\mathrm{out}}{\partial v_\mathrm{g}}\right|_\mathrm{OP} - \bar{I}_\mathrm{o} \cdot \left.\frac{\partial \beta}{\partial v_\mathrm{g}}\right|_\mathrm{OP} \right] \cdot \tilde{V}_\mathrm{g}$$

$$(A.36)$$

$$+ \left[\left.\frac{\partial f^*_\mathrm{out}}{\partial v_\mathrm{o}}\right|_\mathrm{OP} - \bar{I}_\mathrm{o} \cdot \left.\frac{\partial \beta}{\partial v_\mathrm{o}}\right|_\mathrm{OP} \right] \cdot \tilde{V}_\mathrm{o}$$

$$\left. + \left[\left.\frac{\partial f^*_\mathrm{out}}{\partial \alpha}\right|_\mathrm{OP} - \bar{I}_\mathrm{o} \right] \cdot \tilde{\alpha} \right\}.$$

The linearized equations can be obtained by neglecting the higher order terms of perturbation signals and retaining only the linear terms in the *Taylor series expansion* for (A.27), (A.31), and (A.35).

Step 7. State-space model.

A set of the linearized state-space model (in (A.37)) of SRC# in the subresonant mode can be generated from (A.27), (A.31), and (A.35). Additionally, the expression of transfer functions between the input state variables and the interesting state variables are given in (A.44) and (A.46):

$$\begin{bmatrix} \dot{\tilde{x}}_1 \\ \dot{\tilde{x}}_2 \end{bmatrix} = \begin{bmatrix} \left.\frac{\partial f_1}{\partial x_1}\right|_\mathrm{OP} & \left.\frac{\partial f_1}{\partial x_2}\right|_\mathrm{OP} \\ \left.\frac{\partial f_2}{\partial x_1}\right|_\mathrm{OP} & \left.\frac{\partial f_2}{\partial x_2}\right|_\mathrm{OP} \end{bmatrix} \begin{bmatrix} \tilde{x}_1 \\ \tilde{x}_2 \end{bmatrix} + \begin{bmatrix} \left.\frac{\partial f_1}{\partial \alpha}\right|_\mathrm{OP} & \left.\frac{\partial f_1}{\partial v_\mathrm{g}}\right|_\mathrm{OP} & \left.\frac{\partial f_1}{\partial v_\mathrm{o}}\right|_\mathrm{OP} \\ \left.\frac{\partial f_2}{\partial \alpha}\right|_\mathrm{OP} & \left.\frac{\partial f_2}{\partial v_\mathrm{g}}\right|_\mathrm{OP} & \left.\frac{\partial f_2}{\partial v_\mathrm{o}}\right|_\mathrm{OP} \end{bmatrix} \begin{bmatrix} \tilde{\alpha} \\ \tilde{V}_\mathrm{g} \\ \tilde{V}_\mathrm{o} \end{bmatrix},$$

$$\tilde{I}_\mathrm{o} = \left[\frac{1}{\bar{\gamma}} \cdot \left[\left.\frac{\partial f^*_\mathrm{out}}{\partial x_1}\right|_\mathrm{OP} - \bar{I}_\mathrm{o} \cdot \left.\frac{\partial \beta}{\partial x_1}\right|_\mathrm{OP} \right] \frac{1}{\bar{\gamma}} \cdot \left[\left.\frac{\partial f^*_\mathrm{out}}{\partial x_2}\right|_\mathrm{OP} - \bar{I}_\mathrm{o} \cdot \left.\frac{\partial \beta}{\partial x_2}\right|_\mathrm{OP} \right] \right]$$

$$\cdot \begin{bmatrix} \tilde{x}_1 \\ \tilde{x}_2 \end{bmatrix} + \left[\frac{1}{\bar{\gamma}} \cdot \left[\left.\frac{\partial f^*_\mathrm{out}}{\partial \alpha}\right|_\mathrm{OP} - \bar{I}_\mathrm{o} \right] \frac{1}{\bar{\gamma}} \right.$$

$$\cdot \left[\left.\frac{\partial f^*_\mathrm{out}}{\partial v_\mathrm{g}}\right|_\mathrm{OP} - \bar{I}_\mathrm{o} \cdot \left.\frac{\partial \beta}{\partial v_\mathrm{g}}\right|_\mathrm{OP} \right] \frac{1}{\bar{\gamma}} \cdot \left[\left.\frac{\partial f_\mathrm{out}}{\partial v_\mathrm{o}}\right|_\mathrm{OP} - \bar{I}_\mathrm{o} \cdot \left.\frac{\partial \beta}{\partial v_\mathrm{o}}\right|_\mathrm{OP} \right] \right]$$

$$\cdot \begin{bmatrix} \tilde{\alpha} \\ \tilde{V}_\mathrm{g} \\ \tilde{V}_\mathrm{o} \end{bmatrix}.$$

$$(A.37)$$

For derivative of equations f_1 and f_2,

$$\frac{\partial f_i}{\partial x_j} = \left(\frac{\partial}{\partial x_j}\frac{\omega_\mathrm{s}}{\gamma}\right) \cdot \left. f^*_i \right|_\mathrm{OP} + \frac{\omega_\mathrm{s}}{\gamma}\left.\frac{\partial f^*_i}{\partial x_j}\right|_\mathrm{OP}, \qquad (A.38)$$

where

$$\gamma = \alpha + \beta \quad \text{for } i = 1, 2, \ j = 1, 2. \qquad (A.39)$$

With the steady-state operating conditions,

$$\left.\frac{\omega_\mathrm{s}}{\gamma}\right|_\mathrm{OP} \neq 0,$$

$$\left. f^*_i \right|_\mathrm{OP} = 0. \qquad (A.40)$$

Therefore,

$$\frac{\partial f_i}{\partial x_j} = \frac{\omega_\mathrm{s}}{\gamma}\left.\frac{\partial f^*_i}{\partial x_j}\right|_\mathrm{OP}. \qquad (A.41)$$

The same derivation process, which are used to the derivatives of f_1 and f_2 with respect to input states x_1 and x_2 can be used to calculate the derivative of f^*_out in the output equation and the derivatives of f_1 and f_2 with respect to input states α, v_g, and v_o. Additionally, the angle β and its steady-state solution can be calculated by the derivation of the large-signal model in (A.7) as

$$\tan(\omega_\mathrm{rs}\beta) = \frac{-x_1 Z_\mathrm{r}}{(v_\mathrm{g} - v_\mathrm{o} - x_2)} = \tan(\omega_\mathrm{rs}\beta - \pi),$$

$$(A.42)$$

$$\left.\beta\right|_\mathrm{OP} = \frac{\pi}{\omega_\mathrm{rs}} + \frac{1}{\omega_\mathrm{rs}} \cdot \tan^{-1}\left[\frac{-\bar{I}_\mathrm{r} Z_\mathrm{r}}{(\bar{V}_\mathrm{g} - \bar{V}_\mathrm{o} - \bar{V}_\mathrm{Cr})} \right].$$

The derivative of β with respect to input states x_1, x_2, v_g, and v_o at the given operating points is expressed by the following equation:

$$\left.\frac{\partial \beta}{\partial x_1}\right|_{\mathrm{OP}} = \left[\frac{1}{\omega_{\mathrm{rs}}} \cdot \frac{-Z_{\mathrm{r}} \cdot \left(v_{\mathrm{g}} - v_{\mathrm{o}} - x_2\right)}{\left(v_{\mathrm{g}} - v_{\mathrm{o}} - x_2\right)^2 + \left(x_1 Z_{\mathrm{r}}\right)^2}\right]\Bigg|_{\mathrm{OP}}$$

$$= \frac{1}{\omega_{\mathrm{rs}}} \cdot \frac{-Z_{\mathrm{r}} \cdot \left(\overline{V}_{\mathrm{g}} - \overline{V}_{\mathrm{o}} - \overline{V}_{\mathrm{Cr}}\right)}{\left(\overline{V}_{\mathrm{g}} - \overline{V}_{\mathrm{o}} - \overline{V}_{\mathrm{Cr}}\right)^2 + \left(\overline{I}_{\mathrm{r}} Z_{\mathrm{r}}\right)^2},$$

$$\left.\frac{\partial \beta}{\partial x_2}\right|_{\mathrm{OP}} = \left[\frac{1}{\omega_{\mathrm{rs}}} \cdot \frac{-x_1 Z_{\mathrm{r}}}{\left(v_{\mathrm{g}} - v_{\mathrm{o}} - x_2\right)^2 + \left(x_1 Z_{\mathrm{r}}\right)^2}\right]\Bigg|_{\mathrm{OP}}$$

$$= \frac{1}{\omega_{\mathrm{rs}}} \cdot \frac{-\overline{I}_{\mathrm{r}} Z_{\mathrm{r}}}{\left(\overline{V}_{\mathrm{g}} - \overline{V}_{\mathrm{o}} - \overline{V}_{\mathrm{Cr}}\right)^2 + \left(\overline{I}_{\mathrm{r}} Z_{\mathrm{r}}\right)^2},$$

$$\left.\frac{\partial \beta}{\partial v_{\mathrm{o}}}\right|_{\mathrm{OP}} = \left[\frac{1}{\omega_{\mathrm{rs}}} \cdot \frac{-x_1 Z_{\mathrm{r}}}{\left(v_{\mathrm{g}} - v_{\mathrm{o}} - x_2\right)^2 + \left(x_1 Z_{\mathrm{r}}\right)^2}\right]\Bigg|_{\mathrm{OP}}$$

$$= \frac{1}{\omega_{\mathrm{rs}}} \cdot \frac{-\overline{I}_{\mathrm{r}} Z_{\mathrm{r}}}{\left(\overline{V}_{\mathrm{g}} - \overline{V}_{\mathrm{o}} - \overline{V}_{\mathrm{Cr}}\right)^2 + \left(\overline{I}_{\mathrm{r}} Z_{\mathrm{r}}\right)^2},$$

$$(\text{A.43})$$

$$\left.\frac{\partial \beta}{\partial v_{\mathrm{g}}}\right|_{\mathrm{OP}} = \left[\frac{1}{\omega_{\mathrm{rs}}} \cdot \frac{x_1 Z_{\mathrm{r}}}{\left(v_{\mathrm{g}} - v_{\mathrm{o}} - x_2\right)^2 + \left(x_1 Z_{\mathrm{r}}\right)^2}\right]\Bigg|_{\mathrm{OP}}$$

$$= \frac{1}{\omega_{\mathrm{rs}}} \cdot \frac{\overline{I}_{\mathrm{r}} Z_{\mathrm{r}}}{\left(\overline{V}_{\mathrm{g}} - \overline{V}_{\mathrm{o}} - \overline{V}_{\mathrm{Cr}}\right)^2 + \left(\overline{I}_{\mathrm{r}} Z_{\mathrm{r}}\right)^2},$$

where ω_{rs} $(=\omega_{\mathrm{r}}/\omega_{\mathrm{s}})$ is defined by the ratio between the natural frequency of the resonant tank (ω_{r}) and the switching frequency of SRC# (ω_{s}).

Step 8. Transfer functions.

According to (A.37), the transfer functions between the converter output current (output rectifier current) and input state variables can be obtained:

$$\tilde{I}_{\mathrm{o}}(s) = \begin{bmatrix} g_1(s) & g_2(s) & g_3(s) \end{bmatrix} \begin{bmatrix} \tilde{\alpha} \\ \tilde{V}_{\mathrm{g}} \\ \tilde{V}_{\mathrm{o}} \end{bmatrix}, \qquad (\text{A.44})$$

where

$$g_1(s) = \left.\frac{\tilde{I}_{\mathrm{o}}(s)}{\tilde{\alpha}(s)}\right|_{\tilde{V}_{\mathrm{g}}(s)=0, \tilde{V}_{\mathrm{o}}(s)=0},$$

$$g_2(s) = \left.\frac{\tilde{I}_{\mathrm{o}}(s)}{\tilde{V}_{\mathrm{g}}(s)}\right|_{\tilde{\alpha}(s)=0, \tilde{V}_{\mathrm{o}}(s)=0},$$

$$(\text{A.45})$$

$$g_3(s) = \left.\frac{\tilde{I}_{\mathrm{o}}(s)}{\tilde{V}_{\mathrm{o}}(s)}\right|_{\tilde{\alpha}(s)=0, \tilde{V}_{\mathrm{g}}(s)=0},$$

$$\tilde{f}_s = \tilde{\alpha}(s) \cdot \frac{f_{\mathrm{r}}}{-\pi},$$

and transfer functions between defined internal state variables and input state are

$$\tilde{X} = \begin{bmatrix} \tilde{I}_{\mathrm{r}} \\ \tilde{v}_{\mathrm{Cr}} \end{bmatrix} = \begin{bmatrix} \tilde{x}_1 \\ \tilde{x}_2 \end{bmatrix} = \begin{bmatrix} g_{\mathrm{xu},11} & g_{\mathrm{xu},12} & g_{\mathrm{xu},13} \\ g_{\mathrm{xu},21} & g_{\mathrm{xu},22} & g_{\mathrm{xu},23} \end{bmatrix} \begin{bmatrix} \tilde{\alpha} \\ \tilde{V}_{\mathrm{g}} \\ \tilde{V}_{\mathrm{o}} \end{bmatrix},$$

$$(\text{A.46})$$

where

$$\begin{cases} g_{\mathrm{xu},11} = \left.\frac{\tilde{I}_{\mathrm{r}}}{\tilde{\alpha}}\right|_{\tilde{V}_{\mathrm{g}}(s)=0, \tilde{V}_{\mathrm{o}}(s)=0}, \\[4pt] g_{\mathrm{xu},12} = \left.\frac{\tilde{I}_{\mathrm{r}}}{\tilde{V}_{\mathrm{g}}}\right|_{\tilde{\alpha}(s)=0, \tilde{V}_{\mathrm{o}}(s)=0}, \\[4pt] g_{\mathrm{xu},13} = \left.\frac{\tilde{I}_{\mathrm{r}}}{\tilde{V}_{\mathrm{o}}}\right|_{\tilde{\alpha}(s)=0, \tilde{V}_{\mathrm{g}}(s)=0}, \\[4pt] g_{\mathrm{xu},21} = \left.\frac{\tilde{v}_{\mathrm{Cr}}}{\tilde{\alpha}}\right|_{\tilde{V}_{\mathrm{g}}(s)=0, \tilde{V}_{\mathrm{o}}(s)=0}, \\[4pt] g_{\mathrm{xu},22} = \left.\frac{\tilde{v}_{\mathrm{Cr}}}{\tilde{V}_{\mathrm{g}}}\right|_{\tilde{\alpha}(s)=0, \tilde{V}_{\mathrm{o}}(s)=0}, \\[4pt] g_{\mathrm{xu},23} = \left.\frac{\tilde{v}_{\mathrm{Cr}}}{\tilde{V}_{\mathrm{o}}}\right|_{\tilde{\alpha}(s)=0, \tilde{V}_{\mathrm{g}}(s)}. \end{cases} \qquad (\text{A.47})$$

The details of derivation of the linearized plant model and the expression of elements in matrixes $[A]$, $[B]$, $[C]$, and $[D]$ are shown in (A.37), where the expressions of transfer functions, $g_1(s)$, $g_2(s)$, and $g_3(s)$, in (A.44) can be obtained by (3).

B. Derivation of Output LC Filter

Step 9. Derivation of transfer function of output LC filter.

The output current of SRC# delivered from the output diode rectifier to the MVDC grid has a high-harmonic content at twice the switching frequency and its multiples, and at the moment, only the mean value of the output current in one event is considered as the real output current. An output LC filter has to be placed on the output terminal of the converter in order to decrease as much as possible the current harmonics and mitigate the inrush current during fault. The following equations give the transfer functions of the output LC filter used to derive the harmonic model of the DC turbine converter:

$$g_{f1} = \frac{\tilde{I}_{turb}}{\tilde{I}_{o,Rec}} = \frac{R_C/(R_C + R_I)}{1 + s \cdot ((L_f + R_C R_L C_f)/(R_C + R_I)) + s^2 \cdot ((L_f C_f R_C)/(R_C + R_I))},$$

$$g_{f2} = \frac{\tilde{V}_{o,Rec}}{\tilde{I}_{o,Rec}} = \frac{R_C R_C + R_L (1 + s(L_f/R_I))}{1 + s \cdot ((L_f + R_C R_L C_f)/(R_C + R_I)) + s^2 \cdot ((L_f C_f R_C)/(R_C + R_I))}, \qquad (B.1)$$

$$g_{f3} = \frac{\tilde{I}_{turb}}{\tilde{V}_{turb}} = \frac{(R_C/(R_C + R_I)) \cdot (1 + sR_C C_f)}{1 + s \cdot ((L_f + R_C R_L C_f)/(R_C + R_I)) + s^2 \cdot ((L_f C_f R_C)/(R_C + R_I))},$$

$$g_{f4} = \frac{\tilde{V}_{o,Rec}}{\tilde{V}_g} = \frac{R_C/(R_C + R_I)}{1 + s \cdot ((L_f + R_C R_L C_f)/(R_C + R_I)) + s^2 \cdot ((L_f C_f R_C)/(R_C + R_I))},$$

where R_C and R_L are the parasitic resistances of filters, and filter capacitance and filter inductance are represented by C_f and L_f, respectively. $\tilde{V}_{turb}$ ($=\tilde{V}_{MVDC}$) is the MVDC-side voltage, $\tilde{V}_{o,Rec}$ is the output diode rectifier voltage of SRC#, and $\tilde{V}_g$ ($=\tilde{V}_{LVDC}$) is the LVDC-side voltage of SRC#. The dynamic model of the output LC filter in the control block is given by Figure 7.

C. Control Design of SRC#

Steps 10 and 12. Control design of DC wind turbine converter.

Main specifications for the design of the controller are addressed in articles [47, 48]. Since this is the first paper to address the closed-loop control of the DC wind turbine and then conductor to the susceptibility study of offshore DC wind farm, the paper did not provide specific numbers (/range) for control design of the DC wind turbine in the offshore wind farm. The following descriptions mainly focus on the identification of the coefficient of transfer function $g_c(s)$ of the SRC# turbine converter, and thus it can give the readers a full freedom of control design related to the DC turbine in the offshore wind farms to meet the performance requirements such as stability, zero steady-state error, settling time, overshoot and ringing, and disturbances rejection capability.

According to the structure of the control block in Figure 7, the control deign of SRC# $g_c(s)$ is influenced by the plant model and output LC filter $H(s)$ where the open-loop transfer function $T(s)$ is defined as

$$T(s) = g_C \cdot H(s), \qquad (C.1)$$

where the current delivered to the MVDC grid is related with the switching frequency of the converter by the transfer function $H(s)$ is represented by

$$H(s) = \frac{\tilde{i}_{turb}}{\tilde{f}_s} = \frac{g_1 \cdot g_{f1}}{1 - g_{f2} \cdot g_3}. \qquad (C.2)$$

The design of control loop $g_c(s)$ starts by evaluating the frequency response of $H(s)$. At the resonant frequency ($\approx 100\,\text{Hz}$) of the output LC filter, a peak on the magnitude occurs. The parasitic resistances of the capacitor C_f and inductor L_f can provide a limited damper to attenuate the resonant peak.

Following the specifications mentioned at the beginning of this subsection, the design of the controller starts by adding an integrator to minimize steady-state error with a step input. Thus, the output current will reach the reference value with a minimized tracking error after the transient.

Second, a pole around the resonant frequency of the output LC filter is added to lower the resonant magnitude peak influence of the filter.

Third, a gain K is necessary to set a proper crossover frequency f_c which is always lower than the resonant frequency of the output LC filter; otherwise, the resonant peak will be present in the feedback loop and the phase margin of the open-loop transfer function $H(s)$ will be less than $0°$, leading to the instability of the closed-loop system. Finally, the generic expression of the transfer function of the designed controller g_C can be expressed as

$$g_C = \frac{K}{s \cdot (1 - (s/\omega_p))}, \qquad (C.3)$$

where K is the gain (Hz/A) of the controller and ω_p is a pole (rad/s) close to the resonant frequency of the output LC filter. In order to improve the transient response of the system and to reject harmonic disturbances, generally, the angular crossover frequency ω_c ($=2\pi \cdot f_c$) of $H(s)$ should be as high as possible. Typically, the selected crossover frequencies f_c for all the operating points, are always less than 10% of the switching frequency f_s ($f_c < 0.1\,f_s$) in order to minimize the influence from the switching harmonics [48]. The crossover frequency of the open-loop transfer function $H(s)$, which can be assumed equal to the bandwidth of the system, characterizes the rapidity of the feedback system. Therefore, the higher the crossover frequency, the faster the response of the system is.

Nomenclature

L_r:	Inductor in the resonant tank
C_r:	Capacitor in the resonant tank
i_r:	Resonant inductor current
v_{Cr}:	Resonant capacitor voltage
v_g:	Input voltage of the resonant tank referred to the secondary side of the medium-frequency transformer
v_o:	Output voltage of the resonant tank

V_{LVDC}: Low-voltage DC

V_{MVDC}: Medium-voltage DC

v_{turb}: Output terminal voltage of the DC wind turbine converter

i_{turb}: Output current of the DC wind turbine converter $(= i_{\text{Lf}})$

$i_{\text{o,Rec}}$: Output current of the diode rectifier of the DC wind turbine converter

$v_{\text{o,Rec}}$: Output voltage of the diode rectifier of the DC wind turbine converter $(= v_{\text{Cf}})$

L_{f}: Inductor in the output filter

C_{f}: Capacitor in the output filter

f_s: Switching frequency of the series resonant converter defined by $= \omega_s/2\pi$

ω_{r}: Natural resonant frequency of the tank defined by $\omega_{\text{r}} = 1/\sqrt{L_{\text{r}}C_{\text{r}}}$.

Conflicts of Interest

The authors declare that they have no conflicts of interest.

References

[1] Y.-H. Chen, C. G. Dincan, R. J. Olsen, M. C. Schimmelmann, P. C. Kjær, and C. L. Bak, "Studies for characterisation of electrical properties of DC collection system in offshore wind farms," in *Proceedings of CIGRÉ General Session 2016, B4-301*, Paris, France, August 2016.

[2] T. Christ, S. Seman, and R. Zurowski, "Investigation of DC converter nonlinear interaction with offshore wind power park system," in *Proceedings of EWEA Off-Shore 2015*, pp. 1/4–4/4, Copenhagen, Denmark, March 2015.

[3] C. Meyer and R. W. De Doncker, "Design of a three-phase series resonant converter for offshore DC grids," in *Proceedings of 42nd IAS Annual Meeting; Industry Applications Conference*, pp. 216–223, September 2007.

[4] C. Meyer, *Key components for future offshore dc grids*, Ph.D. dissertation, Institute for Power Electronics and Electrical Drives, RWTH Aachen University, Aachen, Germany, 2007.

[5] S. Vogel, "*Investigation of DC collection network for offshore wind farms*," M.S. thesis, Department of Wind Energy, Technical University of Denmark, Kongens Lyngby, Denmark, 2014.

[6] V. Vorperian and S. Cuk, "A complete DC analysis of the series resonant converter," in *Proceedings of 13th Annual Power Electronics Specialists Conference (PESC'82)*, pp. 85–100, Cambridge, MA, USA, June 1982.

[7] A. F. Wittulski and R. W. Erickson, "Steady-state analysis of the series resonant converter," *IEEE Transactions on Aerospace and Electronic Systems*, vol. AES-21, no. 6, pp. 791–799, 1985.

[8] R. U. Lenke, J. Hu, and R. W. De Doncker, "Unified steady-state description of phase-shift-controlled ZVS-operated series-resonant and non-resonant single-active-bridge converters," in *Proceedings of IEEE Energy Conversion Congress and Exposition (IEEE-ECCE)*, pp. 796–803, IEEE, San Jose, CA, USA, September 2009.

[9] G. Ortiz, H. Uemura, D. Bortis, J. W. Kolar, and O. Apeldoorn, "Modeling of soft-switching losses of IGBTs in high-power high-efficiency dual-active-bridge DC/DC converters," *IEEE Transactions on Electron Devices*, vol. 60, no. 2, pp. 587–597, 2013.

[10] C. G. Dincan, P. Kjær, Y. Chen et al., "Design of a high power, resonant converter for DC wind turbines," *IEEE Transactions on Power Electronics*, 2018.

[11] P. C. Kjær, C. L. Bak, P. F. M. Da Silva, Y. H. Chen, and C. G. Dincan, "Power collection and distribution in medium voltage DC networks," 2015, http://www.dcc.et.aau.dk.

[12] C. Dincan, P. C. Kjær, Y. H. Chen, S. Munk-Nielsen, and C. L. Bak, "High power, medium voltage, series resonant converter for DC wind turbines," *IEEE Transactions on Power Electronics*, vol. 33, no. 9, pp. 7455–7465, 2018.

[13] D. Jovcic, "Step-up DC-DC converter for megawatt size applications," *IET Power Electronics*, vol. 2, no. 6, pp. 675–685, 2009.

[14] D. Jovcic, "Bidirectional, high-power DC-transformer," *IEEE Transactions on Power Delivery*, vol. 25, no. 4, pp. 2276–2283, 2010.

[15] W. Chen, A. Q. Huan, C. Li, G. Wang, and W. Gu, "Analysis and comparison of medium voltage high power DC/DC converters for offshore wind energy systems," *IEEE Transactions on Power Electronics*, vol. 28, no. 4, pp. 2014–2023, 2013.

[16] A. Parastar and J. K. Seok, "High gain resonant switched capacitor cell based DC/DC converter for offshore wind energy systems," *IEEE Transactions on Power Electronics*, vol. 30, no. 2, pp. 644–656, 2015.

[17] L. Max, *Design and control of A DC grid for offshore wind farms*, Ph.D. dissertation, Department of Energy and Environment, Chalmers University of Technology, Göteborg, Sweden, 2009.

[18] R. Lenke, *A contribution to the design of isolated DC-DC converters for utility applications*, Ph.D. dissertation, Institute for Power Electronics and Electrical Drives, RWTH Aachen University, Aachen, Germany, 2012.

[19] K. Park and Z. Chen, "Analysis and design of a parallel-connected single active bridge DC-DC converter for high-power wind farm applications," in *Proceedings of 15th European Conference on Power Electronics and Applications (EPE)*, vol. 110, Lille, France, September 2013.

[20] D. Duji, F. Kieferndorf, and F. Canales, "Power electronic transformer technology for traction applications—an overview," in *Proceedings of 7th International Power Electronics and Motion Control (PEMC)*, vol. 16, no. 1, p. 5056, Daegu, Korea, June 2012.

[21] M. Steiner and H. Reinold, "Medium frequency topology in railway applications," in *Proceedings of European Conference on Power Electronics and Applications (EPE)*, Aalborg, Denmark, September 2007.

[22] H. Hoffmann and B. Piepenbreier, "Medium frequency transformer in resonant switching dc/dc-converters for railway applications," in *Proceedings of 2011 14th European Conference on Power Electronics and Applications*, pp. 1–8, Birmingham, UK, August-September 2011.

[23] L. Heinemann, "An actively cooled high power, high frequency transformer with high insulation capability," in *Proceedings of APEC Seventeenth Annual IEEE Applied Power Electronics Conference and Exposition*, vol. 1, Los Cabos, Mexico, 2002.

[24] J. W. Kolar and G. I. Ortiz, "Solid state transformer concepts in traction and smart grid applications schedule/outline," in

Proceedings of 15th International Power Electronics and Motion Control Conference EPE-PEMC, pp. 1–166, Novi Sad, Serbia, September 2012.

[25] C. Dincan, P. Kjaer, Y. H. Chen, S. Munk-Nielsen, and C. L. Bak, "Analysis of a high power, resonant DC-DC converter for DC wind turbines," *IEEE Transactions on Power Electronics*, vol. 33, no. 9, pp. 7438–7454, 2017.

[26] Report 9639–01–R0, *MVDC Technology Study: Market Opportunities and Economic Impact*, TNEI Services Ltd., Manchester, UK, 2015.

[27] M. Barnes and A. Beddard, "Voltage source converter HVDC links–the state of the art and issues going forward," *in Energy Procedia*, vol. 24, pp. 108–122, 2012.

[28] F. F. Da Silva and C. L. Bak, *Electromagnetic Transients in Power Cables*, Springer, London, UK, 2013.

[29] *Cigré WG B4.57: Guide for the Development of Models for HVDC Converters in a HVDC Grid*, CIGRE Technical Brochure, Paris, France, 2014.

[30] Manitoba HVDC Research Centre, *USER'S GUIDE on the Use of PSCAD*, Manitoba HVDC Research Centre, Division of Manitoba Hydro International Ltd., Manitoba, ON, Canada, 2018.

[31] J. Beerten, S. D'Arco, and J. A. Suul, "Cable model order reduction for HVDC systems interoperability analysis," in *11th IET International Conference on AC and DC Power Transmission*, pp. 1–10, Birmingham, UK, February 2015.

[32] C. L. Bak, *Power System Technical Performance Issues Related to the Application of Long HVAC Cable*, WG C4.502s, Cigré Technical brochure, Paris, France, 2013.

[33] J. B. Glasdam, L. H. Kocewiak, J. Hjerrild, and C. L. Bak, "Control system interaction in the VSC-HVDC grid connected offshore wind power plant," in *Proceedings of Cigré Symposium 2015*, pp. 142–149, Lund, Sweden, 2015.

[34] L. Kocewiak, *Harmonics in large offshore wind farms*, Ph.D. dissertation, Aalborg University, Aalborg, Denmark, 2012.

[35] V. Preciado, "Harmonics in a wind power plant," in *Proceedings of IEEE Power and Energy Society General Meeting*, pp. 1–5, Denver, CO, USA, July 2015.

[36] K. M. Hasan, K. Rauma, A. Luna, J. I. Candela, and P. Rodriguez, "Harmonic resonance study for wind power plant," in *Proceedings of International Conference on Renewable Energies and Power Quality (ICREPQ'12)*, Santiago de Compostela, Spain, March 2012.

[37] Y. Fillion and S. Deschanvres, "Background harmonic amplifications within offshore wind farm connection projects," in *Proceedings of Power Systems Transient (IPST 2015)*, Cavtat, Croatia, June 2015.

[38] F. Mura, C. Meyer, and R. W. De Doncker, "Stability analysis of high-power dc grids," *IEEE Transactions on Industry Applications*, vol. 46, no. 2, pp. 584–592, 2010.

[39] A. Shafiu, A. Hernandez, F. Schettler, J. Finn, and E. Jørgensen, "Harmonic studies for offshore windfarms," in *Proceedings of 9th IET International Conference on AC and DC Power Transmission (ACDC 2010)*, pp. 1–6, London, UK, October 2010.

[40] B. Badrzadeh, M. Gupta, N. Singh, A. Petersson, L. Max, and M. Hogdahl, "Power system harmonic analysis in wind power plants—Part I: study methodology and techniques," in *Proceedings of Industry Applications Society Annual Meeting (IAS)*, Las Vegas, NV, USA, October 2012.

[41] PSCAD, *PSCAD/EMTDC On-Line Help System*, HVDC Research Centre, Manitoba, ON, Canada, 2013.

[42] R. J. King and T. A. Stuart, "Small-signal model for the series resonant converter," *IEEE Transactions on Aerospace and Electronic Systems*, vol. AES-21, no. 3, pp. 301–319, 1985.

[43] X. Wang, F. Blaabjerg, and W. Wu, "Modeling and analysis of harmonic stability in an AC power-electronics-based power system," *IEEE Transactions on Power Electronics*, vol. 29, no. 12, pp. 6421–6432, 2014.

[44] B. Gustavsen and A. Semlyen, "Rational approximation of frequency domain responses by vector fitting," *in IEEE Transactions on Power Delivery*, vol. 14, no. 3, pp. 1052–1061, 1999.

[45] J. R. Marti, "Accuarte modelling of frequency-dependent transmission lines in electromagnetic transient simulations," *IEEE Transactions on Power Apparatus and Systems*, vol. PAS-101, no. 1, pp. 147–157, 1982.

[46] Y. H. Chen, C. G. Dincan, P. C. Kjær, C. L. Bak, X. Wang, and C. E. Imbaquingo, "Model-based control design of series resonant converter based on the discrete time domain modelling approach for DC wind turbine," *Journal of Renewable Energy*, vol. 2108, Article ID 7898679, 18 pages, 2018.

[47] C. Imbaquingo, N. Isernia, E. Sarrà, and A. Tonellotto, "Modelling DC collection of offshore wind farms for harmonic susceptibility study," May 2017, https://www.dcc.et.aau.dk/digitalAssets/390/390654_aau-2017-carlos-dc-harmonics-on-src-converter.pdf.

[48] R. W. Erickson and D. Maksimovic, *Fundamentals of Power Electronics*, Springer Science & Business Media, Berlin, Germany, 2007.

10

Priority Control Strategy of VSC-MTDC System for Integrating Wind Power

Wen-ning Yan,[1] Ke-jun Li,[1] Zhuo-di Wang,[1] Xin-han Meng,[1] and Jianguo Zhao[2]

[1]*School of Electrical Engineering, Shandong University, Jinan 250061, China*
[2]*State Grid of China Technology College, Jinan 250002, China*

Correspondence should be addressed to Ke-jun Li; lkjun@sdu.edu.cn

Academic Editor: Alfredo Vaccaro

For the obvious advantages in integrating wind power, multiterminal HVDC transmission system (VSC-MTDC) is widely used. The priority control strategy is proposed in this paper considering the penetration rate of wind power for the AC grid. The strategy aims to solve the problems of power allocation and DC voltage control of the DC system. The main advantage of this strategy is that the demands for wind power of different areas can be satisfied and a power reference for the wind power trade can also be provided when wind farms transmit active power to several AC grids through the DC network. The objective is that power is well distributed according to the output power of wind farm with the demand of AC system and satisfactory control performance of DC voltage is obtained.

1. Introduction

Since wind resources especially offshore wind resources are rich in China, large scale integration of wind power transmission technology has drawn a lot of attention in recent years [1, 2]. Compared to the disadvantages of traditional AC transmission, such as the limitation of transmission power due to the charging current, the voltage stability problems for long distance, and the need for a large amount of reactive power compensation equipment, flexible DC transmission technology, especially high voltage direct current based on voltage source converter (VSC-HVDC), has increasingly become the preferred way for remote wind power transmission with the advantages of lower harmonic content, the stability of AC bus voltage, large transmission capacity, and so on [3–5].

Two-terminal VSC-HVDC can realize power transmission from point to point with the active and reactive power being controlled independently. The application of fully controlled devices makes the system have the ability to work in passive inverting mode to supply isolated system. Moreover, it can also play the role of STATCOM for reactive power compensation under some special conditions [6].

Unfortunately, two-terminal system still has some deficiencies, especially worse stability and reliability under some fault conditions. Compared with the two-terminal system, multiterminal HVDC flexible technology (VSC-MTDC) has multiple converter stations to realize the flexibility of power dispatch control and better economical efficiency. Therefore, it can be used in the integration of multiple wind farms in different regions [7, 8].

In recent years, many scholars have done extensive researches on multiterminal HVDC system control strategies for integration of wind farms. The widely used doubly fed induction generator (DFIG) in offshore wind farms has been analyzed in the literature [9, 10], and the variable speed constant frequency maximum power tracking control is achieved. The DC voltage control strategy suitable for wind farm connected with VSC-MTDC is proposed in [11] to ensure the stable operation of VSC-MTDC for wind power integration. The voltage margin control as an improved constant DC voltage control is proposed in [12] and the DC voltage is controlled at a new value by reserved converter station once the power limitation is exceeded. The widely accepted voltage drop control for VSC-MTDC is proposed in the literature [13], in which different converter

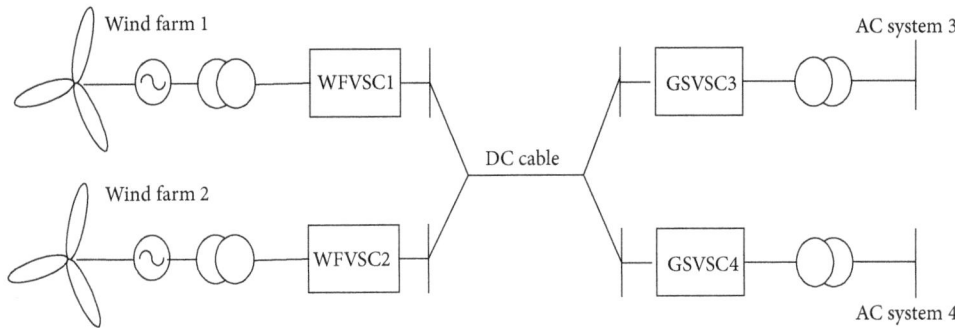

FIGURE 1: Four-terminal VSC-MTDC system connecting wind farms.

stations cooperate to achieve the control of DC voltage and the dispatch of active power at the same time. In [14], the voltage droop control is improved and the droop constants are adapted in relation with the AC system stability. Literature [15] presents control strategy for multiterminal system based on the improved active power characteristics; the problems of active power imbalance and DC voltage instability during the dominant station overload periods are effectively solved; the measures for active power balance are analyzed during the grid system faults in [16] and DC voltage control strategy based on the regulation characteristics of DC voltage and current of VSC-MTDC transmission systems for offshore wind farm is proposed. In [17], a coordinated control strategy of VSC-MTDC named master-auxiliary is proposed by combining the advantages of the voltage margin and voltage droop control; the reliability and stability of the DC network are ensured by master and auxiliary converter stations in normal state or by the APC in abnormal conditions.

The researches above ensure the stability and reliability of the DC system under normal and fault conditions. But the different ratio of wind power penetration on the total load of AC system is not included properly, while in practice the ratio should be controlled in a proper range so that the stability of AC system can be achieved. The amount of transmitting active power of station is related to their DC voltage level; however, DC voltage level is determined once the control strategy is selected; then the transmitting power cannot be adjusted according to different demand of AC system connected to station.

Based on the analysis above, a priority control strategy is proposed in this paper. In this strategy, the VSC stations connected to AC systems which have strong demands of wind power have the highest priority over other stations, so it will receive all the output power of wind farms when the output amount is less than the set value of the connected AC system. While in the high output stage of wind farms, the wind power beyond the set value will be delivered to each AC system in accordance with an adjustable power transmission ratio, and the ratio is in relation to the dependence of each AC system.

Section 2 of this paper gives a brief introduction of the VSC-MTDC system. In Section 3, the characteristic curve of active power and DC voltage for the grid side converters

(GSVSCs) is improved and the priority control strategy is introduced in detail. The controller combining voltage droop and PI control is designed. Finally, the simulation results under uncertain wind output condition and different demand of AC grid are presented to validate the effectiveness of the proposed control strategy.

2. Transmission System Structure of VSC-MTDC

Four-terminal wind power transmission system is established as shown in Figure 1 containing two rectifying converter stations and the two inverting converter stations. Two wind farms working as typical distributed power are connected to system by the wind farm side converter stations (WFVSCs), and the WFVSCs collect the output power of wind farms in different places to the DC system and then through the remote transmission with DC cable; finally the total wind power of the system is delivered into the AC systems in different places by the two grid side converter stations (GSVSCs).

Nowadays the technology of variable speed constant frequency is widely used in wind power generation; wind turbine (in the case of DFIG) with variable speed constant frequency can realize the decoupling of the rotational speed of the generator and the power grid frequency and reduce the mutual influence between the wind power generation and power grid. The control of DFIG will no longer be described in detail considering many existing mature researches.

The primary task of wind farm side converters (WFVSCs) is working as a slack node to transfer the power of wind farms out instantaneously and ensure the stability of bus voltage amplitude and phase in public node of wind farms; hence the control strategy with constant AC voltage and active power control will be applied to WFVSCs. The grid side converters (GSVSCs) using priority control strategy proposed in this paper aim to maintain a stable DC voltage and in the meantime the allocation of the active power into the grid according to the output power of wind farms and requirements of different AC system is achieved.

The converter stations in Figure 1 have the same structure and the topology of their structure when connected to AC system is as shown in Figure 2.

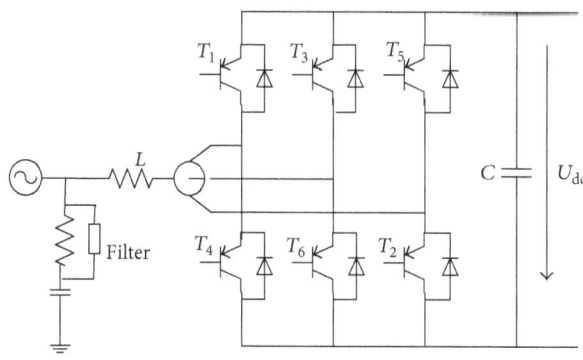

FIGURE 2: Topology diagram of VSC.

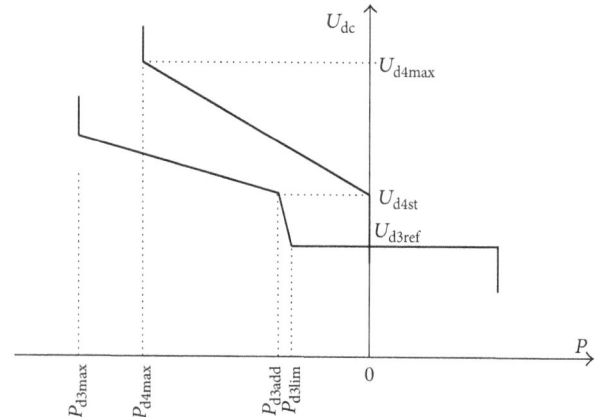

FIGURE 3: Improved characteristic curve of active power and direct voltage.

The mathematical model for VSC in two-phase rotating d-q coordinate system is shown as follows:

$$L\frac{di_d}{dt} = u_{sd} - S_d U_{dc} - R i_d + \omega L i_q,$$

$$L\frac{di_q}{dt} = u_{sq} - S_q U_{dc} - R i_q + \omega L i_d, \qquad (1)$$

$$C\frac{dU_{dc}}{dt} = \frac{3}{2}\left(S_d i_d + S_q i_q\right) - i_L,$$

where ω is the angular velocity of rotation; S_d and S_q are switching function in synchronous rotating d-q coordinate system; u_{sd}, u_{sq} are q- and d-axis component of the grid voltage, respectively; i_d and i_q are d- and q-axis component of the grid current, respectively; i_L is the DC line current; U_{dc} represents the DC voltage; L is the equivalent inductance; R stand for the equivalent loss; C is capacitor in DC side of converter.

3. Priority Control Strategy

As an important index to measure the AC grid which accepts wind power, wind power penetration level refers to the ratio that the wind power capacity accounts for the total capacity of the network load and it shows the dependence of grid on the wind power. The output power of wind farms is unpredictable due to its nature of intermittence and uncertainty. All the generated wind power needs to be transmitted to the AC system (high wind power penetration level) which has priority over others through GSVSC to meet its demand during the period when the total power output is less than the required minimum wind power injection volume (set by the system). With the increase of the power output of wind farms, the system switches to the mode that two grid side converter stations deliver the extra output power beyond the set value to AC systems by adjustable power transmission ratio.

3.1. Improved Characteristic Curve of Active Power and DC Voltage. In order to maintain the normal operation of AC system with high dependence on wind power under uncertain output power of wind farms, the characteristic curve

of active power and DC voltage for the grid side converter (GSVSCs) is improved as shown in Figure 3 where U_{dc3ref} represents the set DC voltage value of the 3rd GSVSC under constant DC voltage control mode; U_{dc4set} represents the trigger voltage for the 4th GSVSC in droop control mode; P_{d3max} and P_{d4max} are maximum transmission power of two GSVSCs, respectively; U_{dc4max} represents the maximum DC voltage of the 4th GSVSC during normal operation; P_{d3lim} represents the set value of transmission power of the 3rd GSVSC under the constant DC voltage mode.

The improved curve is based on the original characteristics [15, 18] and the slope is enhanced in the part of DC voltage curve; thus the working state of GSVSCs is divided into two stages. By the time when the total output power of wind farms reaches the required minimum limitation of AC system with high dependence on wind power (the 3rd AC system), the 3rd GSVSC works in DC voltage mode and all the output power of wind farms is transported to the 3rd AC system through it; meanwhile the transmission power of the 4th GSVSC is zero; the 3rd GSVSC enters the current limiting mode as soon as the total output power of wind farms exceeds the set value $P_{dc3lim}(I_{dc3lim})$ and DC voltage increases accordingly; once the increasing DC voltage reaches the trigger voltage of the 4th GSVSC, the 3rd and 4th GSVSCs start to work in coordination to dispatch the extra wind power exceeding the set value into AC grids by an adjustable power transmission ratio n.

3.2. Analysis of DC Network. The equivalent circuit of DC grid in steady state of four-terminal VSC-MTDC transmission system is shown in Figure 4. In this figure, $R_i(i = 1, 2, 3, 4, 5)$ represent the equivalent resistance of DC cables, respectively; $U_{dci}(i = 1, 2, 3, 4)$ represents the DC voltage of converter stations, respectively; U_{dca} and U_{dcb} represent the DC voltage of collection point and branch point of DC network, respectively; DC power flows from the wind farms side of converter stations WFVSC1 and WFVSC2 through DC cables to the grid side converter stations GSVSC3 and GSVSC4.

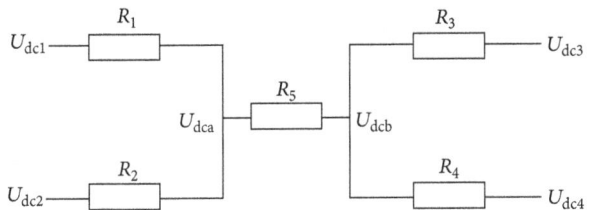

FIGURE 4: DC equivalent circuit of four-terminal MTDC system.

The variables P, U, and P_{ref} of the DC network in the vector form are defined as follows:

$$P = \left[P_{d1} \cdots P_{di} \cdots P_{dn}\right]^T$$

$$U = \left[U_{dc1} \cdots U_{dci} \cdots U_{dcn}\right]^T \quad (2)$$

$$P_{ref} = \left[P_{d1ref} \cdots P_{diref} \cdots P_{dnref}\right]^T.$$

The injected active power from DC nodes into DC network can be expressed as

$$P = U \otimes (YU), \quad (3)$$

where the multiplication symbol represents the calculation that the element of the matrix multiplies bit by bit in accordance with the dimensions; Y represents the admittance matrix of DC network.

The Jacobian matrix is calculated by the partial differential of active power to the DC voltage of each node similar to the definition in AC system as follows:

$$J(k) = \frac{\partial P}{\partial U} = \begin{pmatrix} \dfrac{\partial P_{d1}}{\partial U_{dc1}} \cdots \dfrac{\partial P_{d1}}{\partial U_{dcn}} \\ \vdots \\ \dfrac{\partial P_{dn}}{\partial U_{dc1}} \cdots \dfrac{\partial P_{dn}}{\partial U_{dcn}} \end{pmatrix}. \quad (4)$$

In the above equation, the partial differential is given as

$$\frac{\partial P_{di}}{\partial U_{dck}} = \begin{cases} -Y_{ik}U_{dci} & (k \neq i) \\ -\sum\limits_{j \neq i} Y_{ij}U_{dcj} - 2Y_{ii}U_{dci} & (k = i). \end{cases} \quad (5)$$

The relationship between the change of DC voltage quantity ΔU and the variation of active power ΔP of each node in the network can be expressed as

$$\Delta P = J(k)\Delta U. \quad (6)$$

3.3. Analysis of Working State of GSVSC. As is dispatched in Figure 3, due to the nature of intermittence and uncertainty in wind power output, the working state of two grid side converter stations with priority control strategy can be divided by the boundary point given by the minimum active power limitation of AC system into two stages. The minimum active power limitation P_{d3lim} of AC system is determined

by wind power penetration level, and the two stage working conditions are as follows.

Stage One. The active power output of wind farms is less than the power required by the 3rd AC system with high dependence on wind farms in this stage. The 3rd GSVSC works in a constant DC voltage mode and all the output power of wind farms is transmitted to the 3rd AC system; thus the DC voltage is kept in the setting value of U_{dc3ref} while the transmitted active power of the 4th GSVSC is zero; that is, $I_{dc4} = 0$. In the meantime, the 4th GSVSC can operate in STATCOM mode to provide reactive power support to AC system if needed. The DC voltage of each converter station in this stage can be expressed as

$$\begin{bmatrix} U_{dc1} \\ U_{dc2} \\ U_{dc3} \\ U_{dc4} \end{bmatrix}$$

$$= \begin{bmatrix} R_1 + R_3 + R_5 & R_3 + R_5 & 1 \\ R_3 + R_5 & R_2 + R_3 + R_5 & 1 \\ 0 & 0 & 1 \\ R_3 & R_3 & 1 \end{bmatrix} \begin{bmatrix} \dfrac{P_{d1}}{U_{dc1}} \\ \dfrac{P_{d1}}{U_{dc1}} \\ U_{dc3ref} \end{bmatrix}, \quad (7)$$

where U_{d1}, U_{d2}, and U_{d4} are unknown while P_{d1}, P_{d2}, and U_{d3ref} are known quantity.

Stage Two. The 3rd GSVSC will enter into the current limiting mode when the wind power output is greater than the required minimum active power limit of the GSVSC number 3, furthermore causing the DC voltage to rise rapidly to reach the trigger voltage of the 4th GSVSC; then the working conditions of the two GSVSCs enter stage two and the voltage is given as

$$U_{dc4} = R_3 \frac{P_{dc3lim}}{U_{dc3ref}} + U_{dc3ref}. \quad (8)$$

The trigger voltage U_{dc4st} of the 4th GSVSC should be slightly higher than the U_{dc4} given above considering the system margin and control response. And in the actual calculation the voltage rising volume and the transmission power exceeding the limit part are small before the operation curve enters into the slope control part; hence it can be approximately considered in Figure 3 that $P_{dc3lim} \approx P_{dc3add}$.

Suppose that the slope of active power-DC voltage curve of the two converter stations working in stage two is k_i ($i = 3, 4$); the slope can be expressed as

$$\frac{1}{k_i} = \frac{\Delta P_{di}}{\Delta U_{dci}}. \quad (9)$$

The relationships of the transmission power between the two GSVSCs with droop control are as follows:

$$P_{d3} = P_{d3lim} + \frac{1}{k_3}\Delta U_{dc3} = P_{d3lim} + \frac{1}{k_3}\left(U_{dc3} - U_{dc4st}\right),$$

$$P_{d4} = \frac{1}{k_4}\Delta U_{dc4} = \frac{1}{k_4}\left(U_{dc4} - U_{dc4st}\right). \tag{10}$$

As can be seen in Figure 4, the DC voltage of power branch point can be expressed by the 3rd and 4th GSVSCs as

$$U_{dcb} = U_{dc4} + R_4\frac{P_{d4}}{U_{dc4}} = U_{dc3} + R_3\frac{P_{d3}}{U_{dc3}}. \tag{11}$$

The equation below can be derived by the combination of the above two formulas:

$$\left(P_{d3} - P_{d3lim}\right)\left(k_3 + \frac{R_3}{U_{dc3}}\right) + \frac{P_{d3lim}R_3}{U_{dc3}}$$
$$= P_{d4}\left(k_4 + \frac{R_4}{U_{dc4}}\right). \tag{12}$$

As the resistance is small relative to the DC voltage, thus the smaller ones in the formula can be neglected. The simplified power transmission n for distribution of the extra wind power output which exceeds the set limitation between the two GSVSCs can be approximately given as follows:

$$n = \frac{P_{d3} - P_{d3lim}}{P_{d4}} \approx \frac{k_4}{k_3}. \tag{13}$$

According to the formula above, the transmission power ratio and the slope of characteristic curve are inversely proportional, and thus the change of power transmission ratio can be achieved by adjusting the slope of the droop curve, respectively. If the relationship between slope of the droop control curve and the dependence on wind power of corresponding AC system can be established, then the demand that AC systems with different dependence on wind power have different required transmission power can be achieved.

The DC voltage of each station working in stage two derived from DC network analysis in Section 3.2 and formulas (10) and (12) can be expressed as

$$\begin{bmatrix} U_{dc1} \\ U_{dc2} \\ U_{dc3} \\ U_{dc4} \end{bmatrix}$$
$$= \begin{bmatrix} R_1 + R_{eq} & R_{eq} & \frac{1}{n+1}\left(R_3 - \frac{n}{k_3}\right) & 1 \\ R_2 + R_{eq} & R_{eq} & \frac{1}{n+1}\left(R_3 - \frac{n}{k_3}\right) & 1 \\ \frac{n}{(n+1)k_3} & \frac{n}{(n+1)k_3} & \frac{-1}{n+1}\frac{n}{k_3} & 1 \\ \frac{n}{(n+1)k_4} & \frac{n}{(n+1)k_4} & \frac{-1}{(n+1)k_4} & 1 \end{bmatrix}\begin{bmatrix} \dfrac{P_{d1}}{U_{dc1}} \\ \dfrac{P_{d1}}{U_{dc1}} \\ \dfrac{P_{d3lim}}{U_{dc3ref}} \\ U_{dc4st} \end{bmatrix}. \tag{14}$$

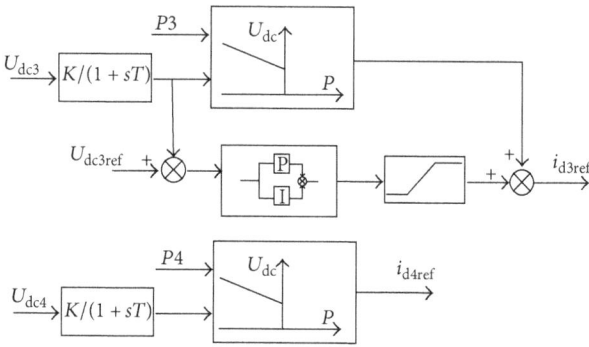

FIGURE 5: Design of the priority controller.

The relationship between the transmission power and DC voltage of the 4th GSVSC is given as

$$P_{d4}$$
$$= \frac{U_{dc4} - U_{dc4st}}{U_{dc4max} - U_{dc4st}}P_{d4max}\left(U_{dc4st} \leq U_{dc4} < U_{dc4max}\right). \tag{15}$$

The DC voltage of each converter station is calculated by formulas (6) and (14) and (15) can provide the reference for determining the DC voltage operation range and the parameters of the controller.

3.4. Design of the Controller. Figure 5 shows the structure diagram of the outer loop controller with priority control strategy. DC voltage is controlled by the 3rd GSVSC with feed forward PI control strategy when the wind power output is less than the minimum power set value; with the wind power output increasing, when the system DC voltage reaches the starting voltage of the 4th GSVSC, two GSVSCs enter into the voltage droop control mode to deliver wind power to AC system in accordance with the adjustable power transmission ratio, where i_{d3ref} and i_{d4ref} are the output signal of the outer loop controller, respectively, which provides the reference values for the inner current control. The measured DC voltage of GSVSCs passes through a first-order low-pass filter which is introduced to prevent the harmonic impact on the control accuracy of the system.

The outer loop DC voltage PI controller in priority controller is of the same structure with the double closed-loop control widely used in many literatures such as [17, 19]. The structure diagram of voltage droop controller for power allocation shown in Figure 6 is analyzed where the parameter e represents the error signal of the controller; k_{p1} and k_{p2} represent the slope of droop curves, respectively; I_{dref} represents the command signal of inner current loop controller; P_{ref} is the transmission power before the curve enters the droop control, which is provided as the reference value for droop control.

In Figure 6, the priority controller works as the droop controller for active power when the slope k_{p1} is set as the constant 1, while it works as droop controller for DC voltage when the slope k_{p2} is the constant 1. The following analysis of the working principle of active power controller is employed

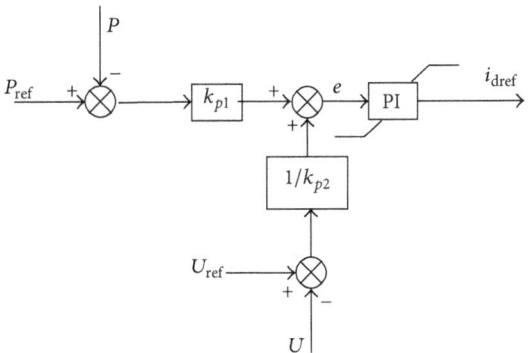

FIGURE 6: Block diagram of DC voltage droop control.

to illustrate the droop control and the influence of active power reference change on the distribution of power between GSVSCs.

The input deviation of PI circuit in the voltage droop controller during steady state is expressed as

$$e = (U_{ref} - U) + k_p (P_{ref} - P) = 0. \tag{16}$$

According to the above formula, the relationship between the active power and DC voltage of station using this controller is as follows:

$$P = P_{ref} + \frac{U_{ref} - U}{k_p}. \tag{17}$$

The relationship between the variation of active power and DC voltage corresponding can be expressed similarly as

$$\Delta P = \Delta P_{ref} + \frac{\Delta U_{ref} - \Delta U}{k_p}. \tag{18}$$

In order to study the influence of active power reference change on the power allocation of each node in this paper, the change quantity of DC voltage reference value is set to zero. The above formula can be rewritten as

$$\Delta P = \Delta P_{ref} - \frac{\Delta U}{k_p}. \tag{19}$$

The relationship given in the above formula can be modified combining formula (5) in chapter 2.2.1 as

$$\Delta P = \left[I + \frac{J(k)^{-1}}{k_p} \right]^{-1} \Delta P_{ref}. \tag{20}$$

Matrix I is unit diagonal matrix with the same dimension of Jacobian matrix $J(k)$.

As can be seen from the above analysis, Jacobian matrix cannot be changed with a determined DC network. The only way to change the distribution of active power between GSVSCs when the power reference value changes (fluctuation of the wind power output) is to change the slope of droop curve, respectively, which coincides with the analysis in formula (12).

4. Simulation Analysis

In order to prove the effect of the proposed control strategy, a simulation system of four-terminal VSC-MTDC grid for wind power transmission is modeled in the electromagnetic transient simulation software PSCAD/EMTDC as shown in Figure 1. The AC systems at the receiving terminals are simulated with ideal AC sources and IEEE standard system with three generators and nine nodes, respectively. Except the above difference, all the parameters of the two simulation models and simulation conditions are the same. Among them, the rated capacity of wind farms which connect WFVSCs is 300 MW, and the rated capacity of GSVSC is 400 MW; reference value for DC voltage of the 3rd GSVSC is 400 kV, the voltage of AC system is 420 kV, and the transformer parameters are 420/230 kV. The calculated trigger voltage of the 4th GSVSC by formula (7) is 403 kV.

4.1. Simulation in Changing the Output of Wind Farms. Assuming that the AC system which is connected to the 3rd GSVSC has stronger dependence on wind power, the minimum input power required is 150 MW. The total wind power output is 130 MW at the initial time and increases to 330 MW at the time of 0.5 seconds; the corresponding change quantity of active power reference value in formula (17) is $\Delta P_{ref} = [100, 100, 0, 0]$; then the wind power output continues to increase to 580 MW at the time of 1 second with an active power reference value change quantity $P_{ref} = [100, 150, 0, 0]$, while at the time of 1.5 seconds the wind power output is reduced to 480; similarly the corresponding active power reference value change quantity $P_{ref} = [-50, -50, 0, 0]$; the power transmission ratio between the 3rd and 4th GSVSCs is $3 : 2$ and remains unchanged. Figure 7 presents the simulation results of active power and DC voltage of each station.

As can be seen from Figure 7(a), when the total power output of the wind farms is less than the minimum active power limit P_{dc3} of AC system before 0.5 seconds, the 3rd GSVSC delivers all the total wind farm output active 127 MW which is less than the set value due to the loss of station and cable to meet its needs, while the transmission power of the 4th GSVSC is zero, and in the meantime the DC voltage is maintained around its reference value. When the total power output of wind farms rises to 330 MW, which is beyond the required limitation set in advance, the 3rd GSVSC enters the current limit mode, which causes the DC voltage to rise rapidly to reach the trigger voltage of the 4th GSVSC. The 3rd and 4th GSVSC start to allocate the extra wind power exceeding the set limitation from 0.5 seconds to 1 second with the transmission power of 118 MW and 76 MW, respectively, which match approximately with the transmission ratio considering the losses as shown in the figure, and the DC voltage is jointly maintained by the two GSVSCs according to droop curve. The transmission power reaches to 396 MW and 178 MW, respectively, when the wind power output increases at 1 second and then decreases at 1.5 seconds when the wind farm output power drops to 338 MW and 139 MW; the above two changes both match the transmission according to the simulation result shown in the figure.

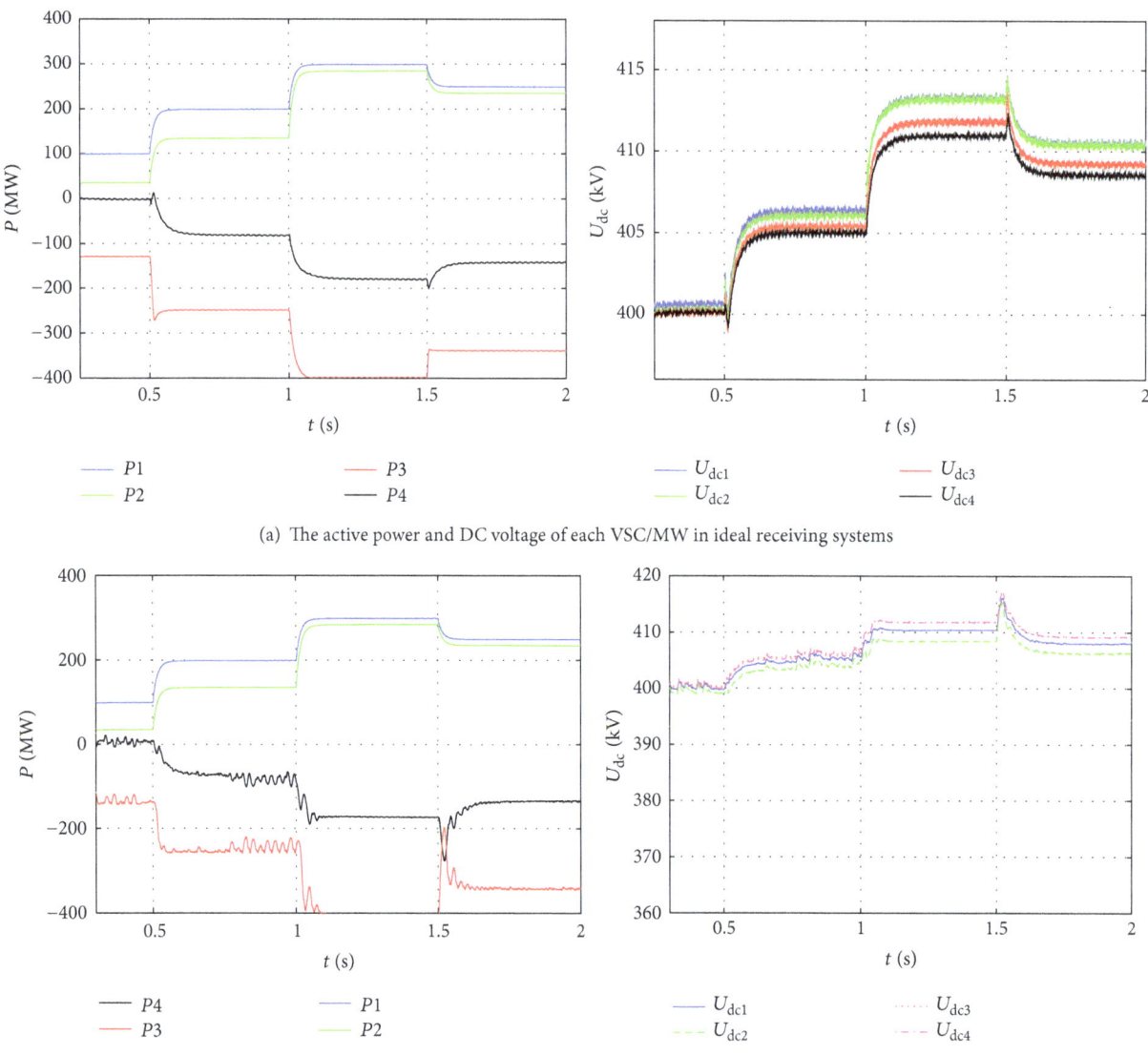

(a) The active power and DC voltage of each VSC/MW in ideal receiving systems

(b) The active power and DC voltage of each VSC/MW in standard receiving systems

FIGURE 7: Simulation of variation of power of wind farm with standard receiving systems.

The results of IEEE standard systems shown in Figure 7(b) are almost the same with Figure 7(a), except for little fluctuation due to the different characteristic between ideal systems with standard systems. And similarly, the active power of the DC system is balanced and the DC voltages are kept in the reasonable range.

The above analysis shows that the control strategy can be applied to both ideal receiving sources and IEEE standard systems according to the similar response to the change of output power of the wind farms, and in the process of changes in the output power from the wind farms, the transmission power of GSVSCs meets the power transmission ratio in allowable range considering the system loss and with smooth transaction. The DC voltage is within the range of normal operation after the transition indicating that the system has good dynamic response.

4.2. Simulation in Changing the Transmission Ratio. The initial output power of wind farms is equal to 130 MW and increases to 430 MW from 0.5 seconds to 1 second with the active power reference value change value $\Delta P_{\text{ref}} = [150, 150, 0, 0]$. Based on the foregoing analysis, the change of transmission power ratio can be achieved by adjusting the slope of the droop curve. The power transmission ratio between the two GSVSCs changes from the initial setting value $3:2$ to $1:1$ at 1 second and then back to the $3:2$ at 1.5 seconds. The simulation results of active power and DC voltage of each station are as shown in Figure 8.

As shown in Figure 8(a), the wind farms power output is less than the minimum set limitation of AC system before 0.5 seconds at the initial moment; therefore, the transmission power transferred by the 3rd GSVSC to the AC power system is 126 MW accounting for the loss. As the total wind farm

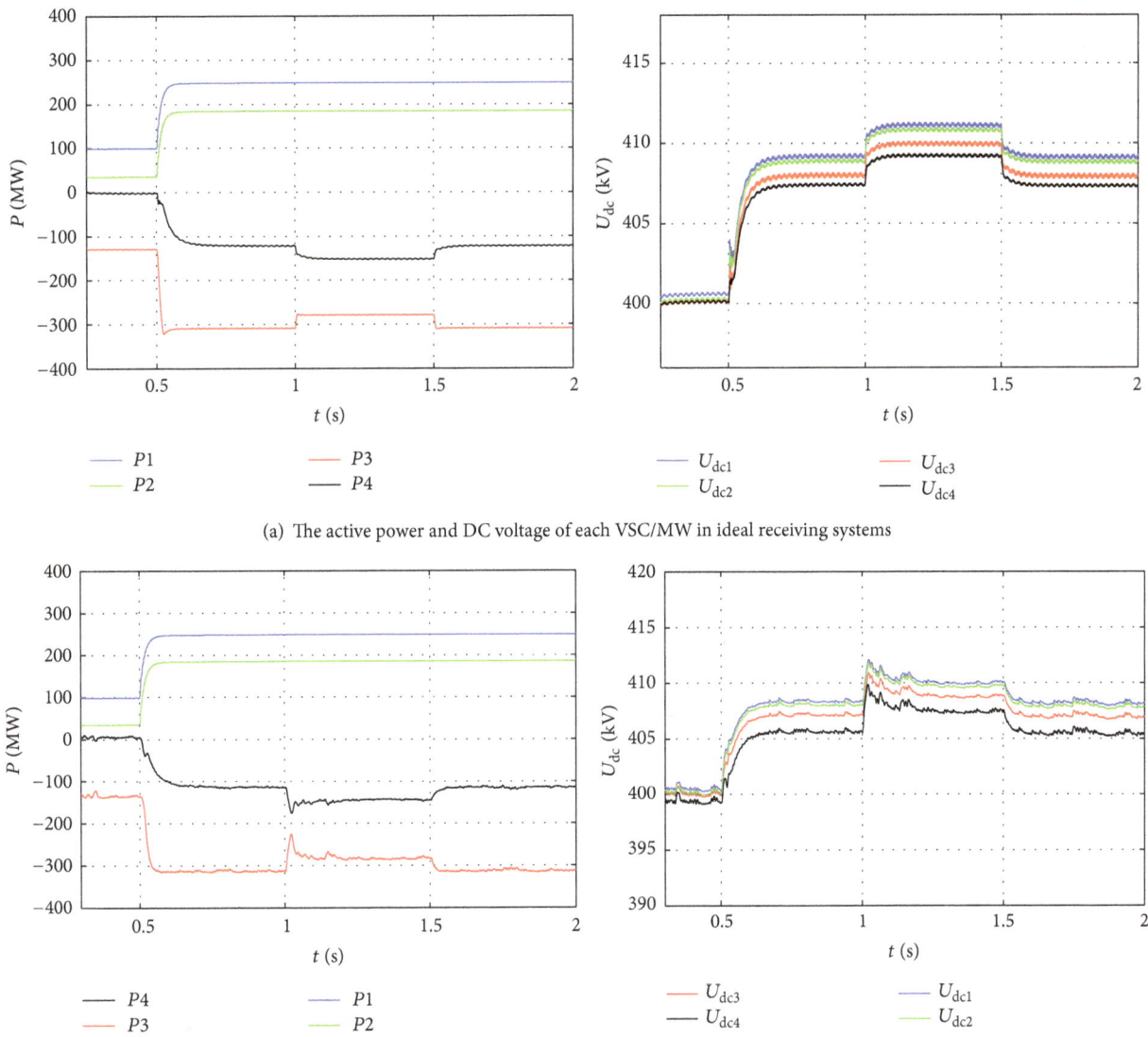

(a) The active power and DC voltage of each VSC/MW in ideal receiving systems

(b) The active power and DC voltage of each VSC/MW in standard receiving systems

FIGURE 8: Simulation of variation of transmission power ratio of GSVSCs.

power output increases to 430 MW during 0.5–1.0 second, the transmission power of the two GSVSCs is 320 MW and 107 MW, respectively, which meets the initial transmission ratio of 3 : 2. The transmission power of the two GSVSCs is 275 MW and 147 MW when the power transmission ratio is adjusted to 1 : 1 during 1.0–1.5 seconds in line with the power transfer ratio, and the DC voltage is jointly maintained at about 410 kV; the power transmission ratio returns to 3 : 2 at 1.5 seconds and the system approximately has the same value of the transmission power and the DC voltage with the aforementioned period of 0.5–1.0 sec. Simulation results show that, within the allowable error range, the actual converter station power distribution ratio consistent with the set transmission power ratio and the DC voltage is maintained within the system reasonable operating range based on droop characteristic curve.

Although the results with IEEE standard system in Figure 8(b) show little fluctuation both in active power and in DC voltage, the normal operation of the system will not be affected for the fact that the little fluctuation can be neglected against the rated power capacity and voltage level, and the working principle of the proposed control strategy is also proved to be effective according to the system response to the change of working states.

4.3. Simulation of Faults in AC System. The initial power output of wind farms is 130 MW before 0.5 seconds while the total output rises to 430 MW with the corresponding change in active power reference value $\Delta P_{\text{ref}} = [150, 150, 0, 0]$. The initial transmission power ratio is 3 : 2 and remains unchanged. A three-phase short circuit fault is applied to AC system which is connected to the 3rd GSVSC at 1 second

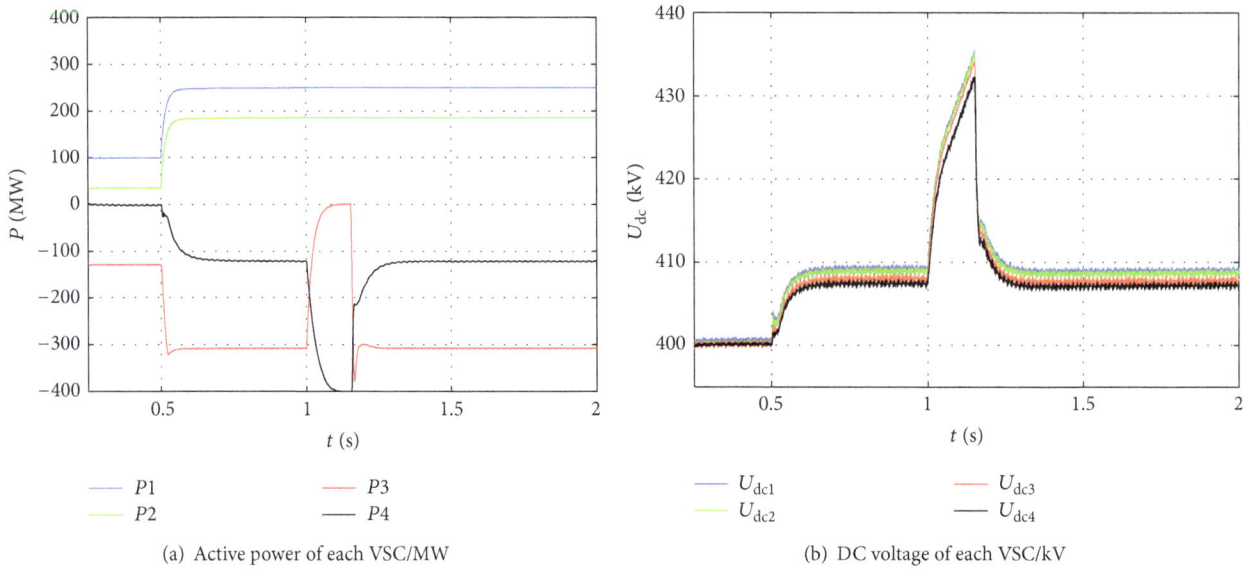

(a) Active power of each VSC/MW

(b) DC voltage of each VSC/kV

FIGURE 9: Simulation of three-phase faults of AC system.

with duration of 0.15 seconds. Figure 9 shows the simulation results of transmission power and DC voltage.

The simulation results of AC system with a serious three-phase short circuit are shown in Figure 9. Before the failure, due to the minimum initial power required, the wind power which is delivered into the AC system by the 3rd GSVSC is 126 MW considering the system loss. When the wind power output increases to 500 MW at 0.5 seconds, the transmission power delivered to AC system is in coordination with the transmission ratio, namely, 305 MW and 118 MW. When the three-phase short circuit is applied at 1 second, the transmission power of the 3rd GSVSC decreases quickly to zero due to the rapid drop of voltage in its AC bus; consequently all the wind power output needs to be transmitted through the 4th GSVSC into the AC system, which is beyond the rated capacity of the station. Thus the 4th GSVSC operates in extreme state with the transmission power of rated 400 MW, the excessive power pushes the DC voltage out of control which is up to about 435 kV, and the abnormal high voltage in actual operation may cause the outage of the system due to equipment limitations or trigger voltage of protection devices.

The situation in which DC voltage is out of control similar to the above can occur when the transmission lines for delivering the wind power out are blocked. In order to avoid this problem caused by the excessive wind power blocked in DC networks, damping resistance can be equipped in the DC side of GSVSC which works in bypass mode during normal operation and consumes excessive power to balance the system when faults occur as shown in [20]. From the point of view of energy conservation, energy storage devices can also be applied in the DC side of WFVSCs such as superconducting magnetic energy storage (SMES) and supercapacitor energy storage devices in [21, 22]. The excessive wind power when the faults occur in DC transmission lines or AC system will be stored in these storage devices and released back to

AC system when faults are cleared. The method mentioned above can also reduce the influence of excessive active power on system and improves the FRT ability of the system.

5. Conclusions

In this paper, the priority control strategy for GSVSC is proposed considering the uncertain output power of wind farms and the different penetration rate of wind power for the AC grid to realize a rational power allocation and smooth DC voltage control. This strategy can provide reference for the wind power trade when wind farms are interconnected with other grids for power supply.

A DC network with four terminals for incorporating wind farms is established. The relationship between the active power and DC voltage of the DC network is analyzed according to the operation condition of the priority controller combined with voltage droop control and PI control. The proposed control strategy under different operating conditions is simulated in PSCAD/EMTDC, and the simulation results show the effectiveness of the system on power distribution and DC voltage control.

Conflict of Interests

The authors declare that there is no conflict of interests regarding the republication of this paper.

References

[1] J. Chen, "Development of offshore wind power in China," *Renewable and Sustainable Energy Reviews*, vol. 15, no. 9, pp. 5013–5020, 2011.

[2] L. Hong and B. Möller, "Offshore wind energy potential in China: under technical, spatial and economic constraints," *Energy*, vol. 36, no. 7, pp. 4482–4491, 2011.

[3] N. B. Negra, J. Todorovic, and T. Ackermann, "Loss evaluation of HVAC and HVDC transmission solutions for large offshore wind farms," *Electric Power Systems Research*, vol. 76, no. 11, pp. 916–927, 2006.

[4] G. Ramtharan, A. Arulampalam, J. B. Ekanayake, F. M. Hughes, and N. Jenkins, "Fault ride through of fully rated converter wind turbines with AC and DC transmission systems," *IET Renewable Power Generation*, vol. 3, no. 4, pp. 426–438, 2009.

[5] T. Lobos, J. Rezmer, T. Sikorski, and Z. Waclawek, "Power distortion issues in wind turbine power systems under transient states," *Turkish Journal of Electrical Engineering & Computer Sciences*, vol. 16, no. 3, pp. 229–238, 2008.

[6] M. S. Carmeli, F. Castelli-Dezza, D. Rosati, G. Marchegiani, and M. Mauri, "MVDC connection of offshore wind farms to the transmission system," in *Proceedings of the International Symposium on Power Electronics, Electrical Drives, Automation and Motion (SPEEDAM '10)*, pp. 1201–1206, June 2010.

[7] R. T. Pinto, P. Bauer, S. F. Rodrigues, E. J. Wiggelinkhuizen, J. Pierik, and B. Ferreira, "A novel distributed direct-voltage control strategy for grid integration of offshore wind energy systems through MTDC network," *IEEE Transactions on Industrial Electronics*, vol. 60, no. 6, pp. 2429–2441, 2013.

[8] N. R. Chaudhuri and B. Chaudhuri, "Adaptive droop control for effective power sharing in multi-terminal DC (MTDC) grids," *IEEE Transactions on Power Systems*, vol. 28, no. 1, pp. 21–29, 2013.

[9] L. Xu and P. Cartwright, "Direct active and reactive power control of DFIG for wind energy generation," *IEEE Transactions on Energy Conversion*, vol. 21, no. 3, pp. 750–758, 2006.

[10] A. D. Hansen and G. Michalke, "Fault ride-through capability of DFIG wind turbines," *Renewable Energy*, vol. 32, no. 9, pp. 1594–1610, 2007.

[11] Y. Xi, A. Qian, H. Jiantao, and A. Yiran, "Application of multipoint DC voltage control in VSC-MTDC system," *Journal of Electrical and Computer Engineering*, vol. 2013, Article ID 257387, 7 pages, 2013.

[12] W. Lu and B.-T. Ooi, "Optimal acquisition and aggregation of offshore wind power by multiterminal voltage-source HVDC," *IEEE Transactions on Power Delivery*, vol. 18, no. 1, pp. 201–206, 2003.

[13] T. Nakajima and S. Irokawa, "A control system for HVDC transmission by voltage sourced converters," in *Proceedings of the IEEE Power Engineering Society Summer Meeting*, vol. 2, pp. 1113–1119, IEEE, Edmonton, Canada, June 1999.

[14] R. L. Hendriks, G. C. Paap, and W. L. Kling, "Control of a multi-terminal VSC transmission scheme for connecting offshore wind farms," in *Proceedings of the European Wind Energy Conference and Exhibition (EWEC '07)*, pp. 565–572, Milan, Italy, May 2007.

[15] J. Ren, K. Li, H. Liu, Y. Sun, and J. Zhao, "Coordinated control strategy of VSC-MTDC system based on improved DC voltage-active power characteristic," *Automation of Electric Power Systems*, vol. 37, no. 15, pp. 133–139, 2013.

[16] L. Xu and L. Yao, "DC voltage control and power dispatch of a multi-terminal HVDC system for integrating large offshore wind farms," *IET Renewable Power Generation*, vol. 5, no. 3, pp. 223–233, 2011.

[17] J. Ren, K. Li, J. Zhao, X. Dong, and X. Zhang, "A multi-point DC voltage control strategy of VSCMTDC transmission system for integrating large scale offshore wind power," in *Proceedings of the IEEE Innovative Smart Grid Technologies—Asia (ISGT Asia '12)*, IEEE, May 2012.

[18] R. Eriksson, J. Beerten, M. Ghandhari, and R. Belmans, "Optimizing DC voltage droop settings for AC/DC system interactions," *IEEE Transactions on Power Delivery*, vol. 29, no. 1, pp. 362–369, 2014.

[19] W. Wang and M. Barnes, "Power flow algorithms for multi-terminal VSC-HVDC with droop control," *IEEE Transactions on Power Systems*, vol. 29, no. 4, pp. 1721–1730, 2014.

[20] G. P. Adam, K. H. Ahmed, S. J. Finney, and B. W. Williams, "AC fault ride-through capability of a VSC-HVDC transmission systems," in *Proceedings of the 2nd IEEE Energy Conversion Congress and Exposition (ECCE '10)*, pp. 3739–3745, Atlanta, Ga, USA, September 2010.

[21] C. Rao, H. Wang, W. Wang et al., "Enhancement of the stable operation ability of large scale wind power system based on the VSC-HVDC embedded in energy storage apparatus," *Power System Protection and Control*, vol. 4, pp. 1–7, 2014.

[22] J. Wang, T. Liu, and X. Li, "Influence of connecting wind farms and energy storage devices to power grid on reliability of power generation and transmission system," *Power System Technology*, vol. 5, pp. 165–170, 2011.

11

New Application's Approach to Unified Power Quality Conditioners for Mitigation of Surge Voltages

Yeison Alberto Garcés Gomez,[1] **Nicolás Toro García,**[2] **and Fredy E. Hoyos**[2,3]

[1]*Universidad Católica de Manizales and Universidad Nacional de Colombia, Colombia*
[2]*Universidad Nacional de Colombia Sede Manizales, Colombia*
[3]*Universidad Nacional de Colombia Sede Medellin, Colombia*

Correspondence should be addressed to Yeison Alberto Garcés Gomez; yagarsesg@unal.edu.co

Academic Editor: Raj Senani

This paper outlines a new approach for the compensation of power systems presented through the use of a unified power quality conditioner (UPQC) which compensates impulsive and oscillatory electromagnetic transients. The newly proposed control technique involves a dual analysis of the UPQC where the parallel compensator is modelled as a sinusoidal controlled voltage source, while the series compensator is modelled as a sinusoidal controlled current source, opposed to the traditional approach where the parallel and series compensators are modelled as current and voltage nonsinusoidal sources, respectively. Also a new compensation algorithm is proposed through the application of the theory of generalized reactive power; this is then compared with the theory of active and reactive instantaneous power, or *pq* theory. The results are presented by means of simulations in MATLAB-Simulink®.

1. Introduction

Currently, one of the main problems with power quality is the increase of electronic devices, which require a high level of waveform voltage quality to operate properly at both residential and industrial levels [1]. These electronic devices are mainly responsible for the deterioration of the power quality acting as nonlinear loads [2].

The capacitor-switching transients, or CST, constitute the most common cause of surge voltage, followed only for lightning in most systems [3]. These transients cause misoperation or faults in devices at both residential and industrial levels; therefore, this problem of power quality has recently gained more attention due to devices that use solid state electronics and are more sensitive to surges than their predecessors [4].

Custom power devices or CPDs used in distribution systems can control power quality problems such as voltage and current harmonics, poor power factor, unbalance at the source, load imbalance, and flicker [5, 6]. The most common CPDs are parallel compensators for current and power factor correction [7, 8] and the series compensators to compensate

harmonic voltage, sag, swell, and flicker. One of the most efficient CPDs consists of parallel and series compensators with a common DC bus, which is called *unified power quality conditioner* or UPQC [1, 9, 10]. This combination allows the simultaneous compensation in the source current side via the parallel compensator and the load voltage side with the series compensator, thus isolating the system of power quality problems generated from the load and the load of problems from the source.

CPDs reference signals generation has been a major research problem. To date, generating reference signals for the compensation of most power quality problems has been widely studied; however, there is no compensation model for voltage surges through CPD. This is because the traditional model of compensation and estimation algorithms requires that the CPD generate a highly distorted wave and wide frequency range. Therefore, the focus on mitigating the transient has been focused on the *surge protective devices* (SPD) or limiter type switch as gas tubes, MOVs (Metal Oxide Varistors), and avalanche junction semiconductor devices [11].

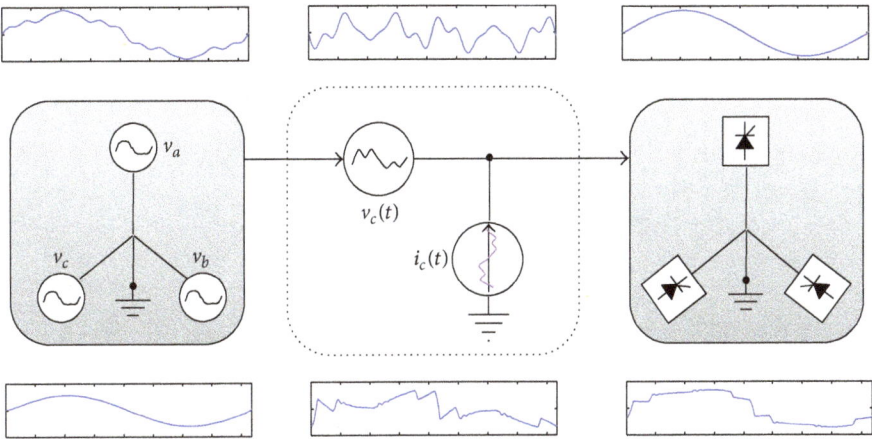

FIGURE 1: Classical scheme of compensation with the UPQC.

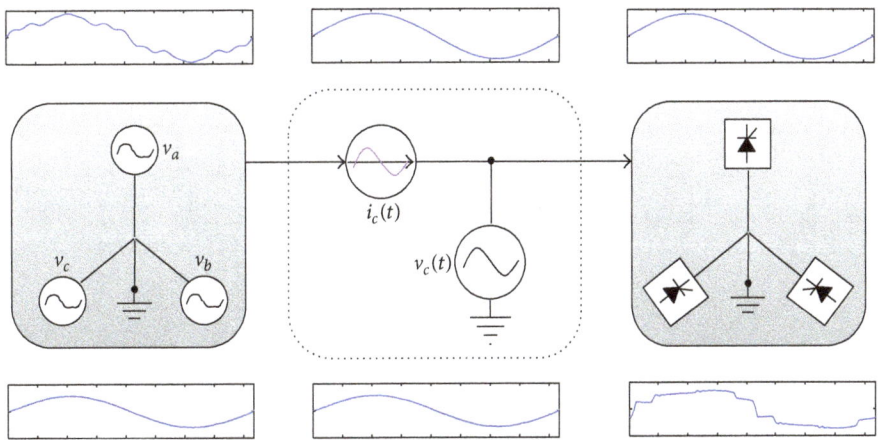

FIGURE 2: Dual scheme of compensation with the UPQC.

The main objective of this research is to develop a control technique and an algorithm for generating a reference signal that permits us to extend the use of CPD to compensate voltage surge.

In order to validate the proposed schematics, a power system compensation was undertaken in a test circuit by means of a simulation. It compensates low power factor, voltage and current harmonics, and oscillatory transient in voltage with a UPQC in dual topology.

2. Unified Power Quality Conditioner (UPQC)

2.1. Principle of Operation. The UPQC is composed of two power inverters in voltage source configuration VSI, connected back-to-back to a single DC bus. Each inverter acts like a controlled voltage and current sources.

In the classic model of compensation, the series compensator was modelled like a voltage controlled source that compensates the power quality problems in the voltage waveform. On the other hand, the parallel compensator acts like a current controlled source that removes the quality problems generated by the load current [9, 10]. This classical model of compensation is shown in Figure 1.

In [12–14], a new scheme of compensation is proposed for the UPQC, where the functions of the series and parallel compensators are invested, so that the series compensator acts like a sinusoidal current source that isolates the source from the power quality issues of the current in the load side. The parallel compensator functions like a sinusoidal voltage source that guarantees that the load is fed with a pure sinusoidal and sag-swell free voltage waveform. This model of compensation is called *dual* scheme of the UPQC and is illustrated in Figure 2.

The main advantage of the dual UPQC is that the waveforms that are needed to compensate the system are purely sinusoidal.

The general circuit of an electric power system with a UPQC is shown in Figure 3. The UPQC is composed of two three-leg VSI power inverters. The DC bus is composed of two capacitors with the middle point connected to GND; this configuration is called *three-leg split capacitor inverter* or *TLSC inverter*. The source has an impedance, R_{rd} and L_{rd}, in each phase.

The load is a nonlinear full controlled AC/CC converter that generates harmonic currents and a low power factor.

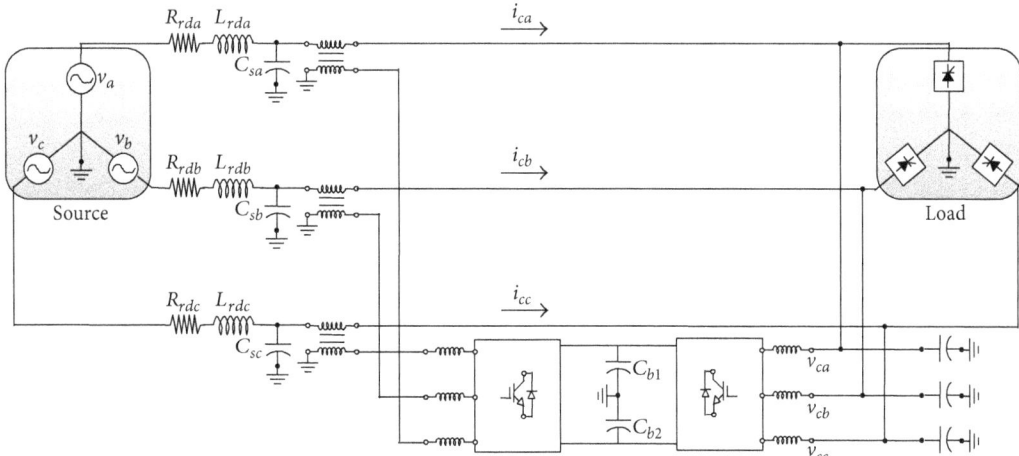

FIGURE 3: Circuit scheme of the UPQC.

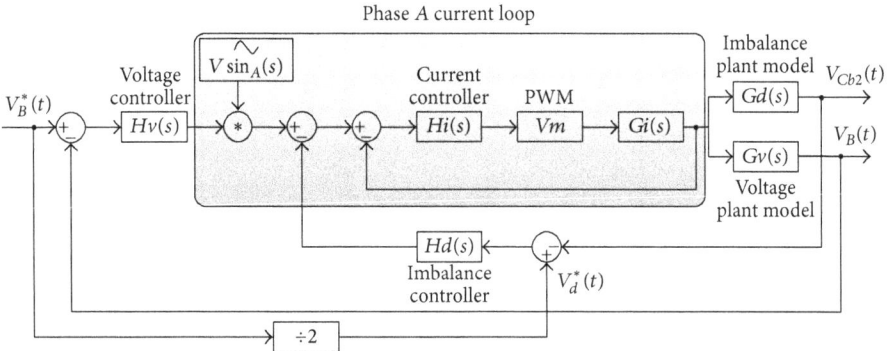

FIGURE 4: Block diagram of the UPQC control loops.

2.2. Control Scheme. The loop control for the UPQC is necessary to guarantee that the inverters generate efficiently the reference signals. The split capacitor structure has the main advantage that it allows us to make an uncoupled analysis of each phase of the series and parallel inverters.

The series compensator control is composed of four control loops. Since the dual model of compensation, the current control scheme proposed in this document is illustrated in Figure 4. The more internal control loop is the sinusoidal current controller of the source; it guarantees that the source current is sinusoidal and in phase with the voltage waveform. The intermediate loop is the imbalance control of the DC bus capacitors voltage, this controller must keep the same voltage in both capacitors, and this control loop acts like a DC current reference that charges or discharges the capacitors according to the difference in the value of the voltage between them.

Finally, the outer control loop in Figure 4 is the DC bus voltage control loop; this control keeps the voltage in the DC terminals of the inverters in a reference value, so that the UPQC can correct the power quality problems in the source and the load and has enough power to compensate sags and swells.

The space state model of the source's current loop is derived from the average model of the left-side inverter [15], so that the one per-phase circuit, referring to the primary of

the coupled transformer, can be simplified in the circuit of Figure 5. Equation (1) is the space state model of the circuit:

$$
\begin{bmatrix} V_B D\left(s\right) \\ 0 \end{bmatrix} = \begin{bmatrix} R_p + sL_{eq\_p} + Z_m & Z_m \\ Z_m & sL_{eq\_s} + R_{eq\_s} + Z_m \end{bmatrix} \\ * \begin{bmatrix} I_{cs\_p}\left(s\right) \\ I_{cs\_s}\left(s\right) \end{bmatrix}. \tag{1}
$$

The parallel compensator control loop has the main objective of providing the load with pure three-phase sinusoidal waveforms, balanced and with nominal value voltage. It has a unique control loop.

Every controller is designed in frequency with the methodology presented in [14]; then the transfer functions are discretized to be compensated with PID classical control schemes.

3. Generalized Reactive Power Theory Applied to Dual UPQC

In 2007 the generalized reactive power formulation applied to poly-phase systems was presented [16, 17]; this was defined later in 2010 as *instantaneous power tensor theory* [18, 19]. This

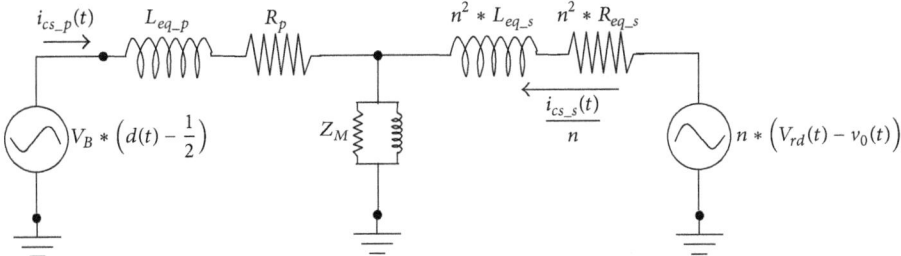

FIGURE 5: Simplified averaged circuit of the series inverter.

formulation is based on the interpretation of the instantaneous voltage and current vectors like first-order tensors to define the components of power from operations with *dyadic* or *tensorial* product.

From the instantaneous vectors of voltage and current, $\vec{u} = \mathbf{u} = [u_1 \ u_2 \ \cdots \ u_4]$ and $\vec{i} = \mathbf{i} = [i_1 \ i_2 \ \cdots \ u_4]$, in [16, 17] the active instantaneous power $p(t)$ and the imaginary instantaneous power $q(t)$ are defined according to (2) and (3), respectively:

$$p(t) = \vec{u} \bullet \vec{i}, \tag{2}$$

$$q(t) = \vec{i}(t) \wedge \vec{u}(t). \tag{3}$$

In the previous equations ((2) and (3)), $\bullet$ and $\wedge$ operators denote the *dot* and the *outer product*, respectively. The outer product is an antisymmetrization of the *dyadic* or *tensorial product* which is denoted by the $\otimes$ operator, so that (3) can be rewritten like

$$q(t) = \vec{i}(t) \wedge \vec{u}(t) = (\vec{i} \otimes \vec{u}) - (\vec{u} \otimes \vec{i}). \tag{4}$$

Furthermore, the active component of the current is defined in

$$\vec{i}_p = \frac{p(t)}{\vec{u}^T \vec{u}} \vec{u} = \frac{(u_1 i_1 + u_2 i_2 + \cdots + u_n i_n)}{\vec{u} \bullet \vec{u}} \begin{bmatrix} u_1 \\ u_2 \\ \vdots \\ u_n \end{bmatrix}. \tag{5}$$

The total current demanded by the load is defined in

$$\begin{aligned} \vec{i} &= \frac{(\vec{i} \otimes \vec{u})}{\vec{u} \bullet \vec{u}} \vec{u} = \frac{[(\vec{u} \otimes \vec{i}) + (\vec{i} \otimes \vec{u}) - (\vec{u} \otimes \vec{i})]}{\vec{u} \bullet \vec{u}} \vec{u} \\ &= \frac{(\vec{u} \otimes \vec{i})}{\vec{u} \bullet \vec{u}} \vec{u} + \frac{[(\vec{i} \otimes \vec{u}) - (\vec{u} \otimes \vec{i})]}{\vec{u} \bullet \vec{u}} \vec{u} \\ &= \frac{(\vec{u} \otimes \vec{i})}{\vec{u} \bullet \vec{u}} \vec{u} + \frac{(\vec{i} \wedge \vec{u})}{\vec{u} \bullet \vec{u}} \vec{u} = \frac{p(t)}{\vec{u} \bullet \vec{u}} \vec{u} + \frac{q(t)}{\vec{u} \bullet \vec{u}} \vec{u}. \end{aligned} \tag{6}$$

Finally, the result is that the current can be decomposed in two components, an active component and a reactive or imaginary component like

$$\vec{i} = \vec{i}_p + \vec{i}_q. \tag{7}$$

From this formulation, the estimation of the reference to parallel compensators of current named *perfect harmonic cancellation* or *PHC* is given by

$$\vec{i}_{\text{ref}} = \vec{i} - \vec{i}_{p\text{-}f}^+, \tag{8}$$

where $\vec{i}_{p\text{-}f}^+$ is the instantaneous vector of direct sequence and fundamental frequency current, given by

$$\vec{i}_{p\text{-}f}^+ = \frac{\text{tr}\left(\overline{\wp}_{ij}\right)}{(1/T) \int_T \left(\vec{v}_f^+ \cdot \vec{v}_f^+\right)} \vec{v}_f^+. \tag{9}$$

Equation (9) is current reference for the UPQC in dual topology because it is at fundamental frequency, positive sequence, and in phase with the fundamental component of the voltage. However, $\vec{v}_f^+$, called *direct sequence and fundamental frequency voltage vector* [18], is calculated by using the Fortescue transformation and the decomposing in Fourier series of a periodical waveform that is inadequate to compensate sags and swells because the decomposition of the expression in these events has shown $k \subset \mathbb{R}$ factor in the amplitude of the voltage reference, and the current reference $\vec{i}_{p\text{-}f}^+$ will be affected as in

$$\vec{i}_{p\text{-}f}^+ = k \frac{\text{tr}\left(\overline{\wp}_{ij}\right)}{(1/T) \int_T \left(\vec{v}_f^+ \cdot \vec{v}_f^+\right)} \vec{v}_f^+. \tag{10}$$

As regards the aforementioned drawback, it is necessary to reconsider the calculation of $\vec{v}_f^+$. That is why we propose the use of $\vec{v}_{f\_u}^+$: that is, the *positive sequence and fundamental frequency unit voltage vector*. To extract this vector we use a phase locked loop in a synchronous reference frame *SRF-PLL* and a unit vector template generator *UVTG* [20] to obtain the three-phase sinusoidal signals in unitary amplitude and later getting

$$v_{f\_u\_a}^+ = \sqrt{2} V_{\text{RMS\_ref}} \sin\left(\hat{\theta}\right),$$

$$v_{f\_u\_b}^+ = \sqrt{2} V_{\text{RMS\_ref}} \sin\left(\hat{\theta} - 120\right), \tag{11}$$

$$v_{f\_u\_c}^+ = \sqrt{2} V_{\text{RMS\_ref}} \sin\left(\hat{\theta} + 120\right).$$

In (11), $V_{\text{RMS\_ref}}$ is the nominal voltage value of the load. The scheme to obtain θ is shown in Figure 6.

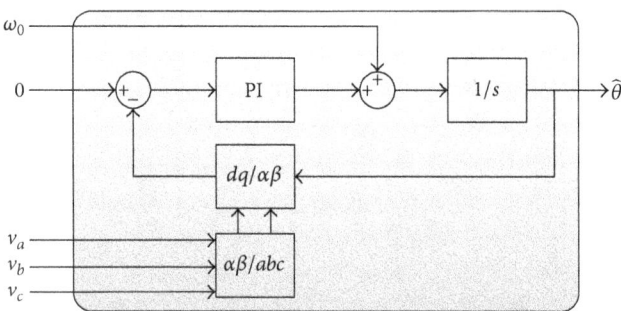

FIGURE 6: SRF-PLL scheme.

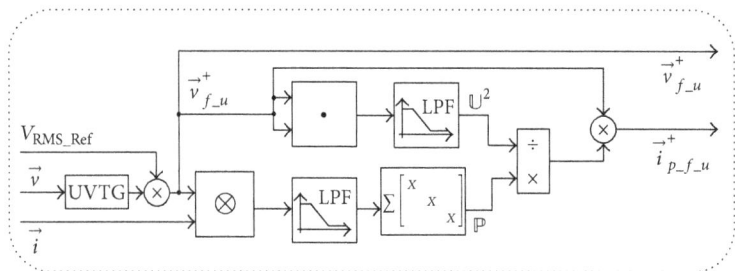

FIGURE 7: Reference signals generator scheme for the iUPQC with SRF-PLL and UVTG.

The expressions in (11) are the voltage reference signals for the *iUPQC*. With (11), (10) can be formulated like in

$$\vec{i}_{p\text{-}f\text{\_}u}^{+} = \frac{\mathbb{P}}{\mathbb{U}^2}\vec{v}_{f\text{\_}u}^{+}, \tag{12}$$

where $\mathbb{U}$ is the average norm of the *positive sequence and fundamental frequency unit voltage vector* and $\mathbb{P}$ is the average active power to fundamental voltage that is provided in

$$\mathbb{P} = \vec{v}_{f\text{\_}u}^{+} \bullet \vec{i}. \tag{13}$$

The vector $\vec{i}_{p\text{-}f\text{\_}u}^{+}$ has the three-phase current reference signals for the unified power quality conditioner; thus, the algorithm for the reference estimation in voltage and current in the *iUPQC* is fully developed in the scheme of Figure 7.

4. Simulation and Results

The simulated iUPQC is composed of a DC bus with split capacitor; the switching pulses are generated by PID discrete controllers compared with 20 kHz triangular signals. The series and parallel compensators have the function of compensating current in the source and voltage in the load, respectively (dual compensation). In addition, by means of the series compensator the DC bus voltage and the DC capacitors imbalance of voltage are controlled. The complete scheme implemented in MATLAB-Simulink is presented in Figure 8.

4.1. Simulation System Parameters

(i) *Source.* $115\,V_{RMS}$, 60 Hz. The harmonic pollution parameters are in Table 1. The source impedance is

TABLE 1: Harmonic distortion in the voltage source per phase.

	$v_{f_{RMS}}$	$v_{5th_{RMS}}$	$v_{7th_{RMS}}$	$v_{11th_{RMS}}$	$v_{13th_{RMS}}$
v_a	115 V	15 V	0 V	5.7 V	5.7 V
v_b	115 V	15 V	15 V	0 V	5.7 V
v_c	108 V	12 V	12 V	5.4 V	0 V

$R_{rd} = 0.04\,\Omega$ and $L_{rd} = 107\,\mu H$. In addition, in the voltage source occurs an oscillatory surge voltage or transient in 4.16 ms.

(ii) *Load.* Composed of a fully controlled thyristors rectified bridge, with $40°$ phase angle, the impedance in DC side is a parallel RC load in series with an inductor; the values are $R = 75\,\Omega$, $C = 10\,\mu F$, and $L = 100\,mH$.

(iii) *Series Inverter.* The series compensator is connected by means of $L = 20\,mH$ reactors and 1 kVA single-phase transformers with $n = 1$. The open circuit and short circuit parameters were measured. At source side a $C_s = 100\,\mu F$ capacitor per phase is placed to mitigate the high frequency of the inverter.

(iv) *Parallel Inverter.* This is connected by $L_p = 650\,mH$ reactors and a high frequency RC filter with a capacitor of $C = 10\,\mu F$.

(v) *DC Bus.* These are two series capacitors with the middle point ground connected: each one is of 6 mF for a total capacitance of 3 mF.

4.2. Results. In Figure 9 results of compensation in the source current are shown. The series compensator enables us to follow the reference of the algorithm in the generalized reactive

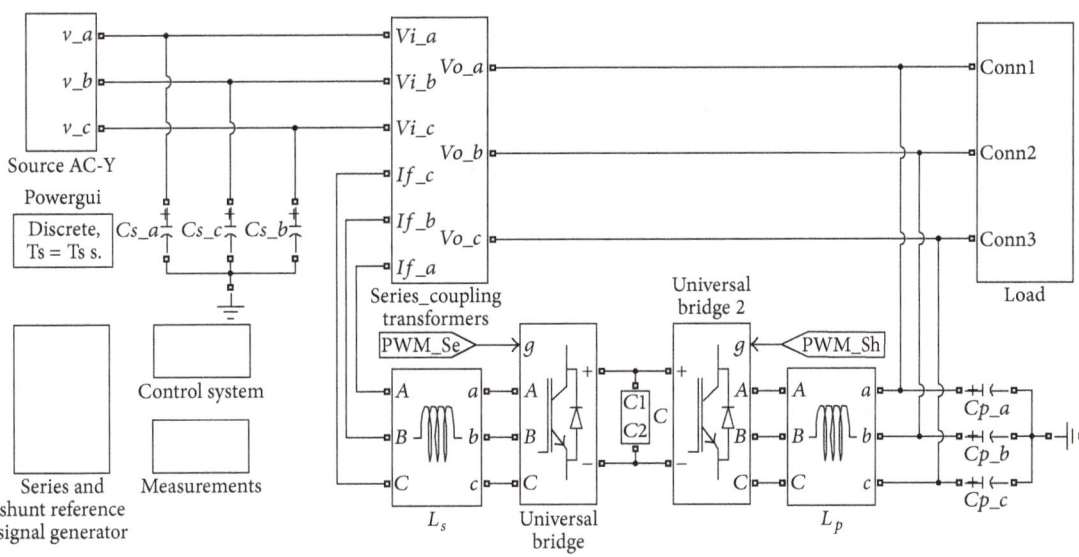

FIGURE 8: MATLAB-Simulink simulation algorithm.

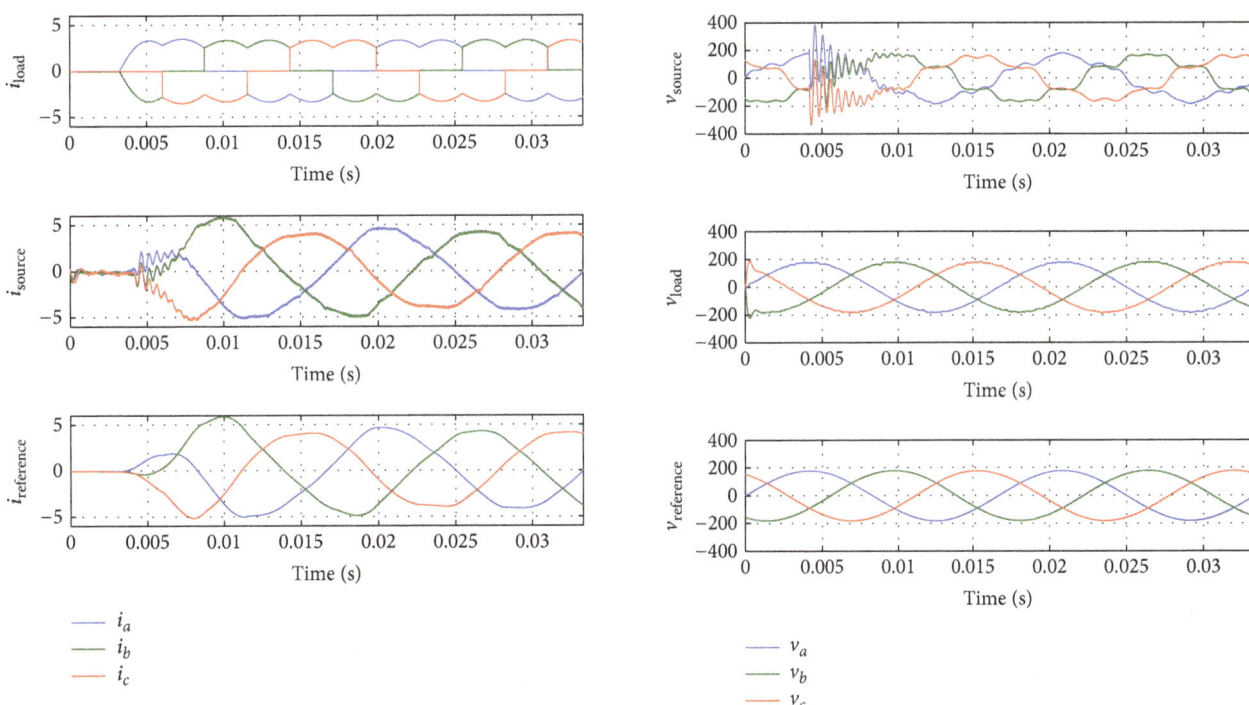

FIGURE 9: Current waveforms before and after compensation and reference signals.

FIGURE 10: Voltage waveforms before and after compensation and reference signals.

power theory frame. Figure 10 shows the results of compensating the load voltage. In the three phases of the source system a surge voltage is generated (voltage corresponding to nonsynchronous closing in a capacitor bank [21]), but these do not affect the load by the action of the iUPQC. In the source current (Figure 9), we show that at the same time of surge voltage there is an oscillation; this is because of the interaction of the surge with the source impedance Z_{rd}.

The frequency results of compensation are summarized in Table 2. The per-phase harmonic distortion index was

evaluated during and after the voltage surge for voltage and for the current before and after compensation THD.

Finally, in Figure 11 we show the comparison of the current signal reference estimation in a particular load case using the generalized reactive power theory with the proposed algorithm and the same reference estimation results of [12] derived with the instantaneous reactive power theory or pq theory.

TABLE 2: Per-phase THD values before and after compensation.

	THD before compensation with surge voltage	THD before compensation without surge voltage	THD after compensation
v_a	40.39%	14.96%	1.24%
v_b	41.88%	19.29%	1.10%
v_c	42.32%	16.51%	1.08%
i_a	NA	30.36%	3.45%
i_b	NA	30.26%	4.46%
i_c	NA	30.33%	3.43%

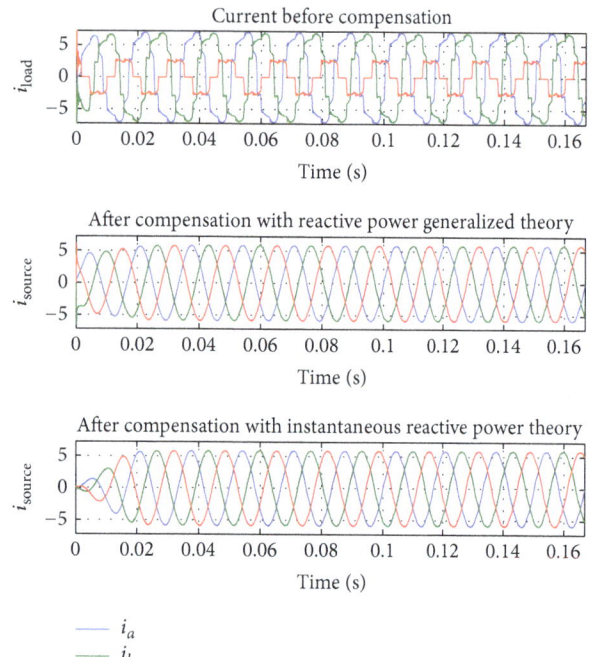

FIGURE 11: Current waveforms for a particular load and signal reference estimated with two different algorithms. Comparison of the proposed algorithm and [12] algorithm.

5. Conclusions

In this paper, we propose a new reference signal algorithm to estimate the references for a UPQC in dual topology using the generalized reactive power theory, and inverting the functioning of the compensators to work like sinusoidal voltage dependent source and sinusoidal current dependent source. This is called the dual of the traditional concept where the dependent sources are highly distorted. The results of the simulations show that the UPQC with the proposed algorithm is able to compensate highly distorted and imbalanced source currents with the series compensator, and at the same time it can compensate sag, swell, unbalance, and voltage harmonics with the parallel compensator. The proposed control approach for the UPQC has improvements in its functioning by the addition of the capability to compensate oscillatory transients and source voltage. Finally, the proposed algorithm is compared with the algorithm to estimate signals references

by means of the instantaneous reactive power theory showing that the new one has a faster response in transients, therefore reaching the steady state in less time than the previously proposed algorithm.

Competing Interests

The authors declare that there is no conflict of interests regarding the publication of this paper.

Acknowledgments

This work was supported by the Universidad Católica de Manizales and the Universidad Nacional de Colombia Sede Manizales with the *Grupo de Investigación en Recursos Energéticos* (GIRE-COL0144229), the *Grupo de Investigación Percepción y Control Inteligente* (PCI), the Grupo de Investigación en Potencia, Energía y Mercados (GIPEM), and The Universidad Nacional de Colombia Sede Medellín (Instrumentación Científica e Industrial-COL0037917) with Research Project HERMES30115.

References

[1] A. Mokhtarpour, H. Shayanfar, and S. M. T. Bathaee, *Reference Generation of Custom Power Devices (CPs)*, INTECH, 2013.

[2] J. L. Afonso, J. G. Pinto, and H. Gonçalves, "Active power conditioners to mitigate power quality problems in industrial facilities," in *Power Quality Issues*, A. Zobaa, Ed., chapter 5, InTech, Rijeka, Croatia, 2013.

[3] J. L. Durán-Gómez and P. N. Enjeti, "A new approach to mitigate nuisance tripping of PWM ASDs due to utility capacitor switching transients (CSTs)," *IEEE Transactions on Power Electronics*, vol. 17, no. 5, pp. 799–806, 2002.

[4] F. D. Martzloff, "A guideline on surge voltages in AC power circuits rated up to 600 V," in *Proceedings of the 3rd International Symposium on Electromagnetic Compatibility*, 1979.

[5] J. Fei, T. Li, F. Wang, and W. Juan, "A novel sliding mode control technique for indirect current controlled active power filter," *Mathematical Problems in Engineering*, vol. 2012, Article ID 549782, 18 pages, 2012.

[6] J. Fei and S. Hou, "Adaptive fuzzy control with supervisory compensator for three-phase active power filter," *Journal of Applied Mathematics*, vol. 2012, Article ID 654937, 13 pages, 2012.

[7] S. S. Patnaik and A. K. Panda, "Particle swarm optimization and bacterial foraging optimization techniques for optimal current harmonic mitigation by employing active power filter,"

Applied Computational Intelligence and Soft Computing, vol. 2012, Article ID 897127, 10 pages, 2012.

[8] Z. Chelli, R. Toufouti, A. Omeiri, and S. Saad, "Hysteresis control for shunt active power filter under unbalanced three-phase load conditions," *Journal of Electrical and Computer Engineering*, vol. 2015, Article ID 391040, 9 pages, 2015.

[9] S. A. Taher and S. A. Afsari, "Optimal location and sizing of UPQC in distribution networks using differential evolution algorithm," *Mathematical Problems in Engineering*, vol. 2012, Article ID 838629, 20 pages, 2012.

[10] R. Dharmalingam, S. S. Dash, K. Senthilnathan, A. B. Mayilvaganan, and S. Chinnamuthu, "Power quality improvement by unified power quality conditioner based on CSC topology using synchronous reference frame theory," *The Scientific World Journal*, vol. 2014, Article ID 391975, 7 pages, 2014.

[11] J. C. Das, *Transients in Electrical Systems: Analysis, Recognition, and Mitigation*, McGraw Hill Professional, New York, USA, 2010.

[12] M. Aredes and R. M. Fernandes, "A unified power quality conditioner with voltage sag/swell compensation capability," in *Proceedings of the Brazilian Power Electronics Conference (COBEP '09)*, pp. 218–224, Bonito, Brazil, October 2009.

[13] B. W. França and M. Aredes, "Comparisons between the UPQC and its dual topology (iUPQC) in dynamic response and steady-state," in *Proceedings of the 37th Annual Conference of the IEEE Industrial Electronics Society (IECON '11)*, pp. 1232–1237, Melbourne, Australia, November 2011.

[14] R. J. M. dos Santos, M. Mezaroba, and J. C. da Cunha, "A dual unified power quality conditioner using a simplified control technique," in *Proceedings of the 11th Brazilian Power Electronics Conference (COBEP '11)*, pp. 486–493, September 2011.

[15] N. Mohan, *First Course on Power Electronics and Drives*, Mnpere, Minneapolis, Minn, USA, 2003.

[16] P. Salmer and R. S. Herrera, "Instantaneous reactive power theory—a general approach to poly-phase systems," *Electric Power Systems Research*, vol. 79, no. 9, pp. 1263–1270, 2009.

[17] R. S. Herrera, P. Salmeron, J. R. Vazquez, and S. P. Litran, "Instantaneous reactive power theory to N wire systems," in *Proceedings of the IEEE International Symposium on Industrial Electronics (ISIE '07)*, pp. 2457–2462, ISIE, June 2007.

[18] A. J. Ustariz, E. A. Cano Plata, and H. E. Tacca, "Instantaneous power tensor theory: improvement and assessment of the electric power quality," in *Proceedings of the 14th International Conference on Harmonics and Quality of Power (ICHQP '10)*, pp. 1–6, IEEE, Bergamo, Italy, September 2010.

[19] A. J. Ustariz, E. A. Cano, and H. E. Tacca, "Tensor analysis of the instantaneous power in electrical networks," *Electric Power Systems Research*, vol. 80, no. 7, pp. 788–798, 2010.

[20] X.-Q. Guo, W.-Y. Wu, and H.-R. Gu, "Phase locked loop and synchronization methods for grid-interfaced converters: a review," *Przeglad Elektrotechniczny*, vol. 87, no. 4, pp. 182–187, 2011.

[21] M. F. Iizarry-Silvestrini and T. E. Vélez-Sepúlveda, *Mitigation of Back-to-Back Capacitor Switching Transients on Distribution Circuits*, Department of Electrical and Computer Engineering, University of Puerto Rico, San Juan, Puerto Rico, USA, 2006.

12

Parameter Extraction of Solar Photovoltaic Modules using Gravitational Search Algorithm

R. Sarjila, K. Ravi, J. Belwin Edward, K. Sathish Kumar, and Avagaddi Prasad

School of Electrical Engineering, VIT University, Vellore, Tamil Nadu, India

Correspondence should be addressed to R. Sarjila; r.sarjila@yahoo.com

Academic Editor: Ahmed M. Soliman

Parameter extraction of a solar photovoltaic system is a nonlinear problem. Many optimization algorithms are implemented for this purpose, which failed in giving better results at low irradiance levels. This article presents a novel method for parameter extraction using gravitational search algorithm. The proposed method evaluates the parameters of different PV panels at various irradiance levels. A critical evaluation and comparison of gravitational search algorithm with other optimization techniques such as genetic algorithm are given. Extensive simulation analyses are carried out on the proposed method and show that GSA is much suitable for parameter extraction problem.

1. Introduction

Precise parameters extraction of solar photovoltaic cells is normally a vital part of a solar photovoltaic (PV) system, which can be interfaced with maximum power point tracker (MPPT) calculations and power electronic converters. This undertaking is vital for the device modelling, characterization, and simulation and for the device quality testing. Maximum power point tracker, ordinarily alluded to as MPPT, is an electronic framework that works the PV modules in a way that permits the modules to deliver all the force they are able to do [1]. MPPT is not a mechanical following framework that "physically moves" the modules to make them point all the more straightforwardly at the sun. MPPT is a completely electronic framework which highlights a brilliant attaching calculation that changes the electrical working purpose of the modules so that the modules can convey most extreme accessible force. Extra power collected from the modules is then made accessible as expanded battery charge current. MPPT calculations are essential in PV applications in light of the fact that the maximum power point (MPP) of a sun oriented board changes with the illumination and temperature, so the utilization of MPPT calculations [2] is required keeping in mind the end goal to acquire the most extreme force from a sun based exhibit. The present scope technique utilizes a range waveform for the PV cluster current

such that the *I-V* normal for the PV exhibit is gotten and upgraded at settled time interims. The most extreme force point voltage can then be processed from the trademark bend at the same interims.

The above addressed problem can be solved efficiently by many prominent techniques. But each prominent technique [3, 4] has its own advantages and limitations. Some of the common drawbacks are neglecting the effects of high level injection and the assumption that the diode was operating in low-level injection while deriving the Shockley equation. These drawbacks can be overcome using the proposed GSA method. In this aspect GSA is found to be superior to other techniques. In parameter extraction, the parameters of the solar panel other than the data given in the manufacturer's data sheet are optimized using an optimization technique. It is done such that the power delivered from the solar panel is maximum even though the solar irradiation and temperature change. These values are then fed into a SIMULINK model which gives the *VI* characteristics of the panel. Output of the SIMULINK model is then compared with the data sheet characteristics. The error has to be less.

2. Gravitational Search Algorithm

In the recent past, different heuristic advancement techniques have been created. Huge numbers of these strategies are

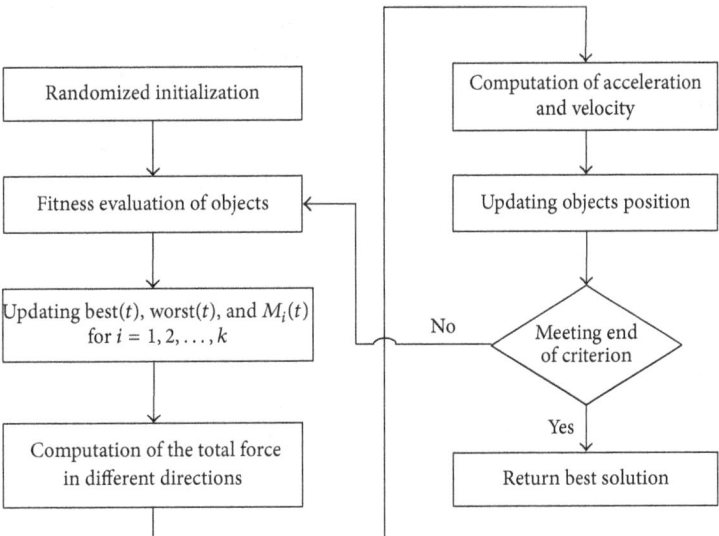

FIGURE 1: Flowchart of gravitational search algorithm.

propelled by swarm practices in nature. In this paper, another advancement calculation in light of the law of gravity and mass connections is presented [5]. There are a few points of interest of GSA that demonstrate its centrality contrasted and other nature enlivened systems. To start with, it requires just two parameters to modify, that is, mass and velocity of particles, and second is has capacity to discover close global optimum solution. The capacity to discover close global optimum solution makes the GSA vary from other nature inspired algorithms. In the proposed calculation, the searcher specialists are an accumulation of masses which connect with each other in light of the Newtonian gravity and the laws of movement. The proposed technique has been contrasted and some understood heuristic pursuit strategies [6]. The acquired results affirm the superiority of the proposed technique in illuminating different nonlinear capacities. Figure 1 shows the flowchart of gravitational search algorithm.

2.1. Algorithm for GSA

Step 1 (initialization).

    Current position: position of particles

    Velocity: velocity

    Force: the gravitational force between the particles

    Acceleration: acceleration

    Mass: mass

    Dim: dimension of test functions

    N: number of particles

    G_o: gravitational constant

    Low, up: search space limits

Step 2. Read the parameters of the PV panel under consideration.

Step 3 (search space identification). Agents that go out of the search space are returned to the boundaries.

Step 4 (evaluation of the population). The objective function is calculated for each iteration.

Step 5 (mass calculation). Mass of each agent is calculated.

    Loop: $i = 1:n$.

Step 6 (force update). Force is updated for each mass.

    Loop: $k = 1:n$

Step 7. Update acceleration and velocity.

    Loop: $i = 1:n$

Step 8. Update agent's position.

Step 9. Repeat Step 4 until the stop criterion is reached.

Step 10. Print the result.

3. Modelling of Photovoltaic Panel

The single diode condition accepts a steady esteem for the ideality factor n. In all actuality the ideality factor is a function of voltage over the device. At high voltage, when the recombination in the device is overwhelmed by the surfaces and the huge regions the ideality factor is near one. However, at lower voltages, recombination in the intersection is overwhelmed and the ideality factor approaches two. The intersection recombination is displayed by including a second diode in parallel with the first and setting the ideality consideration regularly to two. The equivalent circuit for double diode model is shown in Figure 2.

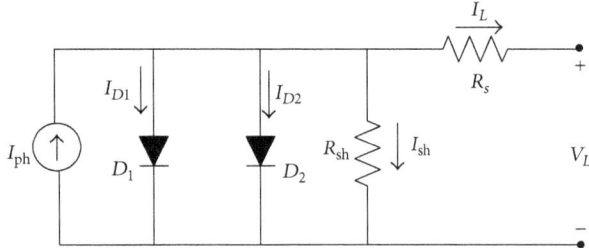

FIGURE 2: Double diode model of solar cell.

In this double diode model, the cell terminal current is calculated as follows:

$$I_L = I_{ph} - I_{D1} - I_{D2} - I_{sh}, \qquad (1)$$

where I_L is terminal current, I_{ph} is cell-generated photocurrent, I_{D1}, I_{D2} are first and second diode currents, and I_{sh} is shunt resistor current.

The diode current equations and leakage current equation are given by

$$I_{D1} = I_{01} \left[\exp\left(\frac{q(V_L + IR_s)}{a_1 kT} \right) - 1 \right]$$

$$I_{D2} = I_{02} \left[\exp\left(\frac{q(V_L + IR_s)}{a_2 kT} \right) - 1 \right] \qquad (2)$$

$$I_{sh} = \frac{V_L + IR_s}{R_{sh}},$$

where R_s and R_{sh} are the series and shunt resistances, respectively; V_L is the terminal voltage; a_1, a_2 are the diffusion and recombination diode ideality factors; k is Boltzmann's constant; q is the electronic charge; and T is the cell absolute temperature in Kelvin.

Thus the expression for cell terminal current is formulated as

$$I_L = I_{ph} - I_{01} \left[\exp\left(\frac{q(V_L + IR_s)}{a_1 kT} \right) - 1 \right]$$

$$- I_{02} \left[\exp\left(\frac{q(V_L + IR_s)}{a_2 kT} \right) - 1 \right] - \frac{V_L + IR_s}{R_{sh}}. \qquad (3)$$

The seven parameters to be estimated that fully describe the I-V characteristics are R_s, R_{sh}, I_{ph}, I_{01}, I_{02}, a_1, and a_2.

3.1. Problem Formulation. The parameters R_s, R_{sh}, I_{ph}, I_{01}, I_{02}, a_1, and a_2 with temperature and irradiance depend on manufacturing tolerance. Such functions have no explicit analytical solutions for either I_L or V_L. The GSA optimization technique [6] is employed to estimate the parameters by minimizing a preselected objective function which is given by

$$f\left(V_L, I_L, I_{ph}, I_{01}, I_{02}, R_s, R_{sh}, a_1, a_2\right) = 0. \qquad (4)$$

The new objective function that sums the individual absolute errors (IAEs) for any given set of measurements is defined as

$$f = \sum_{i=1}^{N} \left| f\left(V_i, I_i, I_{01}, I_{02}, R_s, R_{sh}, a_1, a_2\right)\right|, \qquad (5)$$

where N is the number of data points and I_i and V_i are ith measured current and voltage pair values, respectively.

Objective Function. At maximum power point, the derivative of the power with respect to voltage is equal to zero. That is,

$$\frac{dP}{dV_L} = 0. \qquad (6)$$

The power equation is written as $P = VI_L$. Applying condition for MPP the above equation changes to

$$\frac{dP}{dV_L} = \left[V_L \times \left(\frac{dI_L}{dV_L} \right) \right] + I_L. \qquad (7)$$

In order to obtain the maximum power the term dP/dV_L has to be made zero.

So, RHS is equated to zero.

$$\left[V_L \times \left(\frac{dI_L}{dV_L} \right) \right] + I_L = 0.$$

$$\left(\frac{dI_L}{dV_L} \right) + \left(\frac{I_L}{V_L} \right) = 0. \qquad (8)$$

The objective function to be minimized is

$$J = \left| \frac{dI_L}{dV_L} \right|_{(V_{mp}, I_{mp})} + \frac{I_{mp}}{V_{mp}}$$

$$\left| \frac{dI_L}{dV_L} \right| \qquad (9)$$

$$= \frac{\left(I_{01} \times L_1 \times \exp\left(L_1 \times \left(V_{mpp} + \left(I_{mpp} \times R_s\right)\right)\right) + I_{02} \times L_2 \times \exp\left(L_2 \times \left(V_{mpp} + \left(I_{mpp} \times R_s\right)\right)\right) + G_p\right)}{\left(1 + I_{01} \times L_1 \times R_s \times \exp\left(L_1 \times \left(V_{mpp} + \left(I_{mpp} \times R_s\right)\right)\right) + I_{02} \times L_2 \times R_s \times \exp\left(L_2 \times \left(V_{mpp} + \left(I_{mpp} \times R_s\right)\right)\right) + \left(G_p \times R_s\right)\right) + \left(I_{mpp}/V_{mpp}\right)},$$

where $L_1 = 1/a_1 V_t$; $L_2 = 1/a_2 V_t$; V_{mpp} is Peak Power Voltage; I_{mpp} is Peak Power Current; $G_p = 1/R_p$.

The model parameters extracted are subsequently substituted in the MATLAB/SIMULINK model to plot the I-V

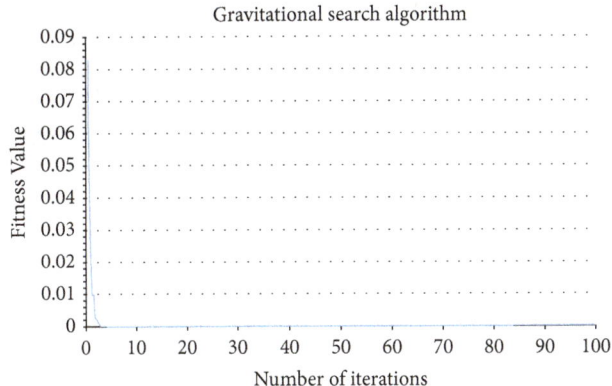

FIGURE 3: Convergence graph for Shell S36.

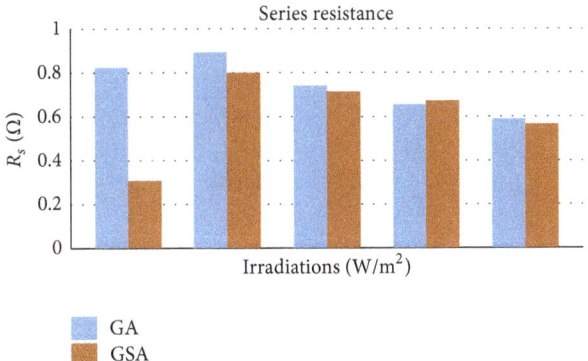

FIGURE 4: Comparison of the optimized series resistance values of Shell S36 under different irradiance conditions.

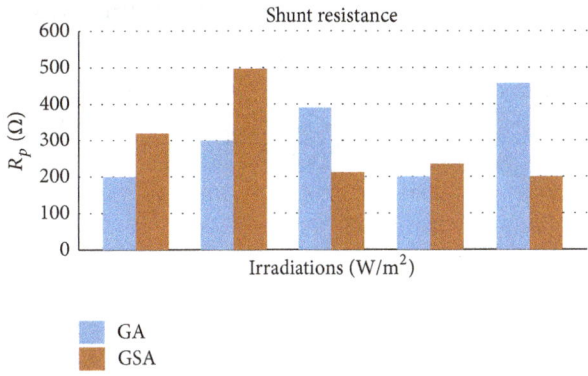

FIGURE 5: Comparison of the optimized shunt resistance values of Shell S36 under different irradiance conditions.

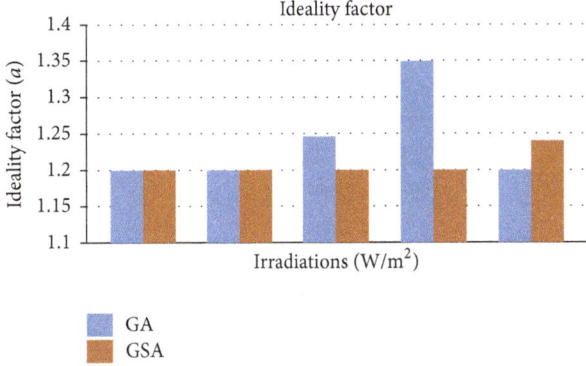

FIGURE 6: Comparison of the optimized ideality factor values of Shell S36 under different irradiance conditions.

characteristics of solar PV modules. Three different panels are considered in this work to validate the efficiency of the proposed method; they are monocrystalline (Shell SP70), thin film (Shell ST40), and multicrystalline (Shell S36). The simulated parameters are compared with the manufacturer's data [7–9].

4. Results and Discussions

Parameter extraction is performed for the 3 above mentioned solar panels. The optimal values of the parameters for respective panels are found out such that the variation of power with respect to voltage is minimum so that the overall power output is maximum. The convergence curves for shells S36 at standard temperature condition of 1000 W/m^2 irradiation and temperature of 25°C are shown in Figure 3.

Comparison of the optimized series resistance, shunt resistance, and ideality factor values of Shell S36 under different irradiance conditions is shown in Figures 4, 5, and 6, respectively.

The simulated characteristics curves of the SIMULINK model for Shell S36 at different irradiations are shown in Figure 7.

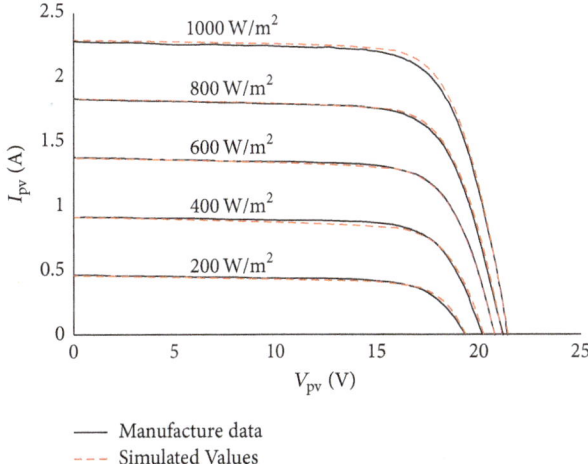

FIGURE 7: I_{pv}-V_{pv} characteristic curves at different irradiation levels for Shell S36.

4.1. Comparison of GA and GSA for Different Irradiations. Optimization results of the algorithms GA and GSA at different irradiation levels for Shell S36, Shell ST40, and Shell SP70 are compared and tabulated in Tables 1, 2, and 3, respectively.

TABLE 1: Comparison of optimization results of GA and GSA at different irradiation levels for Shell S36.

| Parameters | Irradiation in Watt/m^2 | | | | | | | | | |
| | 200 | | 400 | | 600 | | 800 | | 1000 | |
	GA	GSA	GA	GSA	GA	GSA	GA	GSA	GA	GSA
R_s (Ω)	0.825024	0.3069	0.8935	0.7996	0.7429	0.717498	0.656891	0.660802	0.59042	0.59824
R_p (Ω)	200	319.6481	300.098	496.5787	390.029	212.121	200	234.6041	457.478	200
a_1	1.2	1.2	1.2	1.2	1.246	1.2	1.3489	1.2	1.2	1.24
a_2	1	1	1	1	1	1	1	1	1	1
I_{01} (nA)	56.4	56.4	5.23	5.23	10.3	10.2	43.0	4.85	4.73	4.73
I_{02} (nA)	0.147	0.147	0.117	0.117	0.21	0.21	0.0933	0.0933	0.0867	0.0867
I_{pv} (A)	0.46	0.46	0.92	0.92	1.38	1.38	1.84	1.84	2.3	2.3

TABLE 2: Comparison of optimization results of GA and GSA at different irradiation levels for Shell ST40.

| Parameters | Irradiation in Watt/m^2 | | | | | | | | | |
| | 200 | | 400 | | 600 | | 800 | | 1000 | |
	GA	GSA	GA	GSA	GA	GSA	GA	GSA	GA	GSA
R_s (Ω)	0.957967	0.940371	0.995112	0.983382	0.995112	0.995112	0.981427	0.981427	0.991202	0.999
R_p (Ω)	480.9384	200	291.3001	200	295.2102	200	420.3324	200	333.33	200
a_1	1.480938	1.492669	1.2	1.404692	1.2	1.4604	1.2317	1.486804	1.2	1.4956
a_2	1	1	1	1	1	1	1	1	1	1
I_{01} (nA)	52.0	57.2	1.10	22.4	1.05	1.05	1.77	68.6	0.996	72.8
I_{02} (nA)	0.022	0.022	0.0175	0.0175	0.0153	0.0153	0.0139	0.0134	0.0129	0.0129
I_{pv} (A)	0.536	0.536	1.072	1.072	1.608	1.608	2.144	2.144	2.68	2.68

TABLE 3: Comparison of optimization results of GA and GSA at different irradiation levels for Shell SP70.

| Parameters | Irradiation in Watt/m^2 | | | | | | | | | |
| | 200 | | 400 | | 600 | | 800 | | 1000 | |
	GA	GSA	GA	GSA	GA	GSA	GA	GSA	GA	GSA
R_s (Ω)	0.01173	0.066471	0.197458	0.02346	0.250244	0.113392	0.26002	0.183773	0.26002	0.144673
R_p (Ω)	255.132	499.5112	285.435	470.1857	260.016	487.781	200	237.5367	424.2424	385.1417
a_1	1.322	1.2	1.2	1.2	1.2	1.2	1.2	1.2	1.2	1.293255
a_2	1	1	1	1	1	1	1	1	1	1
I_{01} (nA)	11.5	11.5	10.7	10.7	10.2	10.2	9.91	9.91	9.67	9.67
I_{02} (nA)	0.301	0.301	0.24	0.24	0.21	0.21	0.191	0.191	0.177	0.177
I_{pv} (A)	0.94	0.94	1.88	1.88	2.82	2.82	3.76	3.76	4.7	4.7

4.2. Estimation of Absolute Error. To predict the closeness of the results obtained, absolute error is estimated for the proposed and GA method. Absolute error is computed using the following equation:

$$\text{Absolute error} = I_{\text{experimental}} - I_{\text{computed}}. \quad (10)$$

The computed absolute error graph for Shell S36 is presented in Figure 8. To maintain higher level of accuracy and clarity, the results are compared at the same points.

5. Conclusion

In this work, the parameters of the PV module are determined using the proposed GSA. The simulated voltage-current characteristics for three different panels (Shell S36, Shell SP70, and Shell ST40), obtained from the simulation model, are validated with the extracted experimental data. Further, the absolute error curve is plotted for GSA in comparison with GA at different irradiance level (200 to 1000 W/m^2) at 25°C. It is observed that the error obtained in case of GSA is lesser. The extensive simulation results show that the proposed GSA method is superior to the existing GA in terms

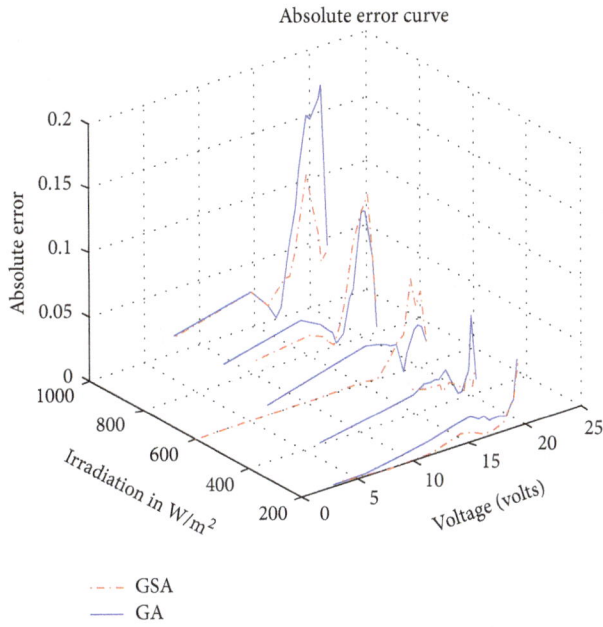

FIGURE 8: Error graph for Shell S36.

of speed of convergence, accuracy, computational efficiency, and consistency of solution.

Competing Interests

The authors declare that there are no competing interests regarding the publication of this paper.

References

[1] D. Verma, S. Nema, A. M. Shandilya, and S. K. Dash, "Maximum power point tracking (MPPT) techniques: recapitulation in solar photovoltaic systems," *Renewable and Sustainable Energy Reviews*, vol. 54, pp. 1018–1034, 2016.

[2] B. Bendib, H. Belmili, and F. Krim, "A survey of the most used MPPT methods: conventional and advanced algorithms applied for photovoltaic systems," *Renewable and Sustainable Energy Reviews*, vol. 45, pp. 637–648, 2015.

[3] R. Sarjila, K. Ravi, J. B. Edward, A. Prasad, and K. S. Kumar, "An AIS based approach for extraction of PV module parameters," in *Information Systems Design and Intelligent Applications*, Advances in Intelligent Systems and Computing, pp. 405–422, Springer, New Delhi, India, 2016.

[4] X. Yuan, Y. Xiang, and Y. He, "Parameter extraction of solar cell models using mutative-scale parallel chaos optimization algorithm," *Solar Energy*, vol. 108, pp. 238–251, 2014.

[5] E. Rashedi, H. Nezamabadi-pour, and S. Saryazdi, "GSA: a gravitational search algorithm," *Information Sciences*, vol. 179, no. 13, pp. 2232–2248, 2009.

[6] R. K. Swain, N. C. Sahu, and P. K. Hota, "Gravitational search algorithm for optimal economic dispatch," *Procedia Technology*, vol. 6, pp. 411–419, 2012.

[7] http://www.proyectodeenergiarenovable.com/descargas/manuales/guia%20para%20el%20bombeo%20de%20agua%20con%20paneles/material%20de%20apoyo/fichas%20tecnicas/siemens/shellsolars36_ukv5.pdf.

[8] http://www.gehrlicher.com/fileadmin/content/pdfs/de/produktarchiv/Shell_ST40.pdf.

[9] http://www.solenerg.com.br/files/SP70.pdf.

13

Short-Term Load Forecasting based on Wavelet Transform and Least Squares Support Vector Machine Optimized by Fruit Fly Optimization Algorithm

Wei Sun and Minquan Ye

Department of Business Administration, North China Electric Power University, Baoding 071000, China

Correspondence should be addressed to Minquan Ye; hdymq2014@163.com

Academic Editor: George S. Tombras

Electric power is a kind of unstorable energy concerning the national welfare and the people's livelihood, the stability of which is attracting more and more attention. Because the short-term power load is always interfered by various external factors with the characteristics like high volatility and instability, a single model is not suitable for short-term load forecasting due to low accuracy. In order to solve this problem, this paper proposes a new model based on wavelet transform and the least squares support vector machine (LSSVM) which is optimized by fruit fly algorithm (FOA) for short-term load forecasting. Wavelet transform is used to remove error points and enhance the stability of the data. Fruit fly algorithm is applied to optimize the parameters of LSSVM, avoiding the randomness and inaccuracy to parameters setting. The result of implementation of short-term load forecasting demonstrates that the hybrid model can be used in the short-term forecasting of the power system.

1. Introduction

Power load forecasting is an important part of management modernization of electric power systems, which has attracted more and more attentions from the academic circle and the practice. Power load forecast with high precision can ease the contradiction between power supply and demand, providing a solid foundation for the stability and reliability of the power grid. It can avoid the waste of resources in the process of grid scheduling and improve the economic benefit. Thus, improving load forecasting methods and the accuracy of prediction constantly is of great important significance to formulate the economic and better power generation plan, reduce spinning reserve capacity, analyse power market demand, and so forth. However, the power load will be influenced by many factors, so features like irregularity and linear independence do exist, which result in the difficulty in making accurate predictions to the power load.

At present, the methods for load forecasting can be divided into two parts: classical mathematical statistical methods and approaches based on artificial intelligence. Most load forecasting theories are based on time series analysis and autoregression models, including vector autoregression model (VAR) and autoregressive moving average model (ARMA) [1–5]. Time series smoothness prediction methods are criticized by researchers for their weakness of nonlinear fitting capability. With the development of the electricity market, the requirement of high accuracy load forecasting is more strict and efficient. So artificial intelligence, which includes neural network [6–8], grey model [9, 10], and support vector machine [11], gains more and more attention from scholars. Due to the strong self-learning, self-adapting ability, and nonlinear change character, artificial neural network is widely used in prediction field. Kandil et al. [12] applied artificial neural network (ANN) for short-term load forecasting without the use of load history and only temperature was used, and it was proved to be effective. After that, BPNN [13] and GRNN [14] are used in improving ANN model for excellent predictive abilities. The shortcomings of neural networks are the requirement of a large amount of data, a long time solving which may cause easily falling into local optimum.

Support vector machine (SVM) is a new statistical learning method [15]. Compared with other machine learning methods, SVM implement the structural risk minimization principle to minimize an upper bound on the generalization error, rather than employing the empirical risk minimization principle to minimize the training error, and it gives SVMs better generative performance. As an extension of SVM, least squares support vector machine (LSSVM) transforms the second optimal inequality constraints problem in original space into equality constraints' linear system in feature space through nonlinear mapping [16], which improved the speed and accuracy of convergence. Different parameters selection in LSSVM (Kernel parameter and penalty factor) would have big influences on the fitting accuracy and generalization ability; inappropriate parameter selection may lead to the limitation of the performance of LSSVM. However, it is possible to employ an optimization algorithm to obtain an appropriate parameter combination. Particle swarm optimization model [17], genetic algorithm model [18], and Firefly Algorithm [19] model are all proposed in parameter optimization for LSSVM; for example, particle swarm optimization finds overall optimal value by following the currently found optimal values. But premature convergence and easily falling into local optimum restrict the application of the particle swarm optimization. Thus, this paper puts forward fruit fly optimization model to optimize the parameters of LSSVM. Fruit fly optimization algorithm was proposed by Pro. Pan [20] in Taiwan in 2012. The FOA has the advantages of being easy to understand due to the shorter program code compared with other optimization algorithms and reaching a better global optimal solution. Li et al. [21] applied the FOA to optimize the parameters of GRNN in order to forecast the annual power load. Zheng et al. [22] proposed a novel fruit fly algorithm for the semiconductor final testing scheduling problem. Pan et al. [23] developed an improved FOA algorithm for continuous function optimization problems.

The wavelet transform (WT) is a recently developed mathematical tool for signal analysis. It has been applied successfully in astronomy, data compression, signal and image processing, earthquake prediction, and so on [12]. The combination of WT and LSSVM is widely used in forecasting fields [24, 25]. This paper proposes WT to reprocess the data and improve its reliability. In order to enhance the accuracy of load forecasting, WT-FOA-LSSVM is put up with, and the examples demonstrate the effectiveness of the model.

The rest of the paper is organized as follows: Section 2 provides some basic theoretical aspects of WT, LSSVM, and FOA and gives a brief description about WT-FOA-LSSVM model; in Section 3, an experiment study is put forward to prove the efficiency of the proposed model; Section 4 is the conclusion of this paper.

2. WT-FOA-LSSVM Model

2.1. Wavelet Transform. Wavelet transform is a mathematical tool for signal analysis which developed in recent decades. It can capture the frequency and location information of the input signal. The basic concept in wavelet transform is to decompose a signal into an approximation component and detail components, in which approximation component is the low-frequency information and contains the important information of the signal. The details are the high-frequency

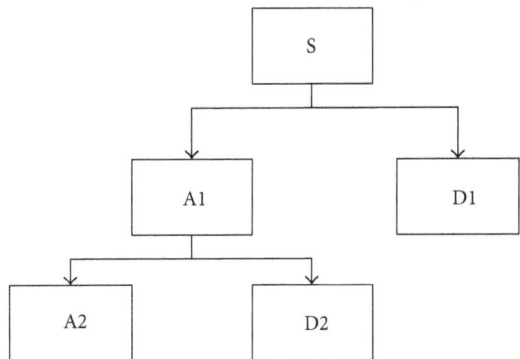

FIGURE 1: Wavelet decomposition.

components which are related to the small-scale space in the signal. Figure 1 is a wavelet decomposition tree showing the decomposition process.

In order to overcome the high redundancy situation which exists in continuous wavelet transform, it captures both frequency and location information in temporal resolution. We propose the discrete wavelet transform (DWT) in this paper, and it is defined as follows:

$$\psi_{a,b}(t) = \psi_{1/2^j, k/2^j}(t) = 2^{j/2} \psi\left(2^j t - k\right), \tag{1}$$

where the scale factor $a = 1/2^j$, $j \in Z$, and the displacement factor $b = k/2^j$, $j \in Z$. In general, $\psi_{j,k}(t)$ represents the discrete transform. Using DWT in the signal $f(x)$, the formula can be shown as

$$\left(W_\psi f\right)(j,k) = \left\langle f, \psi_{j,k}(t)\right\rangle = \int_{-\infty}^{+\infty} f(t) \overline{\psi_{j,k}(t)} \, dt. \tag{2}$$

In this paper, the original load signal is proposed to be decomposed into an approximation component and some detail components. The approximation presents the main fluctuation of the load and the details to contain the spikes and stochastic volatilities. A suitable number of levels can be decided by comparing the similarity between the approximation and the original signal.

2.2. Least Squares Support Vector Machine. LSSVM is an extension of the standard support vector machine (SVM), proposed by Suykens and Vandewalle [26]. It transforms the inequality constraints of traditional SVM into equality constraints and considers sum squares error loss function as the loss experience of the training set, which transforms solving quadratic programming problems into solving linear equations problems [27]. The training set is set as $\{(x_k, y_k) \mid k = 1, 2, \ldots, n\}$, in which $x_k \in R^n$ is the input data and $y_k \in R^n$ is the output data. $\varphi(\cdot)$ is the nonlinear mapping function which transfers the samples into a much higher dimensional feature space $\phi(x_k)$. Establish the optimal decision function in the high-dimensional feature space:

$$y(x) = \omega^T \cdot \varphi(x) + b, \tag{3}$$

where $\varphi(x)$ is mapping function; ω is weight vector; b is constant.

Using the principle of structural risk minimization, the objective optimization function is as follows:

$$\min_{\omega,b,e} (\omega, e) = \frac{1}{2}\omega^T\omega + \frac{1}{2}\gamma \sum_{k=1}^{n} e_k^2. \tag{4}$$

Its constraint condition is

$$y_k = \omega^T\varphi(x_k) + b + e_k, \quad k = 1, 2, \ldots, n, \tag{5}$$

in which γ is the penalty coefficient and e_k is slack variable.

Define the Lagrange function to solve the problem:

$$L(\omega, b, e, \alpha) = \phi(\omega, e)$$
$$- \sum_{k=1}^{n} \left\{ \alpha_k \left[\omega^T\varphi(x_k) + b + e_k - y_k \right] \right\}, \tag{6}$$

where Lagrange multiplier $\alpha_k \in$ R. According to the Karush-Kuhn-Tucker (KKT) conditions, ω, b, e_k, α_k are taken as partial derivatives and required as zero. Consider

$$\omega = \sum_{k=1}^{n} \alpha_k \varphi(x_k),$$

$$\sum_{k=1}^{n} \alpha_k = 0, \tag{7}$$

$$\alpha_k = e_k\gamma,$$

$$\omega^T\varphi(x_k) + b + e_k - y_k = 0.$$

According to (7), the optimization problem can be transformed into solving linear problem, which is shown as follows:

$$\begin{bmatrix} 0 & 1 & \cdots & 1 \\ 1 & K(x_1, x_1) + \dfrac{1}{\gamma} & \cdots & K(x_1, x_l) \\ \vdots & \vdots & \vdots & \vdots \\ 1 & K(x_l, x_1) & \cdots & K(x_l, x_l) + \dfrac{1}{\gamma} \end{bmatrix} \begin{bmatrix} b \\ \alpha_1 \\ \vdots \\ \alpha_l \end{bmatrix}$$
$$= \begin{bmatrix} 0 \\ y_1 \\ \vdots \\ y_l \end{bmatrix}. \tag{8}$$

Solve formula (8) to get α and b; then the LSSVM optimal linear regression function is

$$f(x) = \sum_{k=1}^{l} \alpha_k K(x, x_k) + b. \tag{9}$$

According to Mercer condition, $K(x, x_i) = \varphi(x)^T \cdot \varphi(x_l)$ is kernel function. In this paper, set radial basis function (RBF) as kernel function which is shown in the following equation:

$$K(x, x_k) = \exp\left(-\frac{|x - x_k|^2}{2\sigma^2}\right), \tag{10}$$

where σ^2 is the width of kernel function.

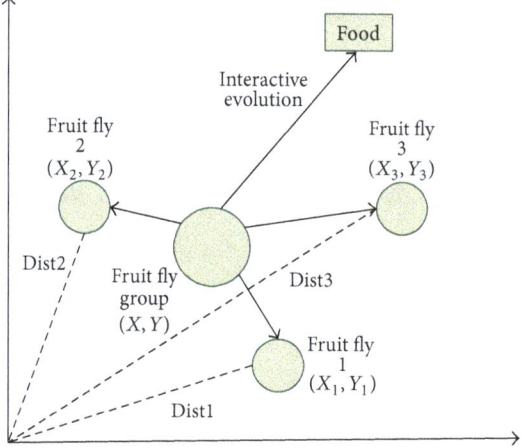

FIGURE 2: Food finding iterative process of a fruit fly swarm.

From the problems of training LSSVM, kernel parameter σ^2 and penalty parameter γ are generally set based on experience, which leads to the existence of randomness and inaccuracy in the application of the LSSVM algorithm. To solve the problem, the paper uses fruit fly optimization algorithm to optimize these two parameters to improve the prediction accuracy of LSSVM.

2.3. Fruit Fly Optimization Algorithm.

Fruit fly optimization algorithm is a kind of intelligent optimization algorithms based on fruit fly foraging behaviours proposed by Pan [20] in 2012. The basic concept of FOA is that fruit fly perceives food concentration according to its position, and then it will move to the site of maximum or minimum concentration by comparing flavor concentration; finally the objective function extreme value can be obtained through repeated iterations of food concentration. Food finding iterative process of fruit fly swarm is shown in Figure 2.

According to the food finding characteristics of fruit fly swarm, the fruit fly optimization algorithm can be divided into following steps:

(1) Randomly initialize the fruit fly swarm location (x_axis, y_axis).

(2) Give the random flight direction and the distance for food finding of an individual fruit fly by using olfactory:

$$X_i = x\text{\_axis} + \text{Random Value}$$
$$Y_i = y\text{\_axis} + \text{Random Value}. \tag{11}$$

(3) Calculate the distance between the origin and each individual fruit fly position (Dist), and then calculate the value of flavor concentration (S); it is the reciprocal of distance:

$$\text{Dist} = \sqrt{X_i^2 + Y_i^2} \tag{12}$$

$$S = \frac{1}{\text{Dist}}. \tag{13}$$

(4) Put the value of flavor concentration S into its fitness function, and then get the flavor concentration of the individual fruit fly location (Smell).

(5) Find out the individual fruit fly with minimal smell concentration among the fruit fly swarm:

$$[\text{best Smell}, \text{best index}] = \max{(\text{Smell})}. \qquad (14)$$

(6) Retain the best flavor concentration and its X, Y coordinates, and then the fruit flies fly to the position by using vision. Enter iterative optimization to repeat steps (2)–(5). When the fitness value reaches target set, or the iterative number reaches the maximal iterative number, the circulation stops. Update the information as follows:

$$\text{Smell best} = \text{best Smell}$$

$$x\_\text{axis} = X \, (\text{best index}) \qquad (15)$$

$$y\_\text{axis} = Y \, (\text{best index}).$$

2.4. The Introduction of WT-FOA-LSSVM. Flowchart of the WT-FOA-LSSVM model is shown in Figure 3, and the detailed processes are as follows.

(1) Data Preprocessing Using DWT. Decompose the load signal into the approximation A1 and the details D1, and select A1 as the training data and testing data.

(2) Initialization of the Fruit Fly Optimization Algorithm Parameters. Fruit fly optimization algorithm parameters contain the initial fruit fly swarm location ($x\_$axis, $y\_$axis), the random flight distant range FR, the population size sizepop, and the maximum iterative number $N_{\max}$. In this paper, we set $x\_$axis = rands(1, 2), $y\_$axis = rands(1, 2), where rands() represents the random number generation function in MATLAB program. In addition, we suppose FR $\subset$ [−10, 10], sizepop = 50, and $N_{\max}$ = 100.

(3) LSSVM Optimized by Fruit Fly Optimization Algorithm. Calculate the distance D_i between the origin and each individual fruit fly i according to formulas (11) and (12) and the smell concentration judgment value S_i according to formula (13). In the FOA-LSSVM program, the parameters $[C, \sigma^2]$ of LSSVM model are represented by $[S(i, 1), S(i, 2)]$, and we set $C = 20 * S(i, 1)$ and $\sigma^2 = S(i, 2)$ for LSSVM model training. According to the electric load forecasting result, the value of fitness function can be calculated. In this paper, we employ the mean absolute percentage error (MAPE) as the fitness function, and the formula is as follows:

$$\text{MAPE} = \frac{1}{n} \sum_{i=1}^{n} \left| \frac{\widehat{y}_i - y_i}{y_i} \right|, \qquad (16)$$

where y_i represents the actual value at period i; $\widehat{y}_i$ is the forecasting value at period i and n is the number of forecasting periods.

If the maximum iterative number $N_{\max}$ = 100 or MAPE < 0.01%, stop the iterative process and output the best values of C and σ^2.

(4) Forecast Using Least Squares Support Vector Machine. Put the optimal parameter value obtained from step (3) in the least squares support vector machine and do the forecast. Finally, get the forecasting load value.

3. Case Studies

3.1. Data Preprocessing. This paper chooses the 24-hour power load data from May 1, 2013, to July 23, 2013, in Shanxi province for model checking. In this paper, we select 1986 pieces of load data from May 1 to July 22 as training set and 24 pieces of load data of July 23 as testing set. In order to eliminate the effects of random fluctuations of load data, we decompose the original load data S into the approximation component A1 and detail component D1 through one-level DWT, as shown in Figure 4.

From Figure 4, it is clear that the major fluctuation of A1 shows high similarity to the original load data S. The detail component D1 is excluded from the original data to ensure the stability of the input data. So, A1 is selected as the input in proposed model.

3.2. Selection of Input. Human activities are always disturbed by many external factors and then the power load is affected. So, some effective features are considered as input features. In this paper, the input features are discussed as follows. The first feature is the highest temperature and the lowest temperature. Temperature is one of these effective features. In [27–29], temperature was considered as an essential input feature and the forecasting results were accurate enough. So, the highest and lowest temperatures are taken into consideration. The second feature is weather conditions. The weather conditions are divided into four types: sunny, cloudy, overcast, and rainy. For different weather conditions, we set different weights: {sunny, cloudy, overcast, and rainy} = {0.9, 0.7, 0.5, 0.2}. The third feature is days type. For different days type, the electric power consumption is different. Figure 5 shows the load data from June 10, 2013, to June 16, 2013, among which Wednesday is dragon festival in China. From Figure 5, we can see that the mean power load of Wednesday is higher than other days and different days type has different curve features. So, we assign values to days type in Table 1.

3.3. Parameters Setting of Comparison Models. In this paper, we introduce five other models, WT-LSSVM, least squares support vector machine optimized by fruit fly optimization algorithm (FOA-LSSVM), least squares support vector machine optimized by particle swarm optimization algorithm (PSO-LSSVM), least squares support vector machine, and the BP neural network, to make a comparison with the proposed model. Referring to some of the relative literature [21, 27], the parameters of the comparison models are set as shown in Table 2.

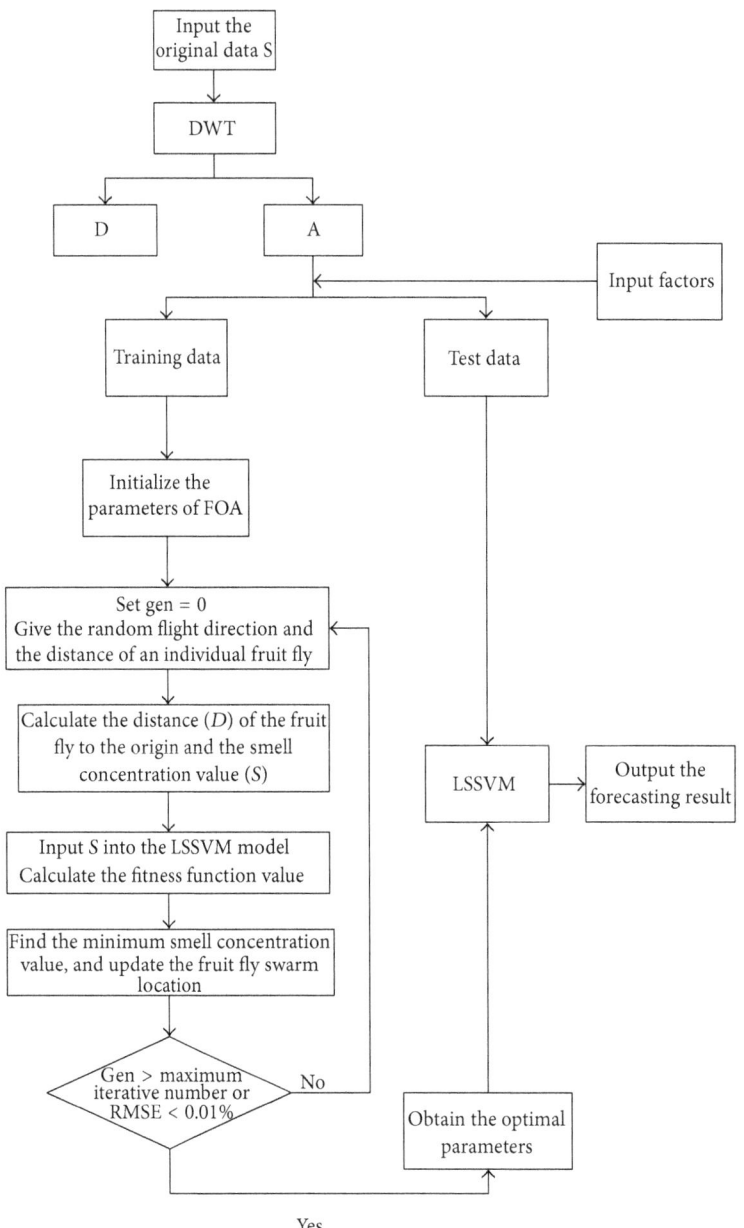

FIGURE 3: Flowchart of the WT-FOA-LSSVM modeling.

TABLE 1: The weights of days type.

	Monday	Tuesday	Wednesday	Thursday	Friday	Saturday	Sunday	Holiday
Weights	1	2	3	4	5	6	7	8

TABLE 2: Parameters of models mentioned in the paper.

Model	Parameters
WT-LSSVM	$\gamma = 40; \sigma^2 = 5$
FOA-LSSVM	Initial-location[rands$(1,2)$, rands$(1,2)$]; FR $= [-10, 10]$; sizepop $= 50$; $N_{max} = 100$
PSO-LSSVM	$C_1 = C_2 = 2$; $N_{max} = 100$; sizepop $= 20$; $w_0 = 0.8$; $w_n = 0.3$
LSSVM	$\gamma = 40; \sigma^2 = 5$
BPNN	$N_{max} = 100$; hidden-layer-node $= [5, 5]$; learning-rate $= 0.1$; goal $= 0.00004$

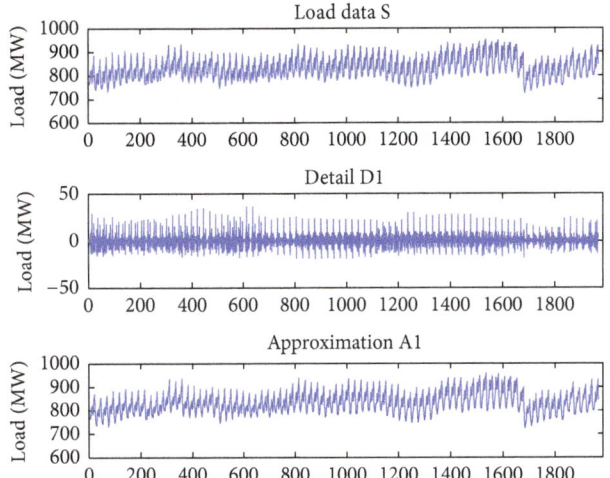

FIGURE 4: Original load signal and its approximation component and detail component decomposed by DWT.

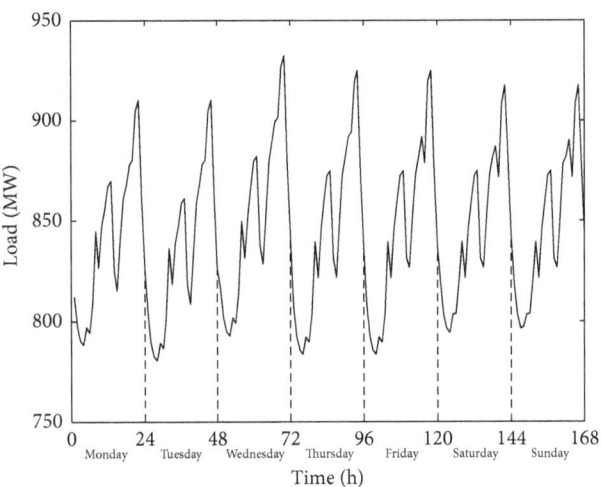

FIGURE 5: Weekly load curve.

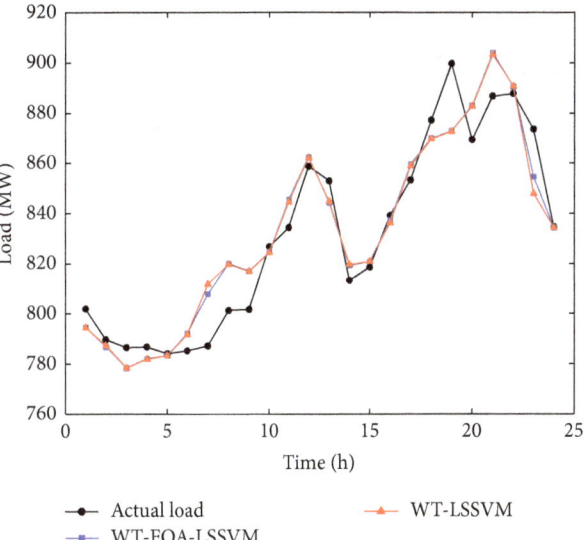

FIGURE 6: Actual load and forecasting results of WT-FOA-LSSVM and WT-LSSVM.

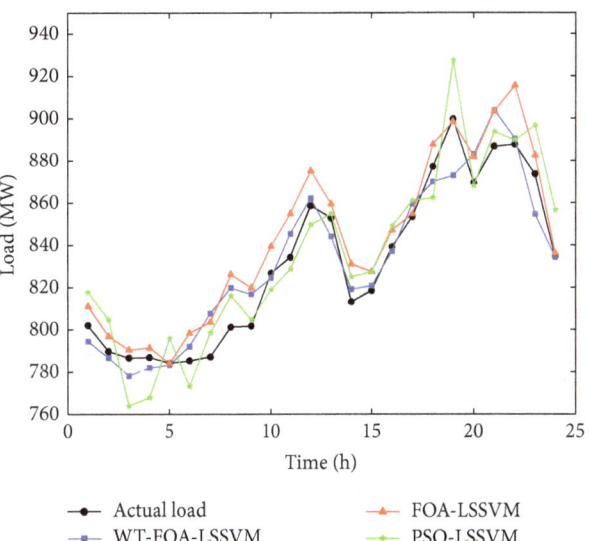

FIGURE 7: Actual load and forecasting results of WT-FOA-LSSVM, FOA-LSSVM, and PSO-LSSVM.

3.4. Model Performance Evaluation.

To examine the performance of model, the relative error (RE), the mean absolute percentage error (MAPE), the mean square error (MSE), and the mean absolute error (MAE) are proposed to measure the forecast accuracy. The formulas are as follows:

$$\text{RE}(i) = \frac{\widehat{y}_i - y_i}{y_i} \times 100\%,$$

$$\text{MAPE} = \frac{1}{n} \sum_{i=1}^{n} \left| \frac{\widehat{y}_i - y_i}{y_i} \right|,$$

$$\text{MSE} = \frac{1}{n} \sum_{i=1}^{n} \left(\widehat{y}_i - y_i \right)^2, \tag{17}$$

$$\text{MAE} = \frac{\sum_{i=1}^{n} \left| \widehat{y}_i - y_i \right|}{n},$$

where y_i represents the actual value at period i; $\widehat{y}_i$ is the forecasting value at period i and n is the number of forecasting periods.

3.5. Analysis of Forecasting Results.

The program in this paper is run in MATLAB R2011b under the XP system. Table 3 shows the short-term electric load forecasting results of the WT-FOA-LSSVM, WT-LSSVM, FOA-LSSVM, PSO-LSSVM, LSSVM, and BPNN models. Figures 6, 7, and 8 present the comparisons of the forecasting results between the proposed model and the others. Figures 9, 10, and 11 show the comparisons of relative errors between the proposed model and the others. The relative error ranges [−3%, 3%] and [−1%, 1%] are always considered as a standard to assess

TABLE 3: Actual load and forecasting results.

Time/h	Actual value/MW	WT-FOA-LSSVM	WT-LSSVM	FOA-LSSVM	PSO-LSSVM	LSSVM	BPNN
1:00	802	794.57	794.48	811.09	817.87	813.47	811.50
2:00	789.71	786.76	787.26	796.86	804.78	798.72	800.40
3:00	786.59	778.32	778.41	790.53	764.12	792.73	785.70
4:00	786.82	782.07	781.90	791.48	768.02	792.93	782.74
5:00	784.16	783.38	783.36	784.04	795.92	796.98	801.99
6:00	785.28	792.11	791.62	798.36	773.31	799.73	790.22
7:00	787.18	807.72	811.68	803.65	798.63	810.33	811.39
8:00	801.29	819.94	819.71	826.31	816.22	830.52	817.20
9:00	801.71	816.90	816.87	819.94	804.59	822.10	812.74
10:00	826.78	824.57	824.37	839.52	819.02	842.58	838.74
11:00	834.31	845.44	844.48	854.89	828.72	857.19	850.32
12:00	858.62	862.27	862.07	875.15	849.75	874.93	873.86
13:00	852.87	844.17	844.70	859.51	854.92	871.70	880.61
14:00	813.21	819.19	819.59	831.13	825.04	829.22	830.73
15:00	818.59	820.88	820.84	827.56	827.45	828.01	825.97
16:00	839.14	837.15	836.17	847.18	848.92	849.39	853.78
17:00	853.36	859.53	858.90	854.69	861.14	868.99	878.91
18:00	877.21	869.92	869.73	887.87	862.37	889.47	899.97
19:00	899.76	872.85	872.72	898.40	927.42	900.42	902.82
20:00	869.36	883.01	882.77	881.83	868.06	888.95	877.78
21:00	886.74	903.93	903.35	903.41	893.90	917.73	897.03
22:00	887.77	890.44	890.68	915.67	889.76	919.97	925.14
23:00	873.6	854.58	847.88	882.64	896.88	884.27	866.82
24:00	834.67	834.29	834.34	836.32	856.53	839.34	846.50

TABLE 4: Models performance evaluations.

Index	Model					
	WT-FOA-LSSVM	WT-LSSVM	FOA-LSSVM	PSO-LSSVM	LSSVM	BPNN
MAPE (%)	1.068	1.111	1.3534	1.414	1.8457	1.674
MSE (MW^2)	130.4913	147.9492	181.4398	191.3687	300.9418	268.3937
MAE (MW)	8.9425	9.297083	11.26083	11.90875	15.3725	13.98458

the performance of a forecasting model [30]. First, based on Table 2 and Figures 9, 10, and 11, the relative errors of short-term load forecasting of WT-FOA-LSSVM are all in the range of [−3%, 3%], and the maximum relative error is 2.6096% at 7:00 and the minimum relative error is −2.991% at 19:00. There exist fourteen points that are in the scope of [−1%, 1%]. Second, the WT-LSSVM has two forecasting points that exceed the relative error range [−3%, 3%], which are 3.1121% at 7:00 and 3.006% at 19:00, and there are fifteen forecasting points in the range of [−1%, 1%]. Third, the FOA-LSSVM has two forecasting points that exceed the relative error range [−3%, 3%], which are 3.1225% at 8:00 and 3.1429% at 22:00, and there are eight forecasting points in the range of [−1%, 1%]. Fourth, the PSO-LSSVM has one forecasting point that exceeds the relative error range [−3%, 3%], which is 3.0743% at 9:00, and there are eight forecasting points in the range of [−1%, 1%]. Fifth, the single LSSVM has three forecasting points that exceed the relative error range [−3%, 3%], which are 3.6484% at 8:00, 3.4946% at 21:00, and 3.627% at 22:00, and there are three forecasting

points in the range of [−1%, 1%]. Sixth, the BPNN has three forecasting points that exceed the relative error range [−3%, 3%], which are 3.0755% at 7:00, 3.2531% at 13:00, and 4.209% at 22:00, and there are seven forecasting points in the range of [−1%, 1%]. However, the comparison models also predict more accurately than the proposed model at some points, such as 7:00 and 19:00.

The mean absolute percentage errors, mean square errors, and mean absolute errors of WT-FOA-LSSVM, WT-LSSVM, FOA-LSSVM, PSO-LSSVM, LSSVM, and BPNN are listed in Table 4. From Table 4, we can conclude that the MAPE of the proposed model is 1.068%, which is smaller than the MAPE of WT-LSSVM, FOA-LSSVM, PSO-LSSVM, LSSVM, and BPNN (which are 1.111%, 1.3534%, 1.414%, 1.8457%, and 1.674%). Additionally, the MSE of the proposed model is 130.4913, which is smaller than the MSE of the comparison models (which are 147.9492, 181.4398, 191.3687, 300.9418, and 268.3937). The MAE of the proposed model is 8.9425, which is smaller than the MAE of the comparison models (which are 9.297083, 11.26083, 11.90875, 15.3725, and 13.98458). As

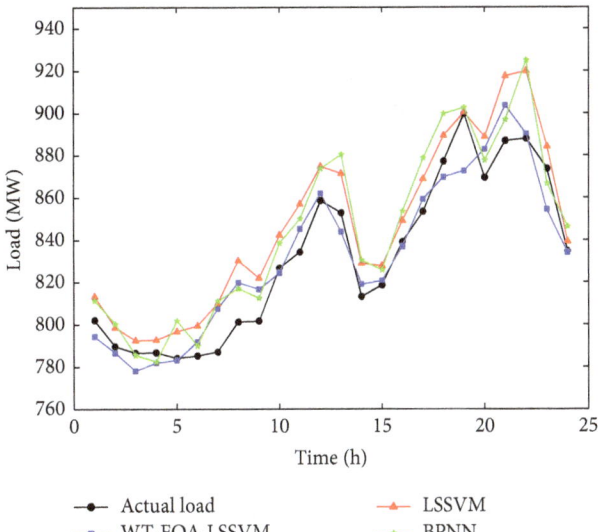

FIGURE 8: Actual load and forecasting results of WT-FOA-LSSVM, LSSVM, and BPNN.

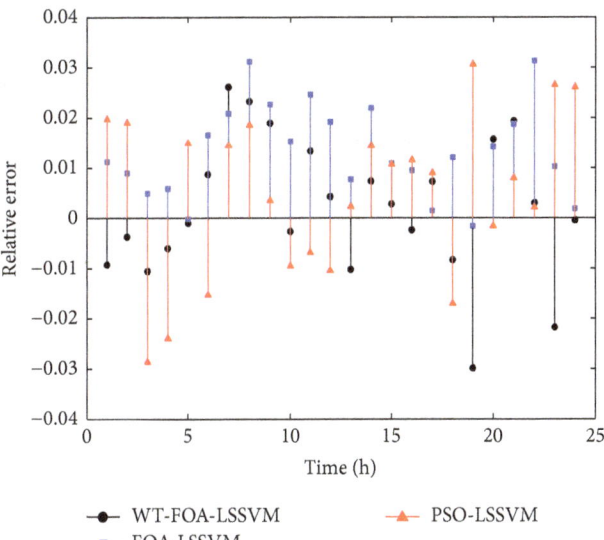

FIGURE 10: Relative errors of WT-FOA-LSSVM, FOA-LSSVM, and PSO-LSSVM.

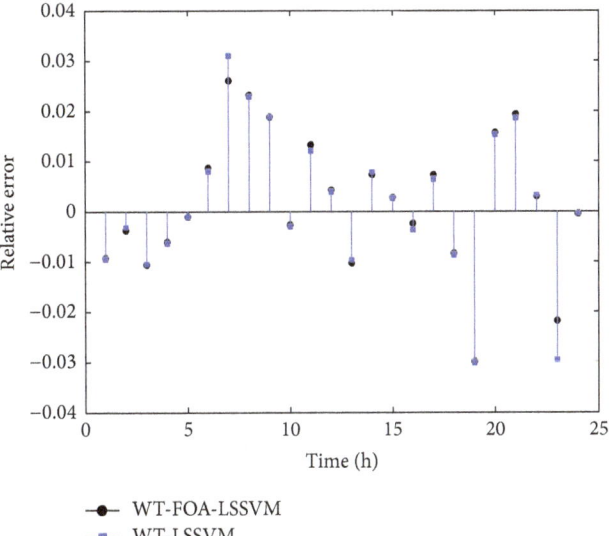

FIGURE 9: Relative errors of WT-FOA-LSSVM and WT-LSSVM.

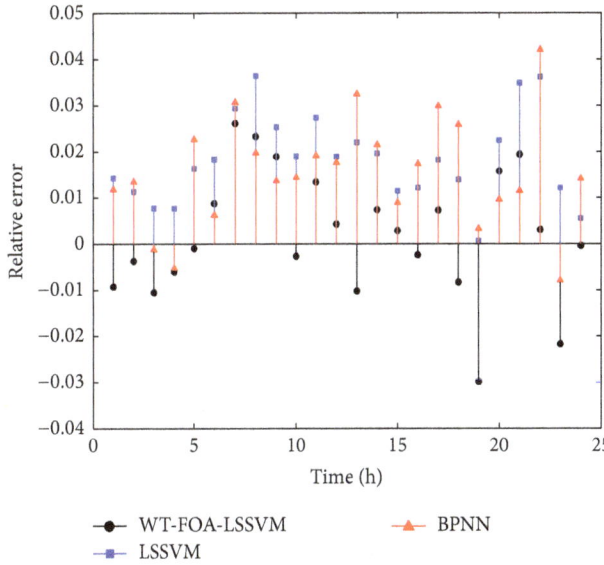

FIGURE 11: Relative errors of WT-FOA-LSSVM, LSSVM, and BPNN.

a result, the MAPE, MSE, and MAE of the WT-FOA-LSSVM are all smaller than those of the WT-LSSVM, so we can conclude that the parameter optimization to LSSVM is essential in the forecasting model. Besides, the MAPE, MSE, and MAE of the WT-LSSVM are all smaller than those of FOA-LSSVM, PSO-LSSVM, LSSVM, and BPNN, indicating the preprocessing of load data is useful for a better performance and higher forecasting accuracy. At the same time, the MAPE, MSE, and MAE of the FOA-LSSVM are all smaller than those of PSO-LSSVM, LSSVM, and BPNN, and it is presented that the optimization result of the fruit fly optimization algorithm is efficient. So, we can conclude that the stability and forecasting accuracy of the proposed model is better than the comparison models, and it is worth of being widely used in the short-term load forecasting.

4. Conclusion

To strengthen the stability and economy of the grid and avoid the waste in grid scheduling, it is essential to improve the forecasting accuracy. Because the short-term power load is always interfered by various external factors with the characteristics like high volatility and instability, the high accuracy of load forecasting should be taken into consideration. Based on the features of load data and the randomness of the LSSVM parameters setting, we propose the model based on wavelet transform and least squares support vector machine optimized by fruit fly optimization algorithm. To validate the proposed model, four other comparison models (FOA-LSSVM, PSO-LSSVM, LSSVM, and BPNN) are employed to compare the forecasting results. Example computation results show that the relative errors of WT-FOA-LSSVM model are all in the range [−3%, 3%], and the MAPE, MSE, and MAE are all smaller than the others. In addition, the fruit fly optimization algorithm is easy to understand and operate, so it is applied widely in parameters optimization. The hybrid model can be effectively used in the short-term load forecasting on power system.

Conflict of Interests

The authors declare that there is no conflict of interests regarding the publication of this paper.

References

[1] J.-F. Chen, W.-M. Wang, and C.-M. Huang, "Analysis of an adaptive time-series autoregressive moving-average (ARMA) model for short-term load forecasting," *Electric Power Systems Research*, vol. 34, no. 3, pp. 187–196, 1995.

[2] H. M. Al-Hamadi and S. A. Soliman, "Short-term electric load forecasting based on Kalman filtering algorithm with moving window weather and load model," *Electric Power Systems Research*, vol. 68, no. 1, pp. 47–59, 2004.

[3] T. Zheng, A. A. Girgis, and E. B. Makram, "A hybrid wavelet-Kalman filter method for load forecasting," *Electric Power Systems Research*, vol. 54, no. 1, pp. 11–17, 2000.

[4] S. S. Pappas, L. Ekonomou, P. Karampelas et al., "Electricity demand load forecasting of the Hellenic power system using an ARMA model," *Electric Power Systems Research*, vol. 80, no. 3, pp. 256–264, 2010.

[5] C.-M. Lee and C.-N. Ko, "Short-term load forecasting using lifting scheme and ARIMA models," *Expert Systems with Applications*, vol. 38, no. 5, pp. 5902–5911, 2011.

[6] B.-L. Zhang and Z.-Y. Dong, "An adaptive neural-wavelet model for short term load forecasting," *Electric Power Systems Research*, vol. 59, no. 2, pp. 121–129, 2001.

[7] S. J. Yao, Y. H. Song, L. Z. Zhang, and X. Y. Cheng, "Wavelet transform and neural networks for short-term electrical load forecasting," *Energy Conversion and Management*, vol. 41, no. 18, pp. 1975–1988, 2000.

[8] H. R. Maier and G. C. Dandy, "Neural networks for the prediction and forecasting of water resources variables: a review of modelling issues and applications," *Environmental Modelling and Software*, vol. 15, no. 1, pp. 101–124, 2000.

[9] J. Kang and H. Zhao, "Application of improved grey model in long-term load forecasting of power engineering," *Systems Engineering Procedia*, vol. 3, pp. 85–91, 2012.

[10] M. Lei and Z. Feng, "A proposed grey model for short-term electricity price forecasting in competitive power markets," *International Journal of Electrical Power & Energy Systems*, vol. 43, no. 1, pp. 531–538, 2012.

[11] H. Nie, G. Liu, X. Liu, and Y. Wang, "Hybrid of ARIMA and SVMs for short-term load forecasting," *Energy Procedia*, vol. 16, part C, pp. 1455–1460, 2012.

[12] N. Kandil, R. Wamkeue, M. Saad, and S. Georges, "An efficient approach for short term load forecasting using artificial neural networks," *International Journal of Electrical Power and Energy Systems*, vol. 28, no. 8, pp. 525–530, 2006.

[13] Z. Xiao, S.-J. Ye, B. Zhong, and C.-X. Sun, "BP neural network with rough set for short term load forecasting," *Expert Systems with Applications*, vol. 36, no. 1, pp. 273–279, 2009.

[14] M. T. Leung, A.-S. Chen, and H. Daouk, "Forecasting exchange rates using general regression neural networks," *Computers & Operations Research*, vol. 27, no. 11-12, pp. 1093–1110, 2000.

[15] V. Vapnik, *The Nature of Statistical Learning Theory*, Springer Science & Business Media, 2000.

[16] M. Mesbah, E. Soroush, V. Azari, M. Lee, A. Bahadori, and S. Habibnia, "Vapor liquid equilibrium prediction of carbon dioxide and hydrocarbon systems using LSSVM algorithm," *The Journal of Supercritical Fluids*, vol. 97, pp. 256–267, 2015.

[17] R. G. Gorjaei, R. Songolzadeh, M. Torkaman, M. Safari, and G. Zargar, "A novel PSO-LSSVM model for predicting liquid rate of two phase flow through wellhead chokes," *Journal of Natural Gas Science and Engineering*, vol. 24, pp. 228–237, 2015.

[18] D. Liu, D. Niu, H. Wang, and L. Fan, "Short-term wind speed forecasting using wavelet transform and support vector machines optimized by genetic algorithm," *Renewable Energy*, vol. 62, pp. 592–597, 2014.

[19] A. Kavousi-Fard, H. Samet, and F. Marzbani, "A new hybrid modified firefly algorithm and support vector regression model for accurate short term load forecasting," *Expert Systems with Applications*, vol. 41, no. 13, pp. 6047–6056, 2014.

[20] W.-T. Pan, "A new fruit fly optimization algorithm: taking the financial distress model as an example," *Knowledge-Based Systems*, vol. 26, pp. 69–74, 2012.

[21] H.-Z. Li, S. Guo, C.-J. Li, and J.-Q. Sun, "A hybrid annual power load forecasting model based on generalized regression neural network with fruit fly optimization algorithm," *Knowledge-Based Systems*, vol. 37, pp. 378–387, 2013.

[22] X.-L. Zheng, L. Wang, and S.-Y. Wang, "A novel fruit fly optimization algorithm for the semiconductor final testing scheduling problem," *Knowledge-Based Systems*, vol. 57, pp. 95–103, 2014.

[23] Q.-K. Pan, H.-Y. Sang, J.-H. Duan, and L. Gao, "An improved fruit fly optimization algorithm for continuous function optimization problems," *Knowledge-Based Systems*, vol. 62, pp. 69–83, 2014.

[24] Y. Zhang, H. Li, Z. Wang, W. Zhang, and J. Li, "A preliminary study on time series forecast of fair-weather atmospheric electric field with WT-LSSVM method," *Journal of Electrostatics*, vol. 75, pp. 85–89, 2015.

[25] H. Shayeghi and A. Ghasemi, "Day-ahead electricity prices forecasting by a modified CGSA technique and hybrid WT in LSSVM based scheme," *Energy Conversion and Management*, vol. 74, pp. 482–491, 2013.

[26] J. A. K. Suykens and J. Vandewalle, "Least squares support vector machine classifiers," *Neural Processing Letters*, vol. 9, no. 3, pp. 293–300, 1999.

[27] W. Sun and Y. Liang, "Least-squares support vector machine based on improved imperialist competitive algorithm in a short-term load forecasting model," *Journal of Energy Engineering*, vol. 141, no. 4, Article ID 04014037, 2015.

[28] R.-A. Hooshmand, H. Amooshahi, and M. Parastegari, "A hybrid intelligent algorithm based short-term load forecasting approach," *International Journal of Electrical Power & Energy Systems*, vol. 45, no. 1, pp. 313–324, 2013.

[29] S. Bahrami, R.-A. Hooshmand, and M. Parastegari, "Short term electric load forecasting by wavelet transform and grey model improved by PSO (particle swarm optimization) algorithm," *Energy*, vol. 72, pp. 434–442, 2014.

[30] D. Niu, Y. Wang, and D. D. Wu, "Power load forecasting using support vector machine and ant colony optimization," *Expert Systems with Applications*, vol. 37, no. 3, pp. 2531–2539, 2010.

14

Power Extraction Control of Variable Speed Wind Turbine Systems based on Direct Drive Synchronous Generator in All Operating Regimes

Youssef Errami [ID],[1] Abdellatif Obbadi,[1] Smail Sahnoun,[1] Mohammed Ouassaid,[2] and Mohamed Maaroufi[2]

[1]Laboratory of Electronics, Instrumentation and Energy, Team of Exploitation and Processing of Renewable Energy, Department of Physics, Faculty of Science, Chouaib Doukkali University, El Jadida, Morocco
[2]Department of Electrical Engineering, Mohammadia School of Engineers, University Mohammed V, Rabat, Morocco

Correspondence should be addressed to Youssef Errami; errami.emi@gmail.com

Academic Editor: Andrea Bonfiglio

Due to the increased penetration of wind energy into the electrical power systems in recent years, the turbine controls are actively occupied in the research. This paper presents a nonlinear backstepping strategy to control the generators and the grid sides of a Wind Farm System (WFS) based Direct Drive Synchronous Generator (DDSG). The control objectives such as Tracking the Maximum Power (TMP) from the WFS, pitch control, regulation of dc-link voltage, and reactive and active power generation at varying wind velocity are included. To validate the proposed control strategy, simulation results for 6-MW-DDSG based Wind Farm System are carried out by MATLAB-Simulink. Performance comparison and evaluation with Vector Oriented Control (VOC) are provided under a wide range of functioning conditions, three-phase voltage dips, and the probable occurrence of uncertainties. The proposed control strategy offers remarkable characteristics such as excellent dynamic and steady state performance under varying wind speed and robustness to parametric variations in the WFS and under severe faults of grid voltage.

1. Introduction

In the past decades, various renewable energy sources have received increasing attention as alternatives of fossil fuels [1, 2]. Among various modern renewable resources, wind power is considered the backbone of renewable power generation. It is regarded as a clean energy resource and it is easily accessible and cost effective. So, wind power penetration greatly increases in the electric power systems and it is anticipated to keep steady growth in the upcoming years [3, 4]. On the other hand, in modern wind power systems, three of the most promising types of wind turbine generators are the Doubly Fed Induction Generator (DFIG), the Direct Drive Synchronous Generator (DDSG) with permanent magnet, and the Squirrel Cage Induction Generator (SCIG) [5–12]. But, the main advantage of DDSG is the absence of gearbox coupling the wind turbine to the DDSG and the dc external excitation current. So, during the past few years direct driven

DDSGs become very attractive in wind energy application because of their high power density, flexible magnet topologies, reduced maintenance, high effectiveness, high operating efficiency, excellent operation performance, and increased reliability [13–19]. In the area of wind power generation technology, Variable Speed Wind Power Generation System (VS-WPGS) has many advantages over fixed speed ones such as better power quality, higher overall efficiency, lower mechanical stress, and increased energy capture. So, it can be controlled to enable the turbine to operate at its maximum coefficient of power and to ensure the Maximum Power Tracking (MPT) ability [20–22]. To control the VS-WPGS based DDSG, full scale power electronic converter systems are generally used as the interface between the VS-WPGS and the electrical network for satisfying the new standards and grid connection requirements [23, 24]. Thus, they allow controlling the wind turbine system, decoupling the DDSG from the power grid, and the VS-WPGS does not need to

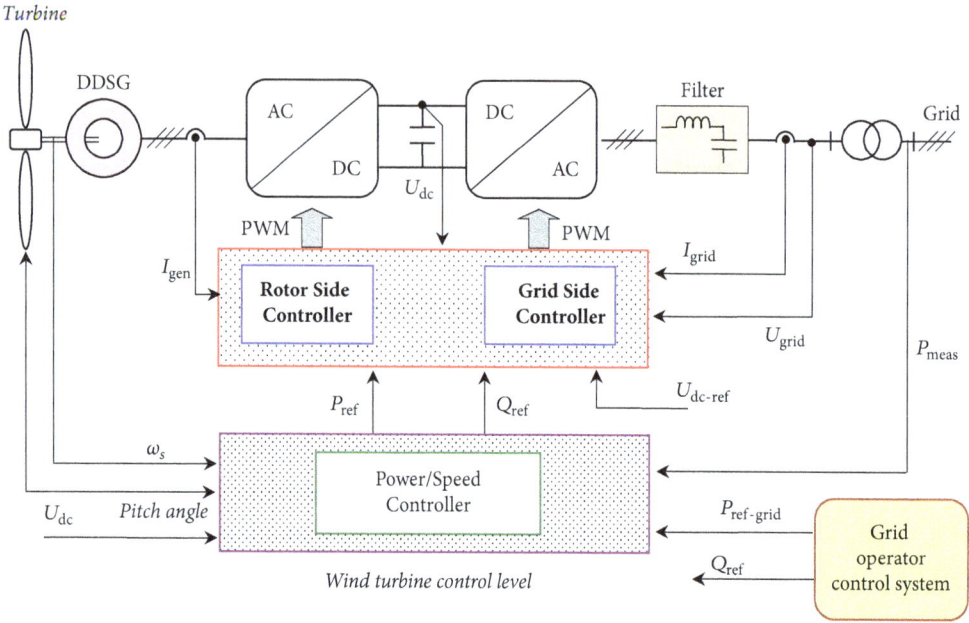

FIGURE 1: Configuration of a VS-WPGS based DDSG.

synchronize its rotational speed with the electrical network frequency [25]. Thus, VS-WPGS is able to attain high efficacy and performance when connected to the electrical network, not only under normal working conditions but also under irregular and faulty network conditions. Figure 1 illustrates the general control structure for VS-WPGS. This consists of a DDSG side and a grid side inverter interconnected through a dc-link system. So, at the DDSG sides, the speed is controlled to ensure the MPT. The grid side converter is employed to regulate the dc-link voltage for a proper transfer of power [13, 19, 26, 27]. Also, the inverter system ought to have the ability to regulate reactive and active power that the VS-WPGS exchange with the electrical network and attain unity power factor (UPF) of total system [28]. Various power electronic device configurations were presented in the literature for VS-WPGS based DDSG [29–31]. On the other hand, to increase the annual energy yield of VS-WPGS, MPT control is essential under the rated wind velocity. Then, by changing the rotor velocity of DDSG according to the varying wind velocity, maximum power is extracted from available wind [21, 22, 32, 33].

Also, variable speed wind turbines frequently employ pitch control system, either active stall or pitch to feather, to reduce the aerodynamic power captured [34, 35].

According to Fault Ride-Through (FRT) requirements, during faults of grid voltage, the WFS must remain connected to the electric network. As an example, Figure 2 presents a diagram of the FRT requirements in some countries. So, these requirements present a considerable challenge to the well-established WFS technologies and their control [36]. Besides, developing effective control approaches WFS will improve the system availability to extract as much power from the wind as possible whereas it is available. On the basic aspect, Vector Oriented Control (VOC) is widely used because of its simple control configuration, ease of design, and economical

cost. In [37, 38], control strategy has been developed for integrated control of wind turbine based PMSG. The works [39, 40] have presented Low Voltage Ride-Through (LVRT) of WECS. But, this technique cannot offer satisfactory control performance if the controlled system is highly nonlinear and uncertain. Moreover, it generally works at particular operating range and it is obligatory to retune the controller if the operating range is modified, whereas the WFS is usually affected by varying the operating condition and outside disturbance.

To overcome the above problems, Direct Control Techniques (DCT) were proposed to substitute the VOC strategy for both rectifier systems and electrical network connection control. Direct power control (DPC) and Direct Torque Control (DTC) of a three-phase converter for fault tolerant DDSG drive were proposed in [29]. The works [19, 41] have proposed TMP algorithm using DTC for DDSG wind turbines. The most important advantages of DCT strategies include low sensitivity to the accuracy of WFS parameter estimation, rapid dynamic response, and easy implementation and they do not necessitate space vector modulation (SVM). But, they can suffer from high torque ripple and variable switching frequency of the power electronic converters and their performances deteriorate at low speed.

On the other hand, modern WFS necessitate efficient and supple technologies that adapt to modifications in load and generation. Also, in order to increase the availability and reliability of WFS, effective strategies are considered crucial means to achieve these goals and to improve the electrical network connected operation capability. Recently, backstepping control approach has received worldwide attention due to its systematic and recursive design methodology for nonlinear control [42]. The stabilization of a virtual control state is the fundamental idea of the backstepping algorithm. This nonlinear control method is principally

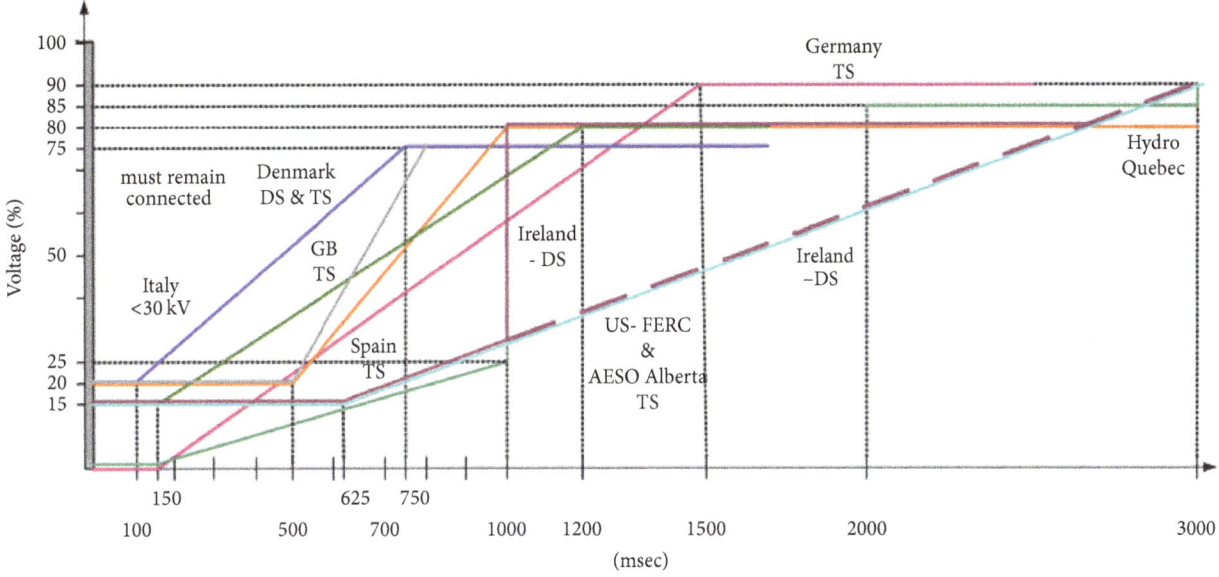

FIGURE 2: National grid codes [36].

interesting due to its capacity to design adaptive controllers for nonlinear systems and to analyze the problem of stability. The advantages of backstepping control are related to the employment of the Lyapunov functions to ensure the robustness and stability of the system, consequently enhancing the performance of system [43]. Also, because of the flexibility to avoid cancellations of suitable nonlinearities and attain the objectives of tracking and stabilization, the backstepping approach has been generally applied in control systems. In the work of [44] backstepping control is developed to deal with the Tracking the Maximum Power (TMP) from the WFS. The work of [45] presents a backstepping control strategy to design a permanent magnet liner synchronous motor servo drive system with uncertainty. Wai et al. [46] proposed the backstepping design for a grid-connected inverter. Ruan et al. [47] proposed an adaptive control strategy using backstepping technique for the Voltage Source Converter (VSC) based High Voltage dc (HVDC) system to develop its dynamic performance. The idea of backstepping approach is to choose recursively some suitable functions of state variables like pseudo-control inputs for lower dimension subsystems of the entire system. The most attractive point of this technique is to employ the virtual control variable to create the original high order system simple. It decomposes a difficult control problem into simpler ones and it chooses recursively suitable functions of state variables which are the virtual control variables. Each backstepping step results in a novel virtual control to deal with a decomposed subsystem problem. So, this virtual control becomes a reference to the next algorithm step for another subsystem. Also, it produces a related error variable that can be stabilized by carefully choosing proper control inputs which can be determined using Lyapunov stability analysis [43]. When the procedure is finished, a true control input for the original control objectives can be generated by summing up the Lyapunov functions associated with each individual design step [48].

This study presents a backstepping control design for the WFS using PMSG, connected to the electrical network. So, a backstepping scheme is developed in the sense of Lyapunov stability theorem to ensure the control objectives of Tracking the Maximum Power (TMP) from the WFS, pitch control, regulation of dc-link voltage, and reactive and active power generation at varying wind velocity. In the first place, mathematical model of WFS is described. In the meanwhile, design of the backstepping technique is provided. Then, simulation studies using MATLAB/Simulink are carried out in Section 3. Finally, some conclusions are summarized in Section 4.

2. Description of the Proposed WFS

This section describes the proposed wind energy conversion system using DDSG and presents its modelling and control strategies.

2.1. System Description. The schematic diagram of the grid-connected DDSG-WTS is shown in Figure 3. It is composed of three wind turbines directly linked to DDSGs connected to the grid throughout a converter, which consists of a grid side inverter (GSI) and a generator-side rectifier (MSR). The MSR control the DDSGs to extract the maximum power from the wind [49]. The GSI sustains a constant dc-link voltage and controls the active and reactive power that the WTS exchanges with the electrical power.

So, the converters allow adapting frequencies and voltages between the DDSG terminals and the grid.

2.2. Wind Turbine Characteristics. According to aerodynamics, the maximum power extracted by the rotor blades is expressed as [50]:

$$P_{w\text{-max}} = \frac{1}{2}\rho_w \pi A_{rw}^2 C_{w\text{-max}}\left(\lambda_{w\text{-}m}, \beta_w\right) v_w^3, \qquad (1)$$

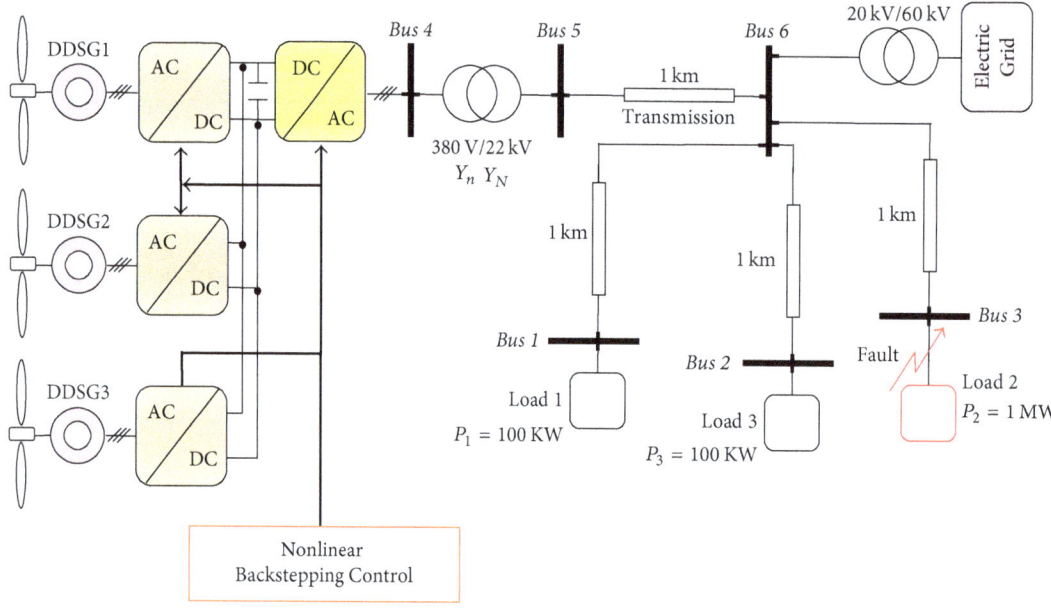

FIGURE 3: Configuration of the VS-WFS.

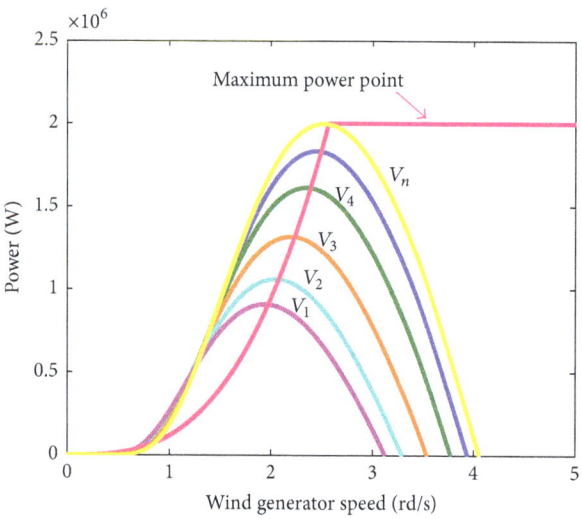

FIGURE 4: Wind turbine power curves at different values of wind speed.

where ρ_w is the air density, A_{rw} is the rotator radius, v_w is the speed of wind, $\lambda_{w\text{-}m}$ is the optimal tip speed ratio (TSR), $C_{w\text{-}\max}(\lambda_m, \beta_w)$ is the maximum value of the performance coefficient of the turbine, and β_w is the blade pitch angle. The tip speed ratio λ_w is determined by the rotor speed ω_r and the wind speed as

$$\lambda_w = \frac{\omega_r A_{rw}}{v_w}.$$ (2)

So, an efficient TMP is fundamental for WFS to develop the energy capture effectiveness from wind and the pitch angle β_w is set as zero during the TMP operation. The optimal TSR strategy is realized by directly regulating the tip speed ratio,

which is calculated by real time using wind velocity and generator speed, to follow a preobtained $\lambda_{w\text{-}m}$. Therefore, one way to modify the captured wind power is to adjust the DDSG speed, which is depicted in Figure 4 [51]. As seen, different turbine velocities are associated with different captured wind power. Consequently, the generator-side rectifiers in Figure 3 regulate the speeds of DDSGs to attain TMP in combination with the pitch control system. Figure 4 depicts the typical wind turbine power-rotor velocity characteristic curves for different wind velocities, where the optimal power curve illustrates the TMP at different conditions of wind speed. The maximum power extracted $P_{w\text{-}\max}$ can be obtained as a function of the shaft velocity:

$$P_{w\text{-}\max} = \frac{\rho_w \pi A_{rw}^5 C_{w\text{-}\max}}{2\lambda_{w\text{-}m}^3} \omega_r^3 = K_{\max} \omega_r^3,$$ (3)

where $K_{\max}$ is the coefficient of rotational velocity for the maximum extracted power through $C_{w\text{-}\max}$ at the wind generator. It should be noted that when the wind velocity is high, the generator velocity and power would exceed the maximum value if the MPPT control is still employed. The pitch control is provided to limit the captured power and the generator speed. So, the pitch angle is set to zero to obtain TMP if the speed is below the maximum speed. But, if the velocity exceeds its rated value, the pitch angle is controlled to limit the captured power to the maximum rated value and the WFS would lead to deviating the TMP searching strategy to prevent overspeeding. The pitch control diagram is shown in Figure 5.

2.3. DDSG Dynamic Modelling. If the rotating reference frame is aligned with the magnetic axis of the rotor, the

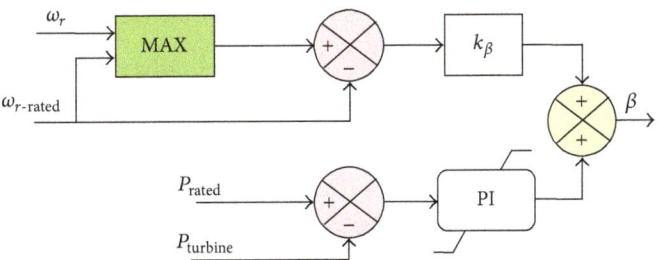

FIGURE 5: Pitch control diagram for generator speed limiting.

electric dynamics equations of a three-phase DDSG can be written in a synchronously rotating dq reference frame as [52]

$$\begin{bmatrix} v_{sd} \\ v_{sq} \end{bmatrix} = \begin{bmatrix} R_s + p \cdot L_{sd} & -\omega_e L_{sq} \\ \omega_e L_{sd} & R_s + p \cdot L_{sd} \end{bmatrix} \begin{bmatrix} i_{sd} \\ i_{sq} \end{bmatrix} + \begin{bmatrix} 0 \\ \omega_e \psi_s \end{bmatrix}, \quad (4)$$

where v_{sq}, v_{sd} are stator terminal voltages in the dq frame (V); i_{sq}, i_{sd} are direct and quadrature currents components in the stator windings (A); L_{sq}, L_{sd} are inductances of the DDSG in the dq frame (H); R_s is resistance of the stator windings (Ω); ψ_s is flux linkage generated by the permanent magnets (Wb).

The electrical angular velocity of the rotor, ω_e, is defined as

$$\omega_e = p_n \omega_r, \quad (5)$$

where p_n is the number of pole pairs of the generator; ω_r is the mechanical angular velocity (rad/sec).

If the DDSG is assumed to have equal q-axis, d-axis in inductances ($L_{sq} = L_{sd} = L_{gs}$), the torque developed by the DDSG is

$$T_{sg} = \frac{3}{2} p_n [\psi_s i_{sq}]. \quad (6)$$

The generator velocity in the one-mass model of the wind turbine systems is given as follows:

$$\frac{d\omega_r}{dt} = \frac{1}{J_s} (T_{sg} - T_{wt} - F_s \omega_r), \quad (7)$$

where J_s is the moment of inertia (Kg·m^2), F_s is the viscous friction coefficient (Nm/rad·sec^{-1}), and T_{wt} is the mechanical torque developed by the wind turbine (WT) (Nm).

2.4. Backstepping Control Design for the MSR. To attain optimum power effectiveness regardless of wind speed variations, the MSR is used to control the DDSG speeds. Control of the MSR permits the generators to regulate the rotational velocities according to the wind variation. On the other hand, the TMP algorithm adopted in this work is the tip speed ratio control. The backstepping control is employed to perform the TMP control of the wind turbine systems. So, according to (2), the reference of the mechanical DDSG speed can be derived as

$$\omega_{r\text{-ref}} = \frac{\lambda_w v_w}{A_{rw}}. \quad (8)$$

For each MSR, the major control objective is to track the velocity reference as

$$\alpha_\omega = \omega_{r\text{-ref}} - \omega_r. \quad (9)$$

The dynamic of the velocity tracking error is expressed as

$$\frac{d\alpha_\omega}{dt} = \frac{d\omega_{r\text{-ref}}}{dt} - \frac{d\omega_r}{dt}. \quad (10)$$

From (6)-(7) and (10), we have

$$\frac{d\alpha_\omega}{dt} = \frac{d\omega_{r\text{-ref}}}{dt} - \frac{1}{J_s}\left(\frac{3}{2}p_n[\psi_s i_{sq}] - T_{wt} - F_s\omega_r\right). \quad (11)$$

So, a first Lyapunov function candidate is defined as [53]

$$\Gamma_1 = \frac{1}{2}\alpha_\omega^2. \quad (12)$$

Differentiating (12) gives us

$$\begin{aligned}\frac{d\Gamma_1}{dt} &= \alpha_\omega \frac{d\alpha_\omega}{dt} \\ &= \alpha_\omega \frac{d\omega_{r\text{-ref}}}{dt} - \frac{\alpha_\omega}{J_s}\left(\frac{3}{2}p_n[\psi_s i_{sq}] - T_{wt} - F_s\omega_r\right).\end{aligned} \quad (13)$$

Equation (13) can be reorganized as

$$\begin{aligned}\frac{d\Gamma_1}{dt} &= -\delta_\omega \alpha_\omega^2 + \frac{\alpha_\omega}{J_s}\left(J_s\delta_\omega\alpha_\omega + T_{wt} - \frac{3}{2}p_n\psi_s i_{sq} + F_s\omega_r \right.\\ &\quad \left. + J_s \frac{d\omega_{r\_ref}}{dt}\right),\end{aligned} \quad (14)$$

where δ_ω is the positive closed-loop feedback constant. According to (14), the backstepping law can be designed as follows:

$$i_{sd\text{-ref}} = 0, \quad (15)$$

$$i_{sq\text{-ref}} = \frac{2}{3p_n\psi_s}\left(J_s\delta_\omega\alpha_\omega + J_s\frac{d\omega_{r\_ref}}{dt} + F_s\omega_r + T_{wt}\right). \quad (16)$$

If the speed error can be made to zero by selecting proper control input, (14) can be simplified to [54]

$$\frac{d\Gamma_1}{dt} = -\delta_\omega \alpha_\omega^2 < 0. \quad (17)$$

Consequently, by Lyapunov stability analysis, the velocity controller is asymptotically stable. To stabilize the currents components i_{sd} and i_{sq}, define the following current errors:

$$\alpha_{sd} = i_{sd\text{-ref}} - i_{sd}$$
$$\alpha_{sq} = i_{sq\text{-ref}} - i_{sq}, \qquad (18)$$

where $i_{sq\text{-ref}}$, $i_{sd\text{-ref}}$ are the current references. Taking the derivative of α_{sd} and α_{sq} with respect to time and using (4), one can obtain

$$\alpha_{sd}\frac{d\alpha_{sd}}{dt} = \alpha_{sd}\left[-\frac{1}{L_{gs}}\left(v_{sd} + L_{gs}\omega_e i_{sq} - R_s i_{sd}\right)\right]$$

$$= -\lambda_{sd}\alpha_d^2 + \frac{\alpha_{sd}}{L_{gs}}\left[-v_{sd} - L_{gs}\omega_e i_{sq} + R_s i_{sd}\right.$$

$$\left. + L_{gs}\lambda_{sd}\alpha_{sd}\right]$$

$$\alpha_{sq}\frac{d\alpha_{sq}}{dt} = \alpha_{sq}\left[\frac{di_{sq\text{-ref}}}{dt} - \frac{di_{sq}}{dt}\right] = -\lambda_{sq}\alpha_{sq}^2 \qquad (19)$$

$$+ \frac{\alpha_{sq}}{L_{gs}}\left[L_{gs}\frac{di_{sq\text{-ref}}}{dt} - v_{sd} + \omega_e\psi_s + R_s i_{sq} + L_{gs}\omega_e i_{sd}\right.$$

$$\left. + \lambda_{sq}L_{gs}\alpha_{sq}\right],$$

where λ_{sd} and λ_{sq} are the positive closed-loop feedback constants. The second Lyapunov function candidate is defined as

$$\Gamma_2 = \Gamma_1 + \frac{1}{2}\alpha_{sq}^2 + \frac{1}{2}\alpha_{sd}^2 = \frac{1}{2}\alpha_\omega^2 + \frac{1}{2}\alpha_{sq}^2 + \frac{1}{2}\alpha_{sd}^2. \qquad (20)$$

Differentiating (20) gives us

$$\frac{d\Gamma_2}{dt} = \alpha_\omega\frac{d\alpha_\omega}{dt} + \alpha_{sd}\frac{d\alpha_{sd}}{dt} + \alpha_{sq}\frac{d\alpha_{sq}}{dt}. \qquad (21)$$

From (14) and (18)–(19), (21) can be reorganized as

$$\frac{d\Gamma_2}{dt} = -\delta_\omega\alpha_\omega^2 + \frac{\alpha_\omega}{J_s}\left(J_s\delta_\omega\alpha_\omega + T_{wt} - \frac{3}{2}P_n\psi_s i_{sq}\right.$$

$$\left. + F_s\omega_r + J_s\frac{d\omega_{r_{ref}}}{dt}\right) - \lambda_{sq}\alpha_{sq}^2 + \frac{\alpha_{sq}}{L_{gs}}\left[L_{gs}\frac{di_{sq\text{-ref}}}{dt}\right.$$

$$\left. - v_{sd} + \omega_e\psi_s + R_s i_{sq} + L_{gs}\omega_e i_{sd} + \lambda_{sq}L_{gs}\alpha_{sq}\right] \qquad (22)$$

$$- \lambda_{sd}\alpha_{sd}^2 + \frac{\alpha_{sd}}{L_{gs}}\left[-v_{sd} - L_{gs}\omega_e i_{sq} + R_s i_{sd}\right.$$

$$\left. + L_{gs}\lambda_{sd}\alpha_{sd}\right].$$

The commands $v_{sq\text{-ref}}$ and $v_{sd\text{-ref}}$ are carried out from (22) as

$$v_{sq\text{-ref}} = \omega_e\psi_s + R_s i_{sq} + L_{gs}\omega_e i_{sd} + L_{gs}\frac{di_{sq\text{-ref}}}{dt}$$

$$+ L_{gs}\lambda_{sq}\alpha_{sq} \qquad (23)$$

$$v_{sd\text{-ref}} = L_{gs}\lambda_{sd}\alpha_{sd} - L_{gs}\omega_e i_{sq} + R_s i_{sd}.$$

Using (16) and (23), (22) can be simplified to

$$\frac{d\Gamma_2}{dt} = -\delta_\omega\alpha_\omega^2 - \lambda_{sd}\alpha_{sd}^2 - \lambda_{sq}\alpha_{sq}^2 \leq 0. \qquad (24)$$

Therefore, by Lyapunov stability analysis, the backstepping regulator is asymptotically stable and the control of the DDSG speeds is attained. The control diagram for individual MSR is depicted in Figure 6.

2.5. Nonlinear dc-link Voltage Regulation. The GSI controls the dc-link voltage and transfers the power from turbine-DDSG to the grid. Also, it contributes to the reactive and active power control of the overall DDSG system [55]. If the reference frame is rotating synchronously with the electrical grid voltage vector, the dynamic model voltage equations of the electrical network connection in grid voltage oriented reference frame (d, q) are represented as follows:

$$\frac{di_{d\text{-grid}}}{dt} = \frac{1}{L_{gf}}\left(e_{gd} - R_{gf}i_{d\text{-grid}} + \omega_g L_{gf}i_{q\text{-grid}} - U_s\right)$$

$$\frac{di_{q\text{-grid}}}{dt} = \frac{1}{L_{gf}}\left(e_{gq} - R_{gf}i_{q\text{-grid}} - \omega_g L_{gf}i_{d\text{-grid}}\right), \qquad (25)$$

where e_{gd}, e_{gq} are the grid side inverter voltage $(d\text{-}q)$ components (V); U_s is the amplitude of the grid voltage (V); R_{gf} (Ω) and L_{gf} (H) are the resistance and inductance of the filter; ω_g is the angular frequency of grid; and $i_{d\text{-grid}}$ and $i_{q\text{-grid}}$ are the d-axis current and q-axis current (A) of the electrical grid, respectively. The active P_g and reactive power Q_g can be calculated as follows [56]:

$$P_g = \frac{3}{2}U_s i_{d\text{-grid}}$$

$$Q_g = \frac{3}{2}U_s i_{q\text{-grid}}. \qquad (26)$$

Equations (26) illustrate that the active and reactive powers are regulated to achieve the required voltage by changing the respective current of the d-q-axis. The dc-link voltage equation can be formulated as in the following equation:

$$C\frac{dV_{dc}}{dt} = i_g - i_{grid}, \qquad (27)$$

where C is the capacitance of the dc-link capacitor (F); V_{dc} is dc-link voltage (V); i_{grid} is the current between the dc-link and the grid; and i_g is the current between the dc-link and the DDSG stators (A). So, the active power along the side of the GSI is

$$V_{dc}i_{grid} = \frac{3}{2}U_s i_{d\text{-grid}}. \qquad (28)$$

From (27)-(28), the dc-link voltage dynamics can be calculated as

$$\frac{dV_{dc}}{dt} = i_g - \frac{1}{C}\frac{3}{2}\frac{U_s}{V_{dc}}i_{d\text{-grid}}. \qquad (29)$$

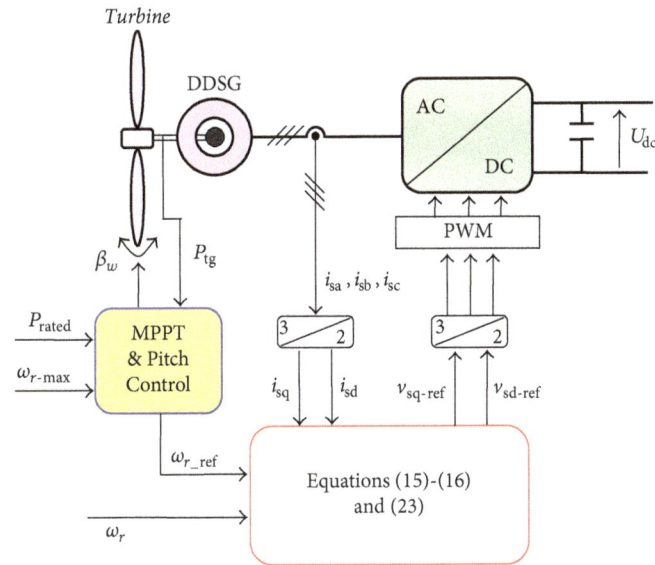

FIGURE 6: Control block diagram of MSR using backstepping approach.

Consequently, to transfer all of the active power generated in turbine-DDSG, the dc-link voltage V_{dc} must be maintained at a constant value using the current control of $i_{d\text{-grid}}$. Equation (29) can be reorganized as

$$V_{dc}C\frac{dV_{dc}}{dt} = P_{Total} - \frac{3}{2}U_s i_{d\text{-grid}}, \qquad (30)$$

where P_{Total} is the generated power of the DDSGs. Equation (30) can be expressed as

$$\frac{C}{2}\frac{dV_{dc}^2}{dt} = P_{Total} - \frac{3}{2}U_s i_{d\text{-grid}}. \qquad (31)$$

The tracking error of the dc-bus voltage is defined as

$$\varepsilon_{dc\text{-link}} = V_{dc-r}^2 - V_{dc}^2. \qquad (32)$$

The dc-link voltage tracking error has the following dynamic:

$$\frac{d\varepsilon_{dc\text{-link}}}{dt} = \frac{dV_{dc-r}^2}{dt} - \frac{2}{C}\left(P_{Total} - \frac{3}{2}U_s i_{d\text{-grid-}r}\right), \qquad (33)$$

where $i_{d\text{-grid-}r}$ is the current reference. Consider a third Lyapunov function candidate:

$$\Phi_{dc\text{-link}} = \frac{1}{2}\varepsilon_{dc\text{-link}}^2. \qquad (34)$$

From (33), differentiating (34) gives us

$$\begin{aligned}\frac{d\Phi_{dc\text{-link}}}{dt} &= \varepsilon_{dc\text{-link}}\frac{d\varepsilon_{dc\text{-link}}}{dt}\\ &= \varepsilon_{dc\text{-link}}\left[\frac{dV_{dc-r}^2}{dt} - \frac{2}{C}\left(P_{Total} - \frac{3}{2}U_s i_{d\text{-grid-}r}\right)\right].\end{aligned} \qquad (35)$$

So, (35) can be reorganized as

$$\begin{aligned}\frac{d\Phi_{dc\text{-link}}}{dt} = &-\lambda_{bus}\varepsilon_{dc\text{-link}}^2 + \varepsilon_{dc\text{-link}}\left[\frac{dV_{dc-r}^2}{dt}\right.\\ &\left. -\frac{2}{C}\left(P_{Total} - \frac{3}{2}U_s i_{d\text{-grid-}r}\right) + \lambda_{bus}\varepsilon_{dc\text{-link}}\right],\end{aligned} \qquad (36)$$

where λ_{bus} is the positive closed-loop feedback constant. According to (36), the backstepping law can be designed as follows:

$$i_{d\text{-grid-}r} = \frac{1}{U_s}\left[\frac{2}{3}P_{Total} - \frac{C}{3}\left(\frac{dV_{dc-r}^2}{dt} + \lambda_{bus}\varepsilon_{dc\text{-link}}\right)\right]. \qquad (37)$$

From (37), (38) can be simplified to

$$\frac{d\Phi_{dc\text{-link}}}{dt} = -\lambda_{bus}\varepsilon_{dc\text{-link}}^2 \le 0. \qquad (38)$$

Consequently, by Lyapunov stability analysis, dc-link voltage controller is asymptotically stable. To stabilize the currents components $i_{d\text{-grid}}$ and $i_{q\text{-grid}}$, define the following current errors:

$$\xi_{d\text{-grid}} = i_{d\text{-grid-}r} - i_{d\text{-grid}} \qquad (39)$$

$$\xi_{q\text{-grid}} = i_{q\text{-grid-}r} - i_{q\text{-grid}}. \qquad (40)$$

The reference $i_{q\text{-grid-}r}$ is calculated from the desired power factor. Based on (37) and (39), (33) can be reorganized as

$$\frac{d\varepsilon_{dc\text{-link}}}{dt} = -\lambda_{dc\text{-link}}\varepsilon_{dc\text{-link}} + \frac{3}{C}U_s\xi_{d\text{-grid}}. \qquad (41)$$

Besides, from (25) and (39)-(40), one can obtain

$$
\xi_{d\text{-grid}} \frac{d\xi_{d\text{-grid}}}{dt} = \xi_{d\text{-grid}} \left[\frac{di_{d\text{-grid-}r}}{dt} - \frac{di_{d\text{-grid}}}{dt} \right]
$$

$$
= -\lambda_{d\text{-grid}} \xi_{d\text{-grid}}^2 + \frac{\xi_{d\text{-grid}}}{L_{gf}} \left[L_{gf} \frac{di_{d\text{-grid-}r}}{dt} - e_{gd} \right.
$$

$$
+ U_s + R_{gf} i_{d\text{-grid}} - L_{gf} \omega_g i_{q\text{-grid}}
$$

$$
\left. + \lambda_{d\text{-grid}} L_{gf} \xi_{d\text{-grid}} \right] \tag{42}
$$

$$
\xi_{q\text{-grid}} \frac{d\xi_{q\text{-grid}}}{dt} = \left[-\frac{di_{q\text{-grid}}}{dt} \right] = -\lambda_{q\text{-grid}} \xi_{q\text{-grid}}^2
$$

$$
+ \frac{\xi_{q\text{-grid}}}{L_{gf}} \left[-e_{gq} + R_{gf} i_{q\text{-grid}} + L_{gf} \omega_g i_{d\text{-grid}} \right.
$$

$$
\left. + \lambda_{q\text{-grid}} L_{gf} \lambda_{q\text{-grid}} \right],
$$

where $\lambda_{d\text{-grid}}$ and $\lambda_{q\text{-grid}}$ are the positive closed-loop feedback constants. Consider the fourth Lyapunov function candidate:

$$
\Phi_{\text{GSI}} = \frac{1}{2} \varepsilon_{\text{dc-link}}^2 + \frac{1}{2} \xi_{q\text{-grid}}^2 + \frac{1}{2} \xi_{d\text{-grid}}^2. \tag{43}
$$

Differentiating (43) gives us

$$
\frac{d\Phi_{\text{GSI}}}{dt} = \varepsilon_{\text{dc-link}} \frac{d\varepsilon_{\text{dc-link}}}{dt} + \xi_{d\text{-grid}} \frac{d\xi_{d\text{-grid}}}{dt}
$$
$$
+ \xi_{q\text{-grid}} \frac{d\xi_{q\text{-grid}}}{dt}. \tag{44}
$$

From (41) and (42), (44) can be reorganized as

$$
\frac{d\Phi_{\text{GSI}}}{dt} = -\lambda_{\text{dc-link}} \varepsilon_{\text{dc-link}}^2 - \lambda_{d\text{-grid}} \xi_{d\text{-grid}}^2 - \lambda_{q\text{-grid}} \xi_{q\text{-grid}}^2
$$

$$
+ \frac{\xi_{d\text{-grid}}}{L_{gf}} \left[\frac{3L_{gf}}{C} U_s \varepsilon_{\text{dc-link}} + L_{gf} \frac{di_{d\text{-grid-}r}}{dt} - e_{gd} \right.
$$

$$
+ R_{gf} i_{d\text{-grid}} - L_{gf} \omega_g i_{q\text{-grid}} + U_s \tag{45}
$$

$$
\left. + \lambda_{d\text{-grid}} L_{gf} \xi_{d\text{-grid}} \right] + \frac{\xi_{q\text{-grid}}}{L_{gf}} \left[-e_{gq} + R_{gf} i_{q\text{-grid}} \right.
$$

$$
\left. + L_{gf} \omega_g i_{d\text{-grid}} + \lambda_{q\text{-grid}} L_{gf} \xi_{q\text{-grid}} \right].
$$

According to (45), the backstepping law can be designed as follows:

$$
v_{d\text{-GSI}} = \frac{3L_{gf}}{C} U_s \varepsilon_{\text{dc-link}} + L_{gf} \frac{di_{d\text{-grid-}r}}{dt} + R_{gf} i_{d\text{-grid}}
$$
$$
- L_{gf} \omega_g i_{q\text{-grid}} + U_s + L_{gf} \lambda_{d\text{-grid}} \xi_{d\text{-grid}} \tag{46}
$$

$$
v_{q\text{-GSI}} = R_{gf} i_{q\text{-grid}} + L_{gf} \omega_g i_{d\text{-grid}} + L_{gf} \lambda_{q\text{-grid}} \xi_{q\text{-grid}}.
$$

TABLE 1: Parameters of the DDSG.

Parameter	Value
P_r: rated power of DDSG	2 (MW)
ω_m: rated mechanical speed	2.57 (rd/s)
R: stator resistance	0.008 (Ω)
L_s: stator d-axis inductance	0.0003 (H)
ψ_f: permanent magnet flux	3.86 (wb)
p_n: pole pairs	60

TABLE 2: Parameters of the turbine.

Parameter	Value
Blade number	3
ρ: the air density	1.08 kg/m^3
A: area swept by blades	4775.94 m^2
$v_{w\text{-}n}$: rated wind speed	12.4 m/s

TABLE 3: System parameters.

Parameter	Value
dc-link voltage reference	2100 V
Grid frequency	50 Hz
Grid phase voltage	660 V
Ratio $X_{\text{Grid}}/R_{\text{Grid}}$	5
dc-link capacitor	38000 μF

Using (37) and (46), (45) can be simplified to

$$
\frac{d\Phi_{\text{GSI}}}{dt} = -\lambda_{\text{dc-link}} \varepsilon_{\text{dc-link}}^2 - \lambda_{d\text{-grid}} \xi_{d\text{-grid}}^2 - \lambda_{q\text{-grid}} \xi_{q\text{-grid}}^2 \tag{47}
$$

$$
\leq 0.
$$

Therefore, by Lyapunov stability analysis, the backstepping regulator is asymptotically stable and the control of the GSI is attained. The control diagram is depicted in Figure 7.

3. Simulation Result Analysis

The performance of the proposed nonlinear control strategy is demonstrated in this section. The controller is tested on the WFS control strategy found in Section 2. The schematic of the overall control scheme of the WFS is illustrated in Figure 7. Numerical simulations are carried out using MAT-LAB/Simulink software to verify WFS performance. System data used are listed in Tables 1, 2, and 3. The performance coefficients of the turbines are evaluated as the following form:

$$
C_w(\lambda_w, \beta_w) = \frac{1}{2} \left(\frac{116}{\lambda_{wc}} - 0.4\beta_w - 5 \right) e^{-(21/\lambda_{wc})}
$$

$$
\frac{1}{\lambda_{wc}} = \frac{1}{\lambda_w + 0.08\beta_w} - \frac{0.035}{\beta_w^3 + 1}, \tag{48}
$$

where C_w reaches the maximum value $C_{w\text{-max}} = 0.4104$ when λ_w is $\lambda_{w\text{-}m} = 8.1$.

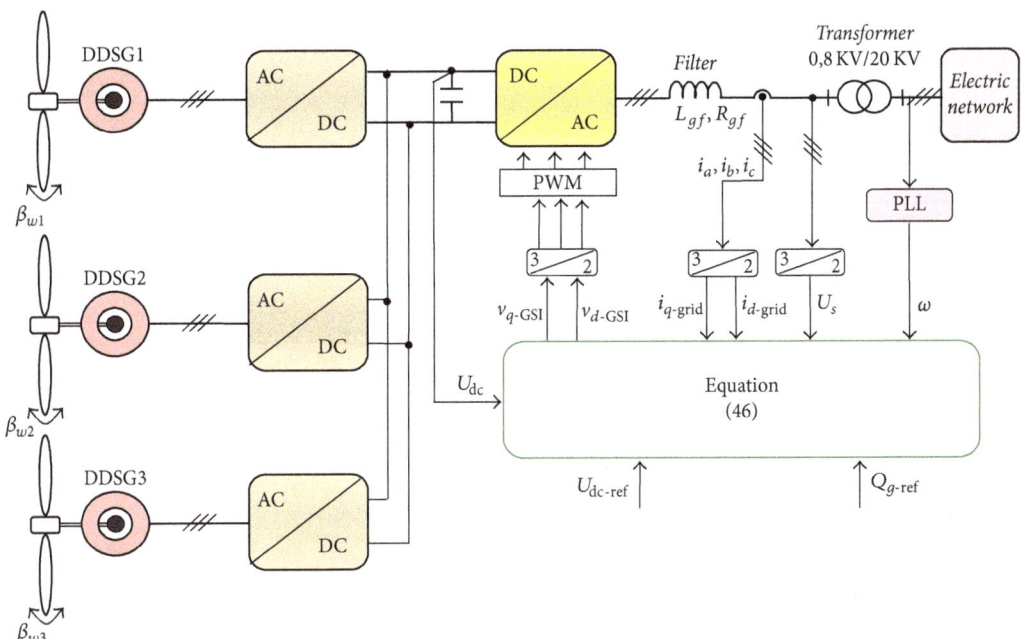

FIGURE 7: Control block diagram of GSI.

Besides, the simulation will be performed in three steps with different objectives. The first scenario consists in operation under changing wind velocities, the second is robustness of proposed control strategy against electrical and mechanical parameter changes, and the third is an AC voltage sag.

3.1. Case 1: Performance Evaluation under Normal Operation of Proposed Control. Figure 8 shows the wind velocities for the different wind turbine systems (a), the individual pitch angle for each wind turbine (b), the performance coefficients of the turbines (c), the speeds of DDSG (d), and the total power extracted by the wind farm (f). It has been noted that the speeds of DDSG are adapted relative to the variation of wind speed and the C_w can be preserved at a maximum value which could yield to TMP by working the turbines at the speed references. The pitch angle controls are activated for velocities above the rated velocity to control the rated power of the DDSGs. If the wind turbines receive wind velocities which are lower than the rated speed, the pitch angles are set as 0° to generate the maximum extracted power. The wind turbine systems operated at theirs maximum performance coefficients and the pitch angle were kept at their optimal values. The controller performs the power maximization according to the wind velocity variation. But, pitch controls are activated to restrict the rotational velocities to below the maximum speed and to limit the powers of turbines when the wind velocities are above the rated speed. In this study, the DDSG rated speed is 2.57 rad/s. So, in high wind velocity regions, the DDSG velocities and extracted power would not exceed the maximum. The generated power of the WFS is shown in Figure 8(f). Figure 8(e) shows wind DDSG1 speed follows the optimal speed provided by the TMP operation as the wind velocity varies. The DDSG1 velocity is able to track its optimal values precisely. Figures 8(g)–8(h)

show the currents of DDSG1, where the generator currents are able to track their optimal values precisely. Figure 9 illustrates the dc-link voltage response. The dc-link voltage controller tried to maintain the constant value so that the dc-link voltage had an approximately constant value of 2100 V. Figure 10 shows the simulation result of reactive power which is controlled to be zero. Figure 11 illustrates a detailed view of the grid currents and phase voltage during the wind variation where the proposed WFS produces sinusoidal currents at electrical network with unity power factor. The simulation results show that the proposed backstepping controllers have good performance and can guarantee acceptable global regulation and tracking performance to extract and convert power under varying wind speeds.

3.2. Case 2: WFS Response Subject to Parameter Uncertainties. In order to verify the robustness of the proposed control strategy against the variation of parameters, it is supposed that the stator inductance L_{gs}, the stator resistor R_s, and the total moment of inertia J_s of the system values are increased by 50% under the condition of nominal values. In Figures 12–15, robustness against WFS parameter variations is tested for the proposed strategy. In the simulation, L_{gs}, R_s, and J_s are increased to 150% of their nominal values, and the other simulation parameters are the same as those in first scenario. The simulation results of the backstepping control and the VOC strategy with and without the mentioned parameter variation are shown in Figures 12–15. The results with and without the parameter variations are marked by "A" and "B," respectively. It can be seen that the backstepping strategy has a fast dynamic response, a reduced settling time, and a lower overshoot. The simulation results confirm that the proposed control approach is robust against uncertainties in the WFS compared with conventional VOC.

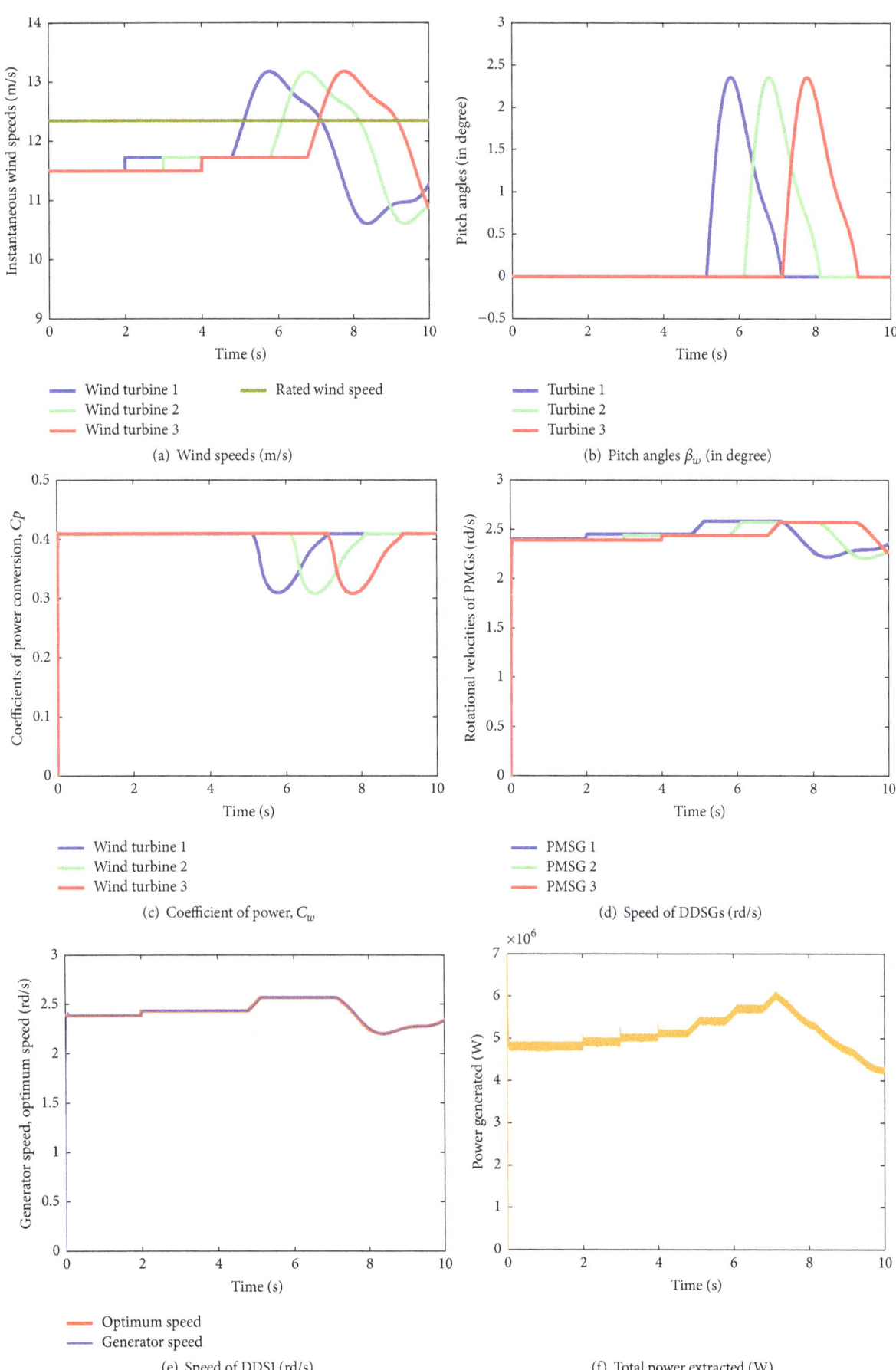

(a) Wind speeds (m/s)

(b) Pitch angles β_w (in degree)

(c) Coefficient of power, C_w

(d) Speed of DDSGs (rd/s)

(e) Speed of DDS1 (rd/s)

(f) Total power extracted (W)

FIGURE 8: Continued.

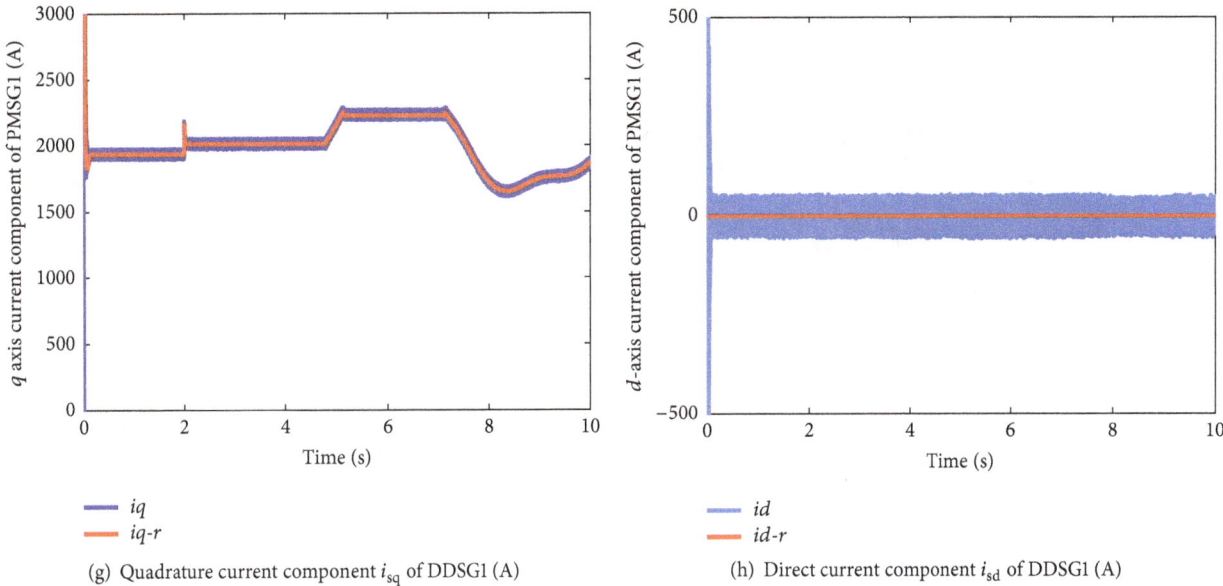

(g) Quadrature current component i_{sq} of DDSG1 (A)

(h) Direct current component i_{sd} of DDSG1 (A)

FIGURE 8: Simulation of the WFS in normal operation and using backstepping strategy.

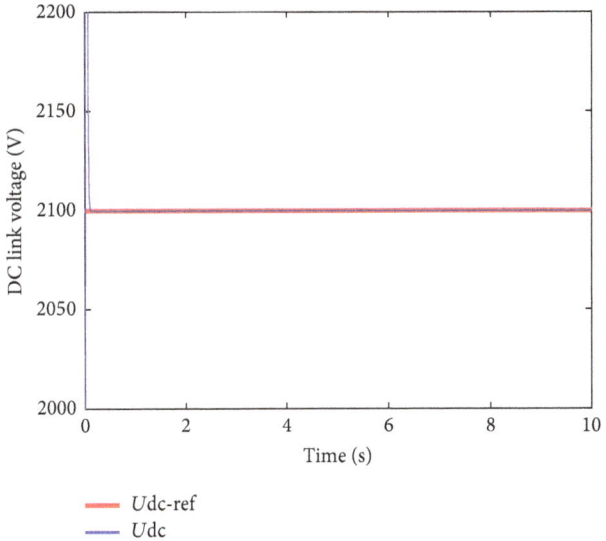

FIGURE 9: dc-link voltage (V).

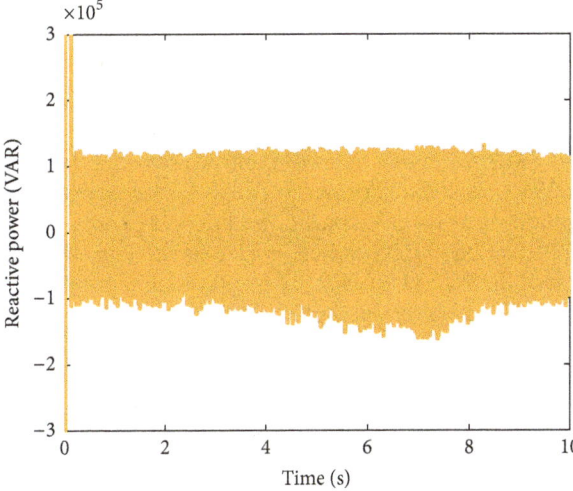

FIGURE 10: Reactive power (VAR).

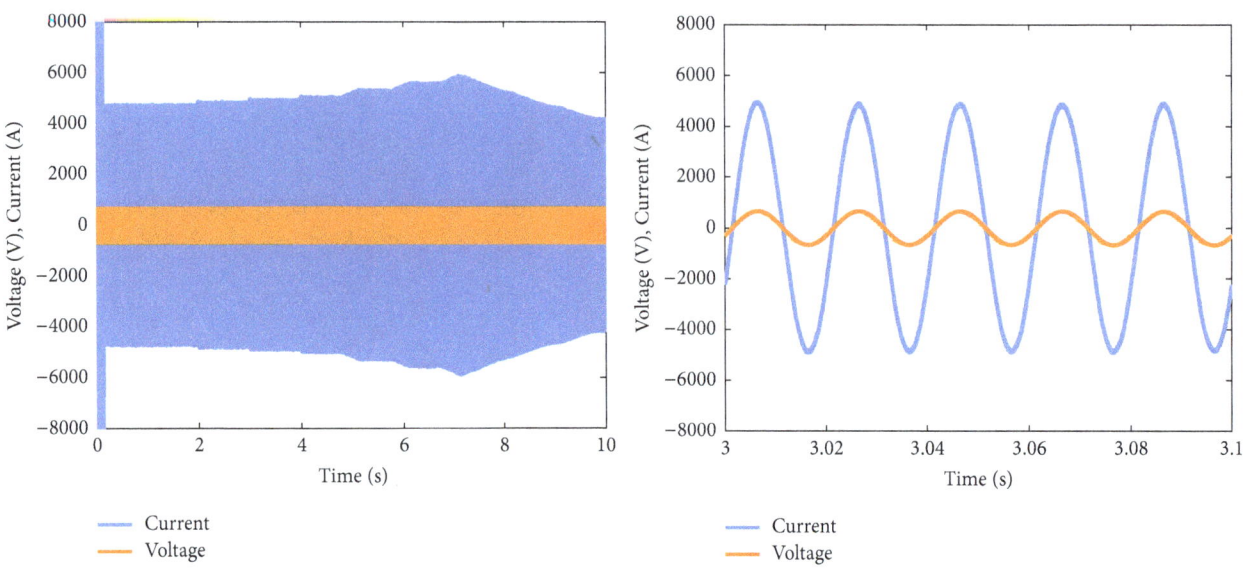

FIGURE 11: Simulation of the three-phase current and voltage of electrical network.

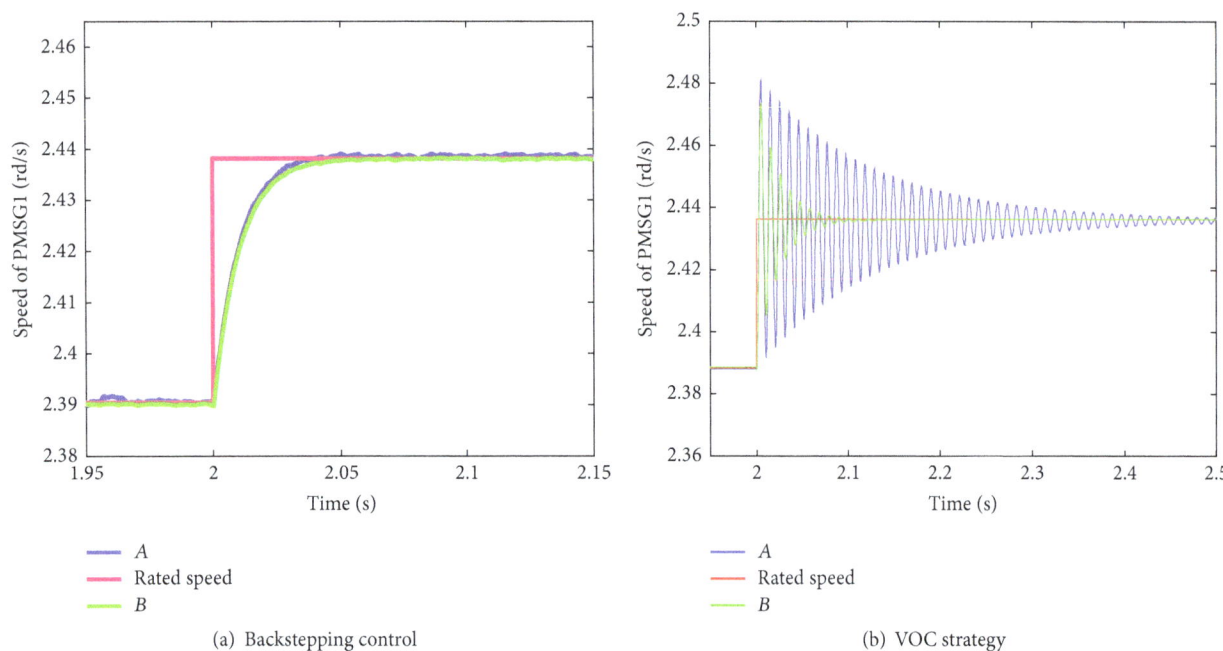

(a) Backstepping control

(b) VOC strategy

FIGURE 12: DDSG1 speed (rd/s).

3.3. Case 3: Simulation under Grid Fault Condition. This simulation example deals with the WFS behaviour during a Three-Phase Grid Fault (3PGF) produced at $t = 5.5$ s and characterised by a voltage drop of 200 ms at Bus 3 (Figure 3). Figures 16 and 17 illustrate the voltage of the WFS at Bus 4 in normal and close zoom views, respectively, for backstepping and VOC strategies. It can be observed that, by applying the backstepping controller, the transient stability of the WFS has been enhanced and the oscillations of the three-phase voltage have been damped quickly when the fault is cleared. Besides, Figure 17 shows that the transient duration with the proposed control is faster than that with the VOC and the three-phase

voltage oscillation amplitude at VOC is evidently larger than that at backstepping control. The results illustrated that the backstepping control strategy can not only reduce the peak value of three-phase voltage but also shorten the transient response time compared with the VOC strategy. Compared to the results depicted in Figure 18, it can be seen that the peak value of frequency deviations can be eliminated during grid voltage drop with the proposed control strategy. Compared with VOC strategy, the peak value of frequency deviations is reduced from 1.9 Hz to less than 0.25 Hz with backstepping strategy. Figure 19 illustrates the dc-link voltage response under proposed control and VOC strategy. The voltage drop

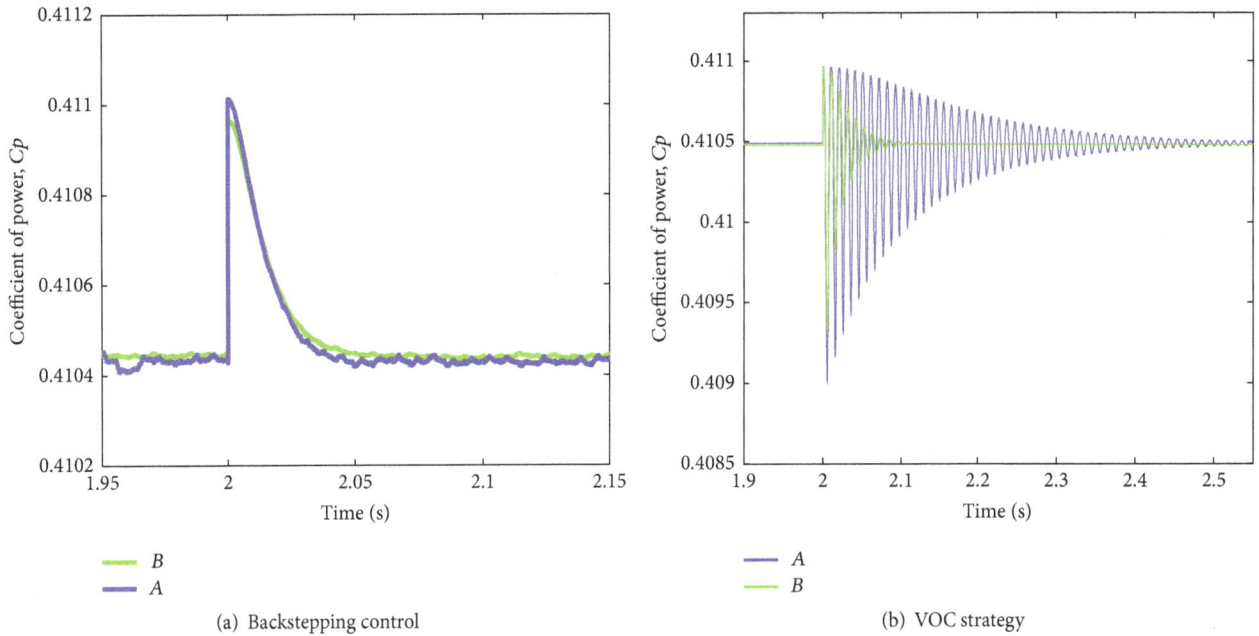

(a) Backstepping control

(b) VOC strategy

FIGURE 13: Coefficient of power.

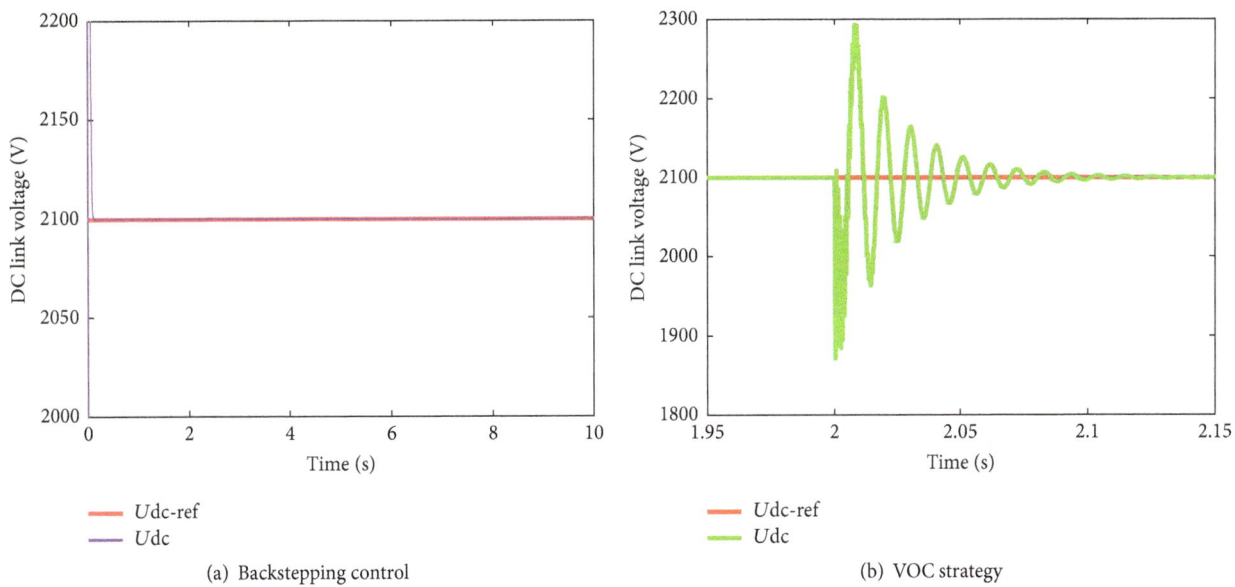

(a) Backstepping control

(b) VOC strategy

FIGURE 14: dc-link voltage (V) (with nominal values).

will not be transmitted to the dc-link system voltage and the converter dc-link voltage is controlled invariable at the reference value 2100 V as in Figure 19.

4. Conclusion

This paper has presented a nonlinear backstepping approach of WFS driven DDSG operating under normal and voltage drop situations. A proposed control has been investigated to deal with problems of simultaneous control of the DDSG speeds and the dc-link voltage to achieve Maximum Power Tracking and pitch control. Simulation research on a 6-MW-DDSG Wind Farm System validates the proposed control strategy. Performance comparison and evaluation with Vector Oriented Control (VOC) are provided under a wide range of functioning conditions, three-phase voltage dips, and the probable occurrence of uncertainties. The main conclusions are summarized as follows: (1) The proposed backstepping strategy can guarantee acceptable global regulation and tracking performance. (2) The proposed control strategy offers remarkable characteristics such as excellent dynamic and steady state performance under varying wind speed and

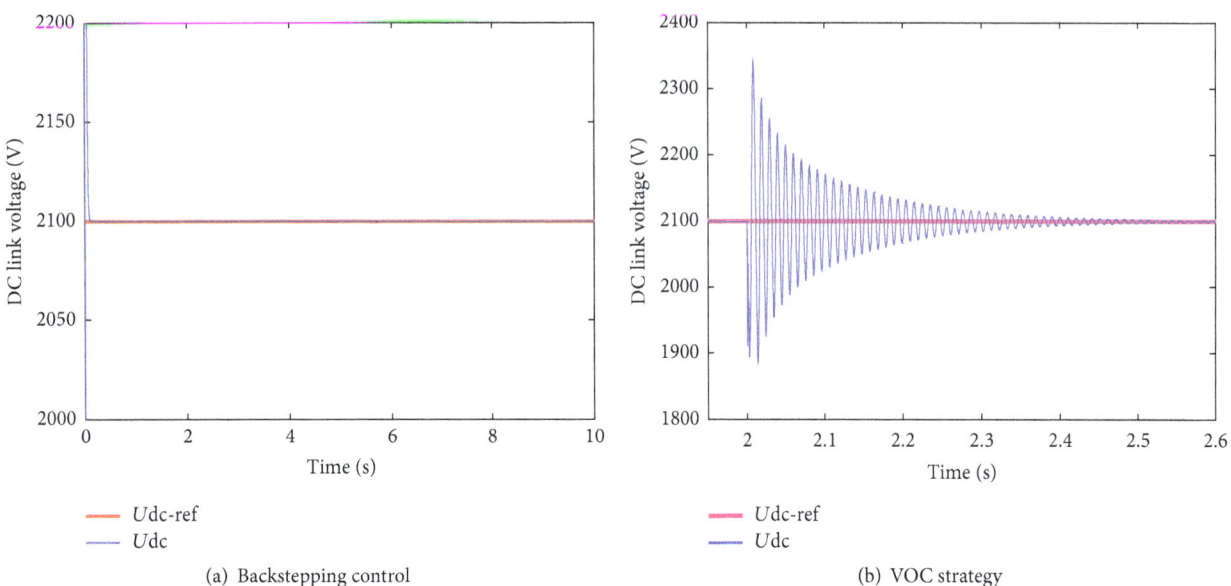

(a) Backstepping control

(b) VOC strategy

FIGURE 15: dc-link voltage (V) (with parameter variations).

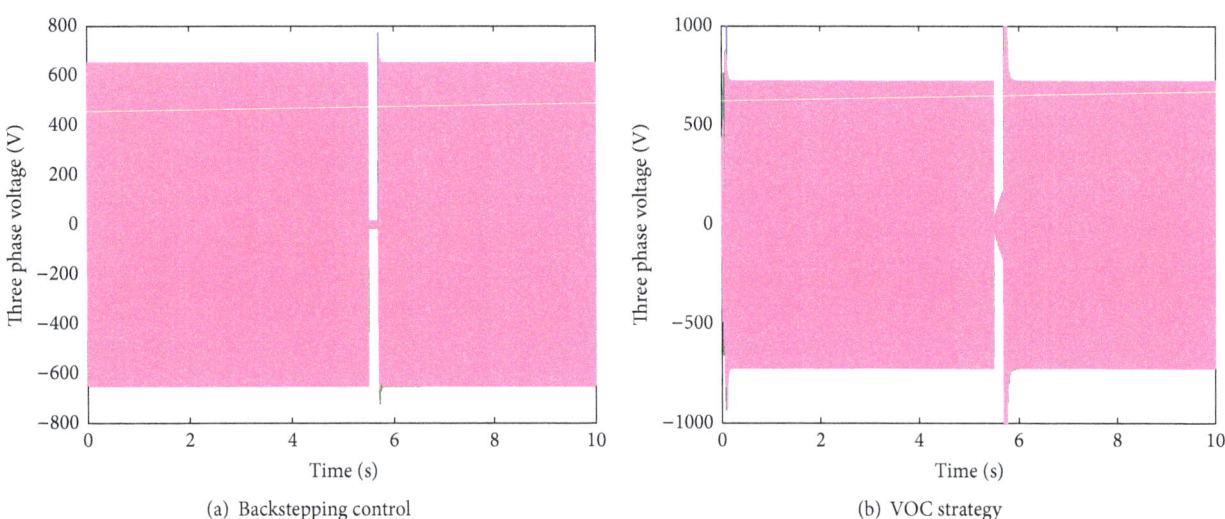

(a) Backstepping control

(b) VOC strategy

FIGURE 16: Waveforms of three-phase voltage at Bus 4.

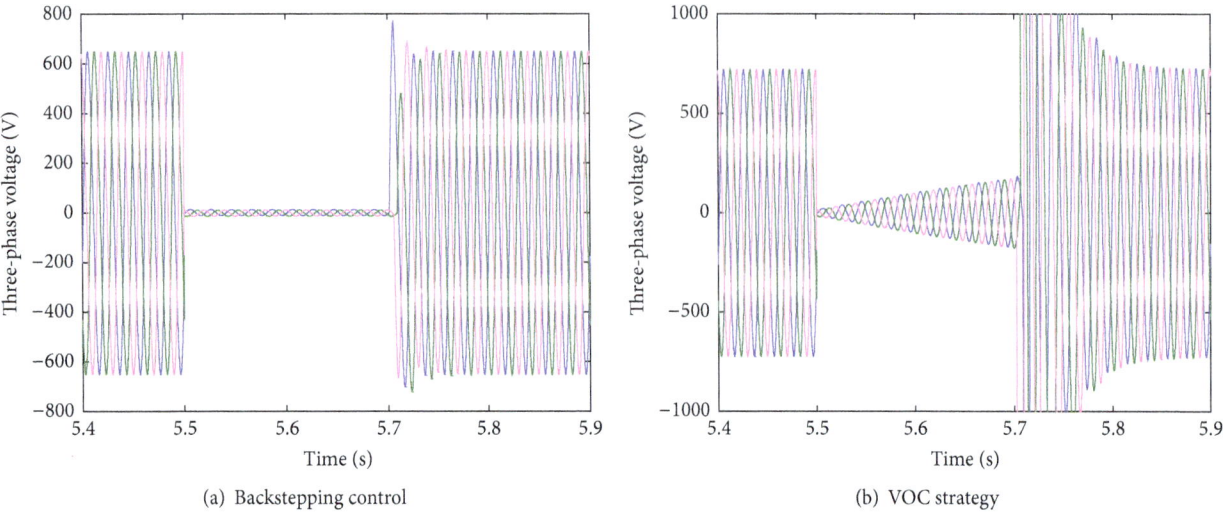

(a) Backstepping control

(b) VOC strategy

FIGURE 17: Zoom-in view of three-phase voltage at Bus 4.

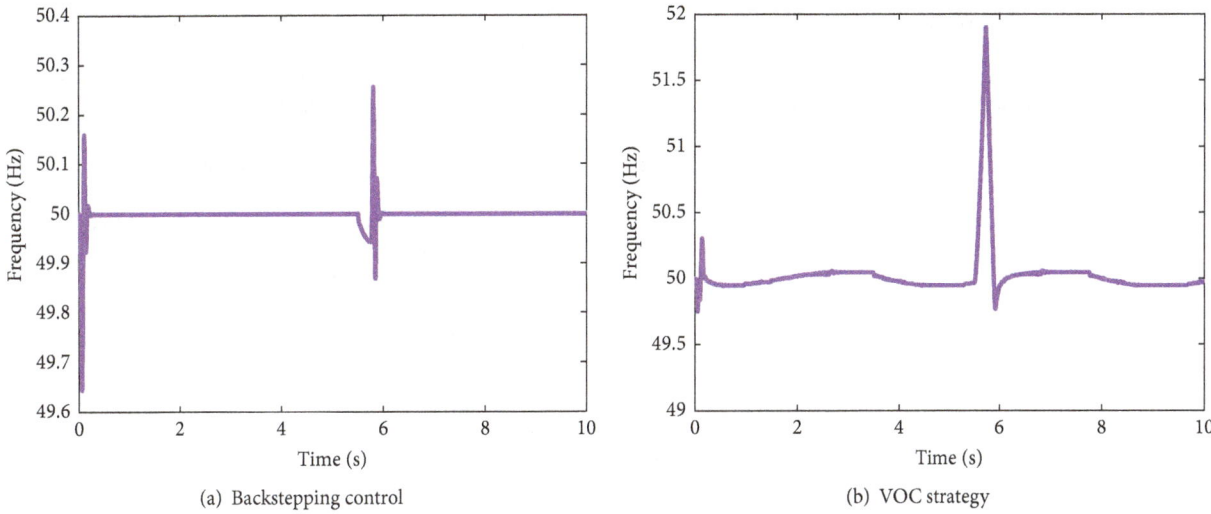

FIGURE 18: Frequency response of the VS-WFS under grid faults.

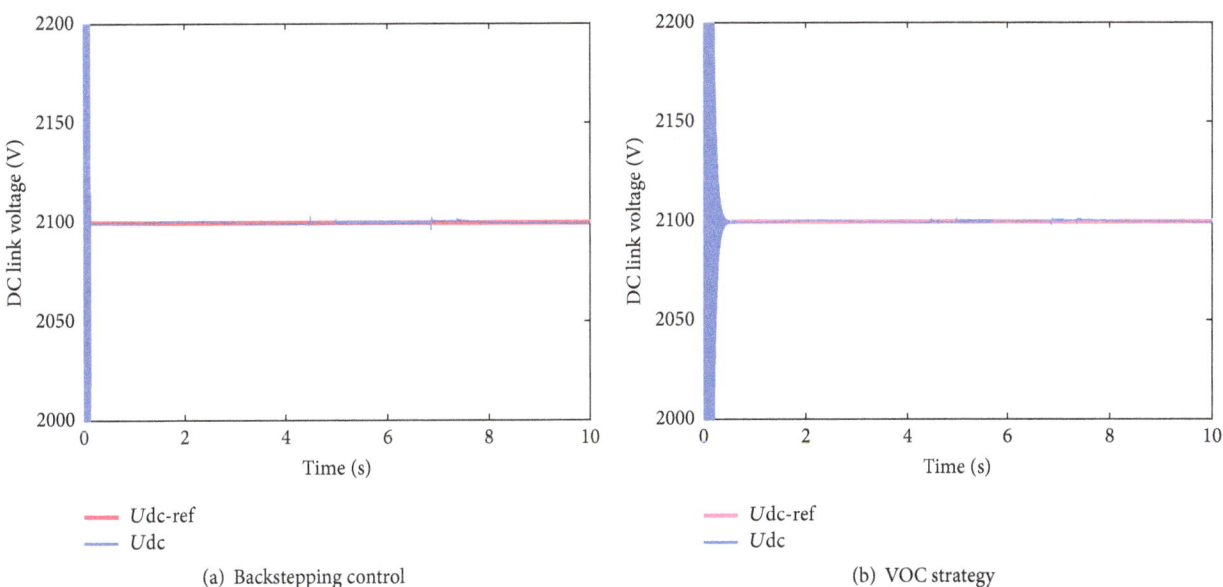

FIGURE 19: dc-link voltage (V).

robustness to parametric variations in the WFS and under severe faults of grid voltage.

Conflicts of Interest

The authors declare that they have no conflicts of interest regarding the publication of this paper.

References

[1] A. Parastar and J.-K. Seok, "High-gain resonant switched-capacitor cell-based DC/DC converter for offshore wind energy systems," *IEEE Transactions on Power Electronics*, vol. 30, no. 2, pp. 644–656, 2015.

[2] P. Yu, W. Zhou, H. Sun, J. Bai, J. Liu, and Y. Liu, "A new pattern wind power suppression system based on hybrid energy storage techniques," *Energy Studies Review*, vol. 4, no. 1, pp. 1–24, 2013.

[3] Y. Errami, M. Ouassaid, and M. Maaroufi, "A performance comparison of a nonlinear and a linear control for grid connected PMSG wind energy conversion system," *International Journal of Electrical Power & Energy Systems*, vol. 68, pp. 180–194, 2015.

[4] F. Luo, K. Meng, Z. Y. Dong, Y. Zheng, Y. Chen, and K. P. Wong, "Coordinated operational planning for wind farm with battery energy storage system," *IEEE Transactions on Sustainable Energy*, vol. 6, no. 1, pp. 253–262, 2015.

[5] A. M. Kassem, "Modeling and control design of a stand alone wind energy conversion system based on functional model predictive control," *Energy Studies Review*, vol. 3, no. 3, pp. 303–323, 2012.

[6] Y. Errami, M. Ouassaid, M. Cherkaoui, and M. Maaroufi, "Variable structure sliding mode control and direct torque control of wind power generation system based on the pm synchronous generator," *Journal of Electrical Engineering*, vol. 66, no. 3, pp. 121–131, 2015.

[7] X. She, A. Q. Huang, F. Wang, and R. Burgos, "Wind energy system with integrated functions of active power transfer, reactive power compensation, and voltage conversion," *IEEE Transactions on Industrial Electronics*, vol. 60, no. 10, pp. 4512–4524, 2013.

[8] M. Benchagra, M. Hilal, Y. Errami, M. Ouassaid, and M. Maaroufi, "Modeling and control of SCIG based variable-speed with power factor control," *International Review on Modelling and Simulations*, vol. 4, no. 3, pp. 1007–1014, 2011.

[9] L. Peng, H. Ma, Y. Li, F. Bruno, and L. Xu, "Reconfiguration of control strategies for high power DFIG wind turbine system with grid fault ride-through capability," in *Proceedings of the 2013 15th European Conference on Power Electronics and Applications, EPE 2013*, fra, September 2013.

[10] K. Nishida, T. Ahmed, S. Mekhilef, and M. Nakaoka, "Cost-effective high-reliability power-conditioning system used for grid integration of variable-speed wind turbine," in *Proceedings of the 1st International Future Energy Electronics Conference, IFEEC 2013*, pp. 530–535, twn, November 2013.

[11] M. Hilal, Y. Errami, M. Benchagra, M. Maaroufi, M. Cherkaoui, and M. Ouassaid, "Doubly fed induction generator wind farm fault ride-through capability," in *Proceedings of the 2012 International Conference on Multimedia Computing and Systems, ICMCS 2012*, pp. 1079–1082, May 2012.

[12] J. Chen, J. Chen, and C. Gong, "On optimizing the aerodynamic load acting on the turbine shaft of pmsg-based direct-drive wind energy conversion system," *IEEE Transactions on Industrial Electronics*, vol. 61, no. 8, pp. 4022–4031, 2014.

[13] O. Alizadeh and A. Yazdani, "A strategy for real power control in a direct-drive PMSG-based wind energy conversion system," *IEEE Transactions on Power Delivery*, vol. 28, no. 3, pp. 1297–1305, 2013.

[14] Y. Errami, M. Ouassaid, and M. Maaroufi, "Optimal Power Control Strategy of Maximizing Wind Energy Tracking and Different Operating Conditions for Permanent Magnet Synchronous Generator Wind Farm," in *Proceedings of the International Conference on Technologies and Materials for Renewable Energy, Environment and Sustainability, TMREES 2015*, pp. 477–490, lbn, April 2015.

[15] S. Alshibani, V. G. Agelidis, and R. Dutta, "Lifetime cost assessment of permanent magnet synchronous generators for mw level wind turbines," *IEEE Transactions on Sustainable Energy*, vol. 5, no. 1, pp. 10–17, 2014.

[16] X.-C. Li, Y.-N. Zhao, and X. Zui-Bing, "Nonlinear control of permanent magnet wind turbine generation (PSMG)," in *Proceedings of the 2013 International Conference on Power, Energy and Control, ICPEC 2013*, pp. 658–661, ind, February 2013.

[17] S. Yang and L. Zhang, "Modeling and control of the PMSG wind generation system with a novel controller," in *Proceedings of the 2013 3rd International Conference on Intelligent System Design and Engineering Applications, ISDEA 2013*, pp. 946–949, chn, January 2013.

[18] S. Zhou, J. Liu, L. Zhou, and Y. Zhu, "Improved DC-link voltage control of PMSG WECS based on feedback linearization under grid faults," in *Proceedings of the 28th Annual IEEE Applied Power Electronics Conference and Exposition, APEC 2013*, pp. 2895–2899, March 2013.

[19] Z. Zhang, Y. Zhao, W. Qiao, and L. Qu, "A space-vector-modulated sensorless direct-torque control for direct-drive PMSG wind turbines," *IEEE Transactions on Industry Applications*, vol. 50, no. 4, pp. 2331–2341, 2014.

[20] C. Huang, F. Li, T. Ding, Z. Jin, and X. Ma, "Second-order cone programming-based optimal control strategy for wind energy conversion systems over complete operating regions," *IEEE Transactions on Sustainable Energy*, vol. 6, no. 1, pp. 263–271, 2015.

[21] G. Buticchi, E. Lorenzani, F. Immovilli, and C. Bianchini, "Active rectifier with integrated system control for microwind power systems," *IEEE Transactions on Sustainable Energy*, vol. 6, no. 1, pp. 60–69, 2015.

[22] F. Bu, Y. Hu, W. Huang, S. Zhuang, and K. Shi, "Wide-speed-range-operation dual stator-winding induction generator DC generating system for wind power applications," *IEEE Transactions on Power Electronics*, vol. 30, no. 2, pp. 561–573, 2015.

[23] F. Blaabjerg and K. Ma, "Future on power electronics for wind turbine systems," *IEEE Journal of Emerging and Selected Topics in Power Electronics*, vol. 1, no. 3, pp. 139–152, 2013.

[24] K. Ma, M. Liserre, and F. Blaabjerg, "Comparison of multi-MW converters considering the determining factors in wind power application," in *Proceedings of the 5th Annual IEEE Energy Conversion Congress and Exhibition, ECCE 2013*, pp. 4754–4761, September 2013.

[25] I. Jlassi, J. O. Estima, S. Khojet El Khil, N. Mrabet Bellaaj, and A. J. Marques Cardoso, "Multiple open-circuit faults diagnosis in back-to-back converters of PMSG drives for wind turbine systems," *IEEE Transactions on Power Electronics*, vol. 30, no. 5, pp. 2689–2702, 2015.

[26] E. Giraldo and A. Garces, "An adaptive control strategy for a wind energy conversion system based on PWM-CSC and PMSG," *IEEE Transactions on Power Systems*, vol. 29, no. 3, pp. 1446–1453, 2014.

[27] Y. Errami, M. Ouassaid, and M. Maaroufi, "Modelling and optimal power control for permanent magnet synchronous generator wind turbine system connected to utility grid with fault conditions," *World Journal of Modelling and Simulation*, vol. 11, no. 2, pp. 123–135, 2015.

[28] T. H. Nguyen and D.-C. Lee, "Advanced fault ride-through technique for PMSG wind turbine systems using line-side converter as STATCOM," *IEEE Transactions on Industrial Electronics*, vol. 60, no. 7, pp. 2842–2850, 2013.

[29] N. M. A. Freire and A. J. M. Cardoso, "A fault-tolerant direct controlled pmsg drive for wind energy conversion systems," *IEEE Transactions on Industrial Electronics*, vol. 61, no. 2, pp. 821–834, 2014.

[30] R. Li and D. Xu, "Parallel operation of full power converters in permanent-magnet direct-drive wind power generation system," *IEEE Transactions on Industrial Electronics*, vol. 60, no. 4, pp. 1619–1629, 2013.

[31] C. Xia, Z. Wang, T. Shi, and Z. Song, "A novel cascaded boost chopper for the wind energy conversion system based on the permanent magnet synchronous generator," *IEEE Transactions on Energy Conversion*, vol. 28, no. 3, pp. 512–522, 2013.

[32] K.-H. Kim, T. L. Van, D.-C. Lee, S.-H. Song, and E.-H. Kim, "Maximum output power tracking control in variable-speed wind turbine systems considering rotor inertial power," *IEEE Transactions on Industrial Electronics*, vol. 60, no. 8, pp. 3207–3217, 2013.

[33] Y. Errami, M. Ouassaid, and M. Maaroufi, "Control of grid connected PMSG based variable speed wind energy conversion system," *International Review on Modelling and Simulations*, vol. 5, no. 2, pp. 655–664, 2012.

[34] H. Polinder, D. Bang, R. P. J. O. M. van Rooij, A. S. McDonald, and M. A. Mueller, "10 MW Wind Turbine Direct-Drive

Generator Design with Pitch or Active Speed Stall Control," in *Proceedings of the IEEE International Conference On Electric Machines & Drives (IEMDC'07)*, vol. 2, pp. 1390–1395, 2007.

[35] C. J. Spruce and J. K. Turner, "Tower vibration control of active stall wind turbines," *IEEE Transactions on Control Systems Technology*, vol. 21, no. 4, pp. 1049–1066, 2013.

[36] K.-H. Kim, Y.-C. Jeung, D.-C. Lee, and H.-G. Kim, "LVRT scheme of PMSG wind power systems based on feedback linearization," *IEEE Transactions on Power Electronics*, vol. 27, no. 5, pp. 2376–2384, 2012.

[37] S. Li, T. A. Haskew, R. P. Swatloski, and W. Gathings, "Optimal and direct-current vector control of direct-driven PMSG wind turbines," *IEEE Transactions on Power Electronics*, vol. 27, no. 5, pp. 2335–2337, 2012.

[38] Y. Errami, M. Maaroufi, and M. Ouassaid, "A MPPT vector control of electric network connected Wind Energy Conversion System employing PM Synchronous Generator," in *Proceedings of the 1st International Renewable and Sustainable Energy Conference, IRSEC 2013*, pp. 228–233, March 2013.

[39] S. Dong, H. Li, and Y. Wang, "Low voltage ride through capability enhancement of PMSG-based wind turbine," in *Proceedings of the International Conference on Sustainable Power Generation and Supply (SUPERGEN 2012)*, pp. 66-66, Hangzhou, China.

[40] S. Alepuz, A. Calle, S. Busquets-Monge, S. Kouro, and B. Wu, "Use of stored energy in PMSG rotor inertia for low-voltage ride-through in back-to-back NPC converter-based wind power systems," *IEEE Transactions on Industrial Electronics*, vol. 60, no. 5, pp. 1787–1796, 2013.

[41] Y. Errami, M. Maaroufi, M. Cherkaoui, and M. Ouassaid, "Maximum power point tracking strategy and direct torque control of permanent magnet synchronous generator wind farm," in *Proceedings of the 2012 International Conference on Complex Systems (ICCS)*, pp. 1–6, Agadir, Morocco, November 2012.

[42] M. Krstic, I. Kanellakopoulos, and P. V. Kokotovic, *Nonlinear and Adaptive*, Control Design Book, John Wiley & Sons, New York, NY, USA, 1995.

[43] G. Foo and M. F. Rahman, "Direct torque and flux control of an IPM synchronous motor drive using a backstepping approach," *IET Electric Power Applications*, vol. 3, no. 5, pp. 413–421, 2009.

[44] N. Khan, S. F. Rabbi, M. J. Hinchey, and M. A. Rahman, "Adaptive backstepping based maximum power point tracking control for a variable speed marine current energy conversion system," in *Proceedings of the 2013 26th IEEE Canadian Conference on Electrical and Computer Engineering, CCECE 2013*, can, May 2013.

[45] C.-S. Ting, Y.-N. Chang, B.-W. Shi, and J.-F. Lieu, "Adaptive backstepping control for permanent magnet linear synchronous motor servo drive," *IET Electric Power Applications*, vol. 9, no. 3, pp. 265–279, 2015.

[46] R.-J. Wai, C.-Y. Lin, W.-C. Wu, and H.-N. Huang, "Design of backstepping control for high-performance inverter with stand-alone and grid-connected power-supply modes," *IET Power Electronics*, vol. 6, no. 4, pp. 752–762, 2013.

[47] S.-Y. Ruan, G.-J. Li, X.-H. Jiao, Y.-Z. Sun, and T. T. Lie, "Adaptive control design for VSC-HVDC systems based on backstepping method," *Electric Power Systems Research*, vol. 77, no. 5-6, pp. 559–565, 2007.

[48] A. Karthikeyan, S. K. Kummara, C. Nagamani, and G. S. Ilango, "Power control of grid connected doubly fed induction generator using Adaptive BackStepping approach," in *Proceedings of the 2011 10th International Conference on Environment and Electrical Engineering, EEEIC.EU 2011*, ita, May 2011.

[49] Y. Errami, M. Ouassaid, and M. Maaroufi, "Modeling and variable structure power control of PMSG based variable speed wind energy conversion system," *Journal of Optoelectronic and Advanced Materials*, vol. 15, no. 11-12, pp. 1248–1255, 2013.

[50] C. Pradhan and C. N. Bhende, "Adaptive deloading of stand-alone wind farm for primary frequency control," *Energy Studies Review*, vol. 6, no. 1, pp. 109–127, 2014.

[51] Y. Errami, M. Hilal, M. Benchagra, M. Ouassaid, and M. Maaroufi, "Nonlinear control of MPPT and grid connected for variable speed wind energy conversion system based on the PMSG," *Journal of Theoretical and Applied Information Technology*, vol. 39, no. 2, pp. 204–217, 2012.

[52] Y. Errami, M. Benchagra, M. Hillal, M. Ouassaid, and M. Maaroufi, "MPPT strategy and direct torque control of PMSG used for variable Speed Wind Energy Conversion System," *International Review on Modelling and Simulations*, vol. 5, no. 2, pp. 887–898, 2012.

[53] J. E. Slotine and W. Li, *Applied Nonlinear Control*, Prentice Hall, Englewood Cliffs, New Jersey, NJ, USA, 1991.

[54] Y. Errami, M. Hilal, M. Benchagra, M. Maaroufi, and M. Ouassaid, "Nonlinear control of MPPT and grid connected for wind power generation systems based on the PMSG," in *Proceedings of the 2012 International Conference on Multimedia Computing and Systems, ICMCS 2012*, pp. 1055–1060, May 2012.

[55] M. Benchagra, M. Hilal, Y. Errami, M. Maaroufi, and M. Ouassaid, "Nonlinear control of DC-bus voltage and power for voltage source inverter," in *Proceedings of the 2012 International Conference on Multimedia Computing and Systems, ICMCS 2012*, pp. 1049–1054, May 2012.

[56] Y. Errami, M. Ouassaid, and M. Maaroufi, "Control of a PMSG based wind energy generation system for power maximization and grid fault conditions," in *Proceedings of the International Conference on Mediterranean Green Energy Forum, MGEF 2013*, pp. 220–229, June 2013.

15

Model Predictive Control Method with Constant Switching Frequency to Reduce Common-Mode Voltage for PMSM Drives

Hao Li,[1] **Shuo Chen,**[1] **Xiang Wu** ⓘ**,**[1] **and Guojun Tan**[1,2]

[1]*School of Electrical and Power Engineering, China University of Mining and Technology, Xuzhou 86221008, China*
[2]*China Mining Drives & Automation Co., Ltd., Xuzhou 221008, China*

Correspondence should be addressed to Xiang Wu; cumtwuxiang@qq.com

Academic Editor: Antonio J. Marques Cardoso

A model predictive control method to reduce the common-mode voltage (MPC-RCMV) with constant switching frequency for PMSM drives is proposed in this paper. Four nonzero VVs are adopted in future control period and the switching sequence is designed to ensure the switching frequency is fixed and equal to the control frequency. By substituting the finite-control nonzero voltage vectors in the current predictive model, a current predictive error space vector diagram is obtained to determine the adopted four VVs. The duty ratio calculating method for the selected four VVs is studied. Compared with the conventional MPC-RCMV method, the current and torque ripples are greatly reduced and the switching frequency is fixed. The simulation and experiment results validate the effectiveness of the proposed method.

1. Introduction

The common-mode voltage (CMV) of the permanent magnet synchronous motor (PMSM) drive system with two-level three-phase voltage source inverter (VSI), as shown in Figure 1, can be approximately calculated according to the voltage between the midpoint of the dc-link capacitor and the neutral point of the three-phase load [1]. The high-frequency CMV can cause many problems including increasing the leakage currents [2], causing damage of the motor shaft [3], and increasing electromagnetic interference [4].

To address this problem, different solutions by improving the control strategy of the PMSM have been studied. In one aspect, many CMV reduction pulse-width modulation (CMVRPWM) strategies have been proposed [5–10]. The research results in [7] suggest that only the active zero-state PWM1 (AZSPWM1) [5] and the near-state PWM (NSPWM) [6] are practical in most cases among the earlier developed CMVRPWM strategies. In addition, the newly developed optimized CMVRPWM strategies with the consideration of the current ripple losses and switching losses minimizing have been proposed in [8]. All of the above CMVRPWM strategies can restrict the amplitude of the

CMV within sixth of the dc-link voltage. Some CMVRPWM strategies are proposed to reduce the low-order components of CMV [11, 12]. A virtual space vector PWM (VSVPWM) for the reduction of CMV in both magnitude and third-order component is studied in [13].

On the other hand, the model predictive control methods [14, 15] to reduce the CMV (MPC-RCMV) have also been proposed in [16–22]. The MPC-RCMV method proposed in [16] adopts only six nonzero voltage vectors (VVs) to calculate the cost function and the CMV amplitude is reduced without utilizing zero VVs. In [17], the optimal VV is selected from three adjacent nonzero VVs, and this MPC-RCMV method needs less compute efforts. However, it has been proven that many spikes may appear in the current ripple with the method in [17]. The problem has been analyzed in [18], and an improved MPC-RCMV method with four candidate VVs is developed to eliminate the current spikes. To eliminate the CMV spikes caused by the dead time, the method proposed in [19] preexcludes, from the candidates for future vectors, those voltage vectors which can increase the common-mode voltage during the dead time. A MPC-RCMV strategy without calculating the cost function is proposed in [20], where the nonzero optimal

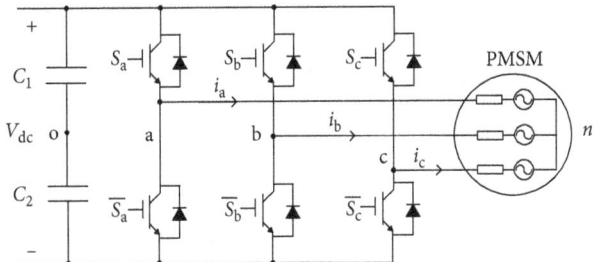

FIGURE 1: PMSM drive system with three-phase two-level VSI.

future VV is determined according to the basis of an inverse dynamics model. A two-vector based MPC-RCMV is proposed in [21], where two nonzero VVs are selected at every control period. The satisfactory load current ripple performance can be yielded; however, it is computationally intensive and the average switching frequency is still lower than $1/T_s$. However, with these above MPC-RCMV strategies, the current and torque ripples may be larger than the conventional field-oriented control (FOC) [23, 24] with the same control frequency ($1/T_s$), mainly owing to that the average switching frequency is far smaller than $1/T_s$. In addition, the switching frequency of these methods are not fixed, which make it hard to analyze the harmonic features and design the filter. Many model predictive controllers with fixed switching frequency have been studied. The fixed switching frequency schemes for finite-control-set MPC are studied in [25, 26]. Modulated MPC strategies [27, 28] have been developed to realize the target of constant switching frequency with the aid of a modulator. However, these fixed switching frequency MPC strategies in [25–28] have not considered the target of CMV reduction, and the amplitude of the CMV reaches to half of the dc-link voltages owing to the adoption of the zero VVs.

In this paper, a MPC-RCMV method with constant switching frequency is proposed. Unlike the conventional MPC-RCMV methods where one or two nonzero VVs are applied in future control period, four nonzero VVs are adopted and the switching sequence is designed to ensure

the average switching frequency is fixed and equal to $1/T_s$. A current predictive error space vector diagram is obtained by substituting the finite-control nonzero VVs in the current predictive model, and according to it, the sector where the coordinate origin locates is calculated to determine the adopted four VVs. The duty ratio calculating method for the selected four VVs is studied. With the proposed method, the current reference can be fast and accurately tracked and the current ripples are much smaller than the conventional MPC-RCMV methods. Compared with the MPC methods with fixed switching frequency studied in [25–28], the proposed method can limit the amplitude of the CMV within sixth of the dc-link voltage. The effectiveness of the proposed method is validated by simulation and experiment.

2. Conventional MPC-RCMV Method

The conventional MPC-RCMV method for PMSM drives is shown in Figure 2. The current predictive model is given in the following equation:

$$
\begin{cases}
i_q(k+1) = \dfrac{T_s}{L_q}\left[-Ri_q - \omega_e\left(L_d i_d + \psi_f\right)\right] + i_q + \dfrac{T_s}{L_q}u_q, \\[2mm]
i_d(k+1) = \dfrac{T_s}{L_d}\left[-Ri_q + \omega_e L_q i_q\right] + i_d + \dfrac{T_s}{L_d}u_d,
\end{cases}
\tag{1}
$$

where L_d and L_q are the direct axis and quadrature axis inductances, ψ_f is the permanent magnet flux linkage, i_d, i_q, u_d, and u_q are the load current and output voltage of the VSI at the kth sampling instant, $i_d(k+1)$ and $i_q(k+1)$ are the predicted direct axis and quadrature axis currents at the $k+1$th sampling instant, R is the stator resistance, T_s is sample period, and ω_e is angular speed.

To compensate the delay caused by the sampling and calculation in digital control, the typical method as proposed in [29, 30] is adopted, i.e., the direct axis and quadrature axis currents at the $k+2$th sampling instant which are predicted as given in Equation (2) are adopted to calculate the cost function as Equation (3).

$$
\begin{cases}
i_q(k+2) = \dfrac{T_s}{L_q}\left[-Ri_q(k+1) - \omega_e\left(L_d i_d(k+1) + \psi_f\right)\right] + i_q(k+1) + \dfrac{T_s}{L_q}u_q(k+1), \\[3mm]
i_d(k+2) = \dfrac{T_s}{L_d}\left[-Ri_d(k+1) - \omega_e L_q i_q(k+1)\right] + i_d(k+1) + \dfrac{T_s}{L_d}u_d(k+1),
\end{cases}
\tag{2}
$$

$$
J = \left|i_q^* - i_q(k+2)\right| + \left|i_d^* - i_d(k+2)\right|.
\tag{3}
$$

For the MPC strategy, the voltage vector (VV) that minimized the cost function is chosen to be applied in the next control period. The difference between the conventional MPC and the MPC-RCMV method is the VV selection in the finite-control set. For the MPC-RCMV method, only nonzero VVs are included in the finite-control set to limit the amplitude of the CMV within sixth of the dc-link

voltage. According to the number of the VVs in the finite-control set, the conventional MPC-RCMV methods can be divided into three categories as shown in Figure 3, i.e., the 6 VV, 3 VV, and 4 VV strategies. In Figure 3, the arrows define the possible changes of the switching states in the next control period. The finite-control set of 6 VV strategy includes all the six nonzero VVs. As shown in

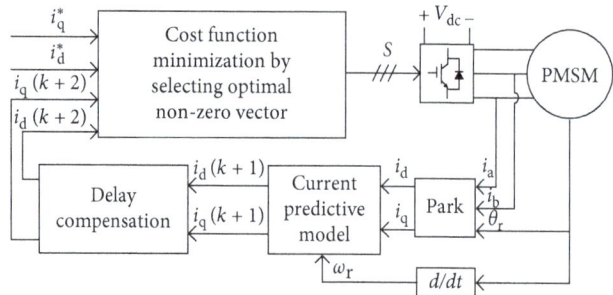

FIGURE 2: The control diagram of the conventional MPC-RCMV method.

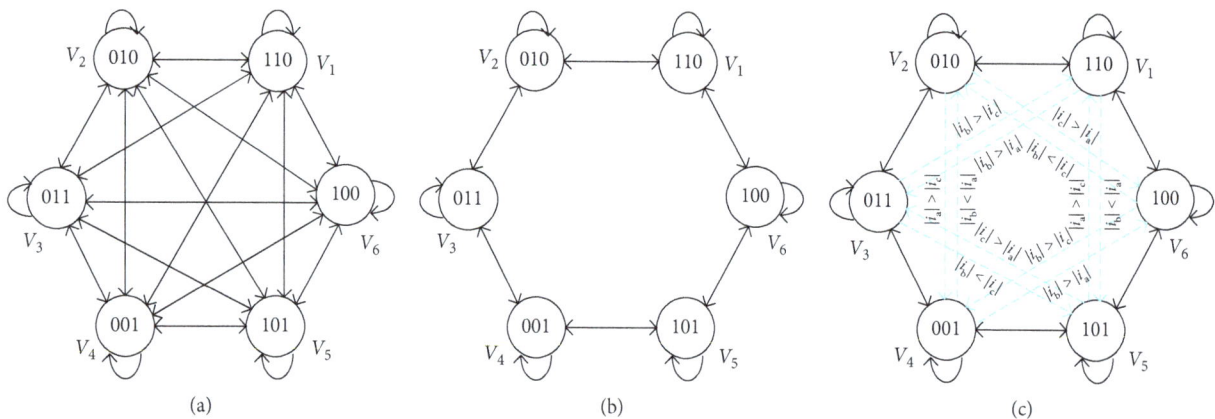

FIGURE 3: Switching state paragraph of the MPC-RCMV: the arrows define the possible changes of the switching states. (a) The 6 VV strategy. (b) The 3 VV strategy. (c) The 4 VV strategy.

Figure 3(a), the optimal VV of the next control sample can be selected as any nonzero VV no matter which VV is applied in the current period. The switching frequency of the 6 VV strategy is relatively high due to too many switches between nonadjacent VVs, e.g., all the power switches turn off or turn on with the switching state changing from 100 to 011. As shown in Figure 3(b), the finite-control set of 3 VV strategy is determined by the VV applied in the current period and the number of the selectable VVs is 3. The VV that applied in the current period and its two adjacent VVs can be selected for the next control period. Compared with the 6 VV strategy, the switching frequency of the 3 VV strategy is limited since at most one switch change is permitted, in addition, the computation amount is reduced by 50% owing to the number of cost function calculation changes from 6 to 3. However, it has been proven that many spikes may appear in the current ripple when 3 VV strategy is utilized due to the lack of enough alternative VVs [16]. To solve this problem, an improved 4 VV strategy is developed in [16] where four VVs are chosen as candidate VVs at the same time as shown in Figure 3(c). Compared with the 3 VV strategy, a nonadjacent VV is included in the finite-control set to ensure that the motor current can be adjusted towards two directions.

It should be noted that there are some problems for the conventional MPC-RCMV methods. Firstly, the switching frequency of these methods is not fixed, which makes the

voltage and current spectrum spread over a wide range of frequencies. Secondly, the average switching frequency is smaller than the control frequency $(1/T_s)$ which should be set small enough considering the computation time, e.g., the results in [16] indicate that the average frequency of the 4 VV strategy is no more than 2 kHz with the control period set as $100\,\mu s$. Thus, the conventional MPC-RCMV methods are not suitable in the fields where the average switching sequence needs to be high enough to realize the control targets such as decrease the torque ripples and the noises.

3. Proposed MPC-RCMV Method with Constant Switching Frequency

The control diagram of the proposed MPC-RCMV method with constant switching frequency is shown in Figure 4. It includes four parts, i.e., current predictive model, delay compensation, voltage vector selection and duty ratio calculation, and switching sequence producing part. Unlike the conventional MPC-RCMV method where only one active VV is applied in a single control period, the proposed method adopts 4 active VVs to keep the switching frequency fixed. The current predictive model part calculates $i_d(k+1)$ and $i_q(k+1)$ by Equation (4) where u_q^{av} and u_d^{av} are the average d- and q-axis voltages in the whole control period. The u_q^{av} and u_d^{av} can be calculated according to the selected

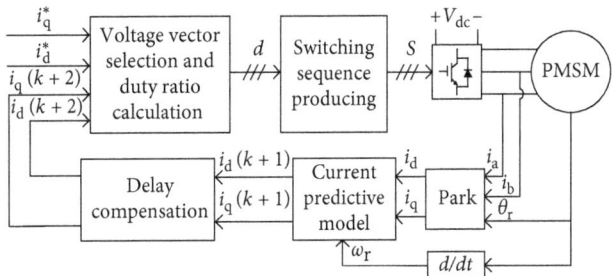

FIGURE 4: The control diagram of the proposed MPC-RCMV method with constant switching frequency.

VVs and their duty ratios which are obtained in the last control period by the proposed MPC-RCMV method.

$$
\begin{cases}
i_q(k+1) = \dfrac{T_s}{L_q}\left[-Ri_q - \omega_e\left(L_d i_d + \psi_f\right)\right] + i_q + \dfrac{T_s}{L_q}u_q^{av}, \\[3mm]
i_d(k+1) = \dfrac{T_s}{L_d}\left[-Ri_q + \omega_e L_q i_q\right] + i_d + \dfrac{T_s}{L_d}u_d^{av}.
\end{cases}
$$
(4)

Then, the direct axis and quadrature axis currents at the $(k+2)$th sampling instant for different active VVs can be calculated by substituting Equation (4) in Equation (2).

The predictive current errors (PCE) are defined as

$$
\begin{cases}
e_{id} = i_d(k+2) - i_d^*, \\
e_{iq} = i_q(k+2) - i_q^*.
\end{cases}
$$
(5)

By combining Equations (2) and (4), the PCE equation can be written as

$$
\begin{cases}
e_{id} = C_d + \dfrac{u_d(k+2)T_s}{L_d}, \\[3mm]
e_{iq} = C_q + \dfrac{u_q(k+2)T_s}{L_q},
\end{cases}
$$
(6)

where C_d and C_q satisfy

$$
\begin{cases}
C_d = \dfrac{T_s}{L_d}\left[-Ri_d(k+1) + \omega_e L_q i_q(k+1)\right] + i_d(k+1) - i_d^*, \\[3mm]
C_q = \dfrac{T_s}{L_q}\left[-Ri_q(k+1) - \omega_e\left[L_d i_d(k+1) + \psi_f\right]\right] + i_q(k+1) - i_q^*.
\end{cases}
$$
(7)

According to Equation (6), the values of PCEs are related with the d- and q-axis voltages, and thus, the position of the PCE in the e_{id}-e_{iq} coordinate system, as shown in Figure 5, are changed by selecting different VVs.

In Figure 5, e_1-e_6 represents the PCE vector by applying the VVs V_1-V_6, and e_0 represents the PCE vector of the zero VVs. Owing to the corresponding d- and q-axis voltages for the zero VVs are both 0, the coordinate of e_0 is (C_d, C_q) according to Equation (6). The variables of C_d and C_q are determined by the VV that applied in the current period and

have no relation with the selected VV in the next period, and thus, they can be viewed as constants in Equation (6).

The reference current tracking problem can be solved by selecting appropriate VVs to make both the e_{id} and e_{iq} become zero, i.e, the current can track its reference by making the current error vector reach the coordinate origin as shown in Figure 5. In fact, the optimal VV selecting process of the conventional MPC-RCMV method can be viewed as determining the nearest PCE from the coordinate origin by minimizing the cost function. As an example, the vector V_1 will be selected for the case shown in Figure 5 with the conventional MPC-RCMV method. Owing to the fact that only single VV is selected during one control period for the conventional MPC-RCMV, the selected optimal VV may not ensure the current error vector reach the coordinate origin. In this part, the MPC-RCMV method with constant switching frequency is proposed by adopting four nonzero VVs in one control period to ensure the average current error approach to 0 within one period for the case shown in Figure 5.

A new coordinate system (e'_{id}-e'_{iq} as shown in Figure 5) is defined by the transformation given in following equation:

$$
\begin{cases}
e'_{id} = e_{id} - C_d = \dfrac{V_{id}T_s}{L_d}, \\[3mm]
e'_{id} = e_{iq} - C_q = \dfrac{V_{iq}T_s}{L_q},
\end{cases}
$$
(8)

The reference current error vector (e^*) is defined as the vector $\overrightarrow{O'O}$, and its coordinate is $(-C_d, -C_q)$. The reference current tracking problem can be solved by synthetizing e^* with the PCE vectors (e_1-e_6 as shown in Figure 5).

The vector selection process can be implemented according to the angle of e^*, e_1, e_2, e_3, e_4, e_5, and e_6 in the e'_{id}-e'_{iq} coordinate system which can be calculated by the following equation:

$$
\theta = a\tan 2\left(e'_{id}, e'_{id}\right).
$$
(9)

The sector (S) is defined as shown in Figure 5 and can be obtained according to Table 1, and then, the selected VVs can be determined. In Table 1, the variable $\theta_i(i = 1, 2, 3, 4, 5,$ and 6) and θ^* represent the angle of e_i and e^*, respectively.

According to Equation (8), the current error vector will reach the position defined in Equation (10) if the voltage vector V_i is adopted with the duty ratio d_i. Equation (10) indicates that the change of the current error vector is proportional to

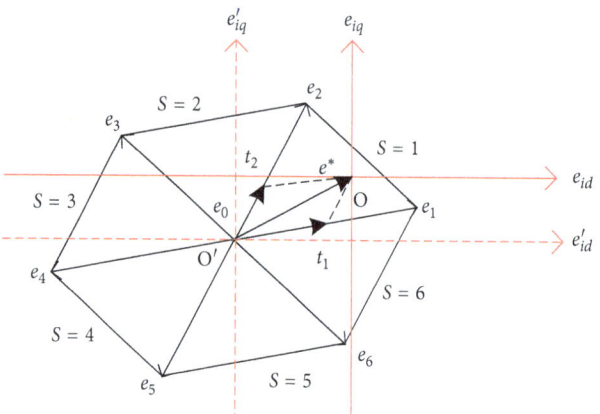

FIGURE 5: Predicted current error space vector diagram.

TABLE 1: Voltage vector selection principle.

The scope of θ^*	S	Selected VVs
$\theta_1 \leq \theta^* < \theta_2$	1	V_1, V_2, V_3, V_6
$\theta_2 \leq \theta^* < \theta_3$	2	V_1, V_2, V_3, V_4
$\theta_3 \leq \theta^* < \theta_4$	3	V_2, V_3, V_4, V_5
$\theta_4 \leq \theta^* < \theta_5$	4	V_3, V_4, V_5, V_6
$\theta_5 \leq \theta^* < \theta_6$	5	V_1, V_4, V_5, V_6
$\theta_6 \leq \theta^* < \theta_1$	6	V_1, V_2, V_5, V_6

the duty ratio of the adopted VV. Accordingly, the reference current tracking can be realized by synthetizing e^* with linear combination of the PCE vector of the selected VVs.

$$\begin{cases} e'_{id} = \dfrac{V_{id}}{L_d} d_i T_s, \\[2ex] e'_{id} = \dfrac{V_{iq}}{L_q} d_i T_s. \end{cases} \tag{10}$$

According to the nearest three-vector principle, e^* can be synthetized by e_0, e_1, and e_2 for the case that S is 1 as shown in Figure 5. Owing to that the zero vectors should be avoided to restrict the CMV amplitude, two opposing PCEs (e_3 and e_6) with equal duty cycle can be utilized to create equivalent e_0. The duty ratios of the adopted VVs can be calculated by Equation (11) if S is 1.

$$\begin{cases} V_{1d}d_1 + V_{2d}d_2 = -\dfrac{C_d L_d}{T_s}, \\[2ex] V_{1q}d_1 + V_{2d}d_2 = -\dfrac{C_q L_q}{T_s}, \\[2ex] d_3 = d_6 = \dfrac{(1 - d_1 - d_2)}{2}, \end{cases} \tag{11}$$

However, the duty ratios calculated by Equation (11) needs to be adjusted as Equation (12) for the case where the reference current error vector (e^*) is outside of the hexagon enclosed by e_1–e_6 as shown in Figure 6.

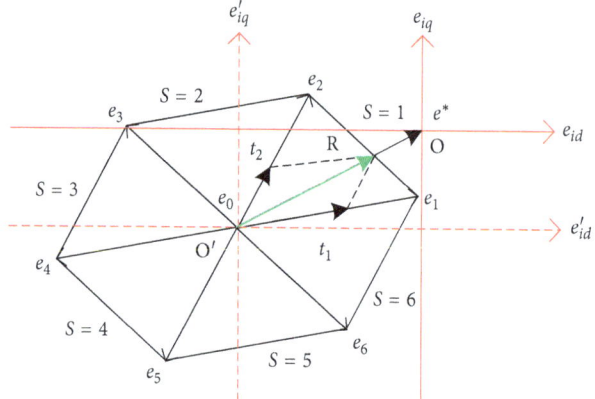

FIGURE 6: Predicted current error space vector diagram.

TABLE 2: Switching sequence design.

S	Switching sequence
1	V_3-V_2-V_1-V_6-V_1-V_2-V_3
2	V_4-V_3-V_2-V_1-V_2-V_3-V_4
3	V_5-V_4-V_3-V_2-V_3-V_4-V_5
4	V_6-V_5-V_4-V_3-V_4-V_5-V_6
5	V_1-V_6-V_5-V_4-V_5-V_6-V_1
6	V_2-V_1-V_6-V_5-V_6-V_1-V_2

$$\begin{cases} d'_1 = \dfrac{d_1}{d_1 + d_2}, \\[2ex] d'_2 = \dfrac{d_2}{d_1 + d_2}, \\[2ex] d'_3 = d'_6 = 0. \end{cases} \tag{12}$$

If e^* locates in other sectors, the duty ratio calculating method is the same as the case where S is 1. After obtaining the duty ratios of the adopted VVs, the switching sequence can be designed as shown in Table 2.

The flowchart of the proposed method is shown in Figure 7(b). The values of $i_q(k+1)$ and $i_d(k+1)$ are firstly calculated by Equation (4), and then, e_{id} and e_{iq} for different

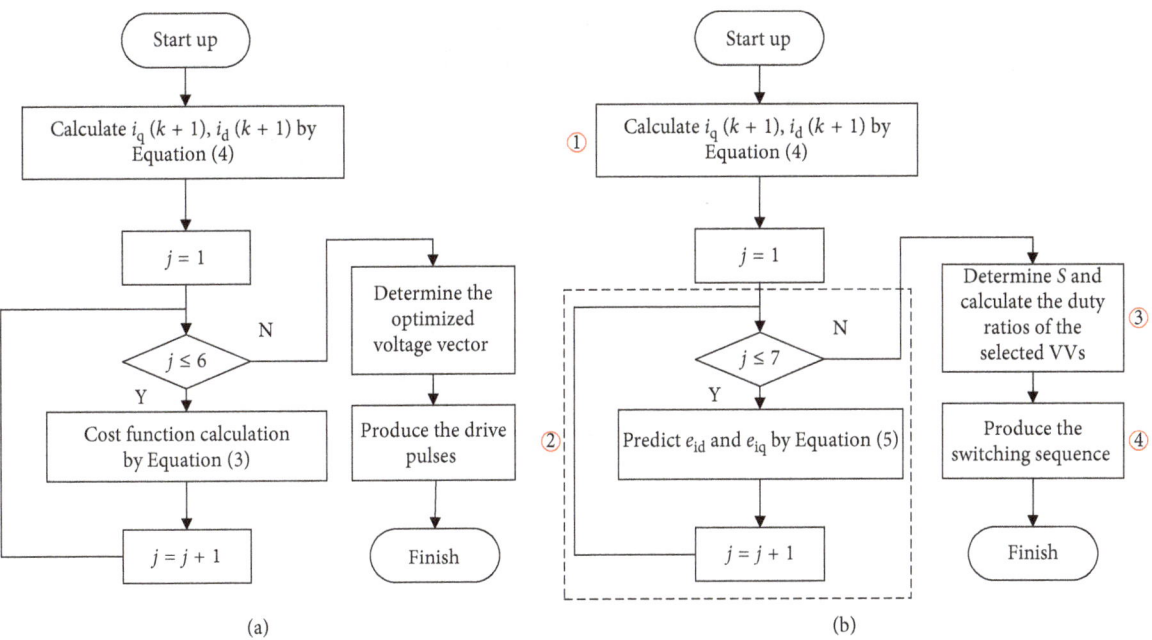

FIGURE 7: Flowchart of the conventional and proposed method. (a) The conventional method. (b) The proposed method.

active VVs can be obtained by Equation (5). The sector and the selected VVs can be determined by Table 1 and the duty ratios of the selected VVs can be calculated by Equations (11) and (12). Finally, the switching sequence can be obtained according to Table 2.

4. Simulation and Experimental Validation

4.1. Simulation Results and Analysis. To validate the effectiveness of the proposed MPC-RCMV with constant switching frequency, the strategy is simulated by Matlab/Simulink and the main parameters are given in Table 3.

To minimize the copper losses of the PMSM, the maximal torque per ampere (MTPA) strategy [31, 32] is adopted and the reference of the d- and q-axis current can be calculated as Equation (13) where i_s^* is the reference of the stator current.

$$\begin{cases} i_d^* = \dfrac{\psi_f}{4(L_q - L_d)} - \sqrt{\dfrac{\psi_f}{16(L_q - L_d)^2} + \dfrac{|i_s^*|^2}{2}}, \\[2ex] i_q^* = \operatorname{sign}(i_s^*)\sqrt{i_s^{*2} - i_d^{*2}}. \end{cases} \quad (13)$$

In the simulation, i_s^* is set as 200 A and 300 A at the interval 0–0.5 s and 0.5–1 s, respectively. The speed of the motor is set as 750 r/min in the simulation.

The line-to-line voltage (V_{ab}), CMV, and three-phase current waveforms for both the conventional 6 VV strategy, and the proposed method are given in Figure 8. The CMV curves in Figure 8 indicate that the amplitude of the CMV has been restricted within sixth of the dc-link voltage with both the conventional and the proposed methods.

TABLE 3: Main parameters of the simulation.

Parameters	Values
Flux linkage	0.225 Wb
D-axis inductor	0.95 mH
Q-axis inductor	2.05 mH
Resistance	0.1 Ω
Pole pairs	4
Sampling time	100 μs

The waveforms of the d- and q-axis current and the electromagnetic torque (T_e) for both the conventional and the proposed methods are shown in Figure 9. It can be seen from Figure 9 that the fluctuations of i_d, i_q, and T_e of the proposed method are far smaller than the conventional method. The comparison results of the peak-to-peak of the fluctuation of i_d, i_q, and T_e are given in Table 4. In the case where T_s of these two methods are both set as 100 μs, the fluctuations of i_d, i_q, and T_e with the proposed method in the case where i_s^* is set 200 A reduce by 95.9%, 97.8%, and 78.2%, respectively. In addition, the fluctuations of i_d, i_q, and T_e in the case where i_s^* is set 300 A reduce by 95.9%, 98.2%, and 74.8%, respectively. In Table 4, the peak-to-peak of the fluctuation of i_d, i_q, and T_e for the conventional method with T_s set as 50 μs is also analyzed, and the fluctuations of i_d, i_q, and T_e are still much bigger than the proposed method with T_s set as 100 μs, which indicate that the control performance of the proposed method is greatly enhanced compared with the conventional method.

The fast Fourier transform analysis results of the phase current for both the conventional and the proposed methods are shown in Figure 10. The results in Figure 10 indicate that the main current harmonics components of the proposed method concentrate on the switching frequency (10 kHz)

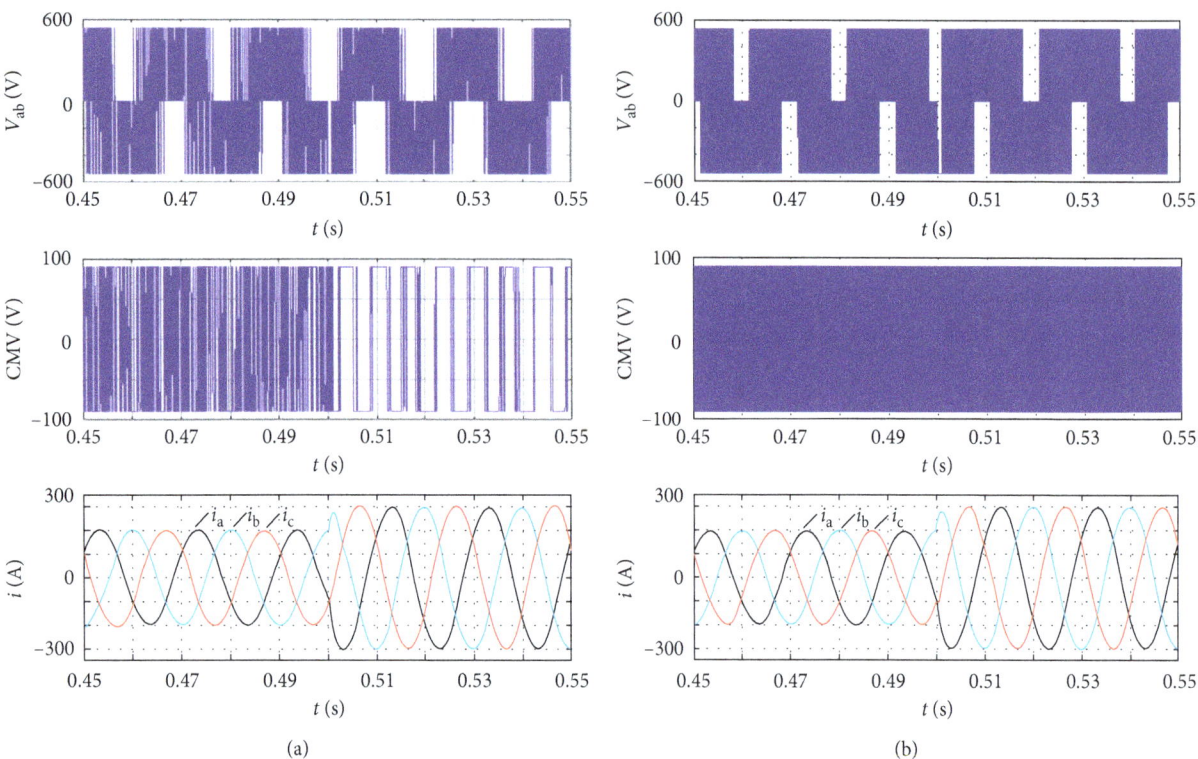

FIGURE 8: The line-to-line voltage (V_{ab}), the common-mode voltage, and three-phase current waveforms. (a) The conventional MPC-RCMV. (b) The proposed MPC-RCMV with constant switching frequency.

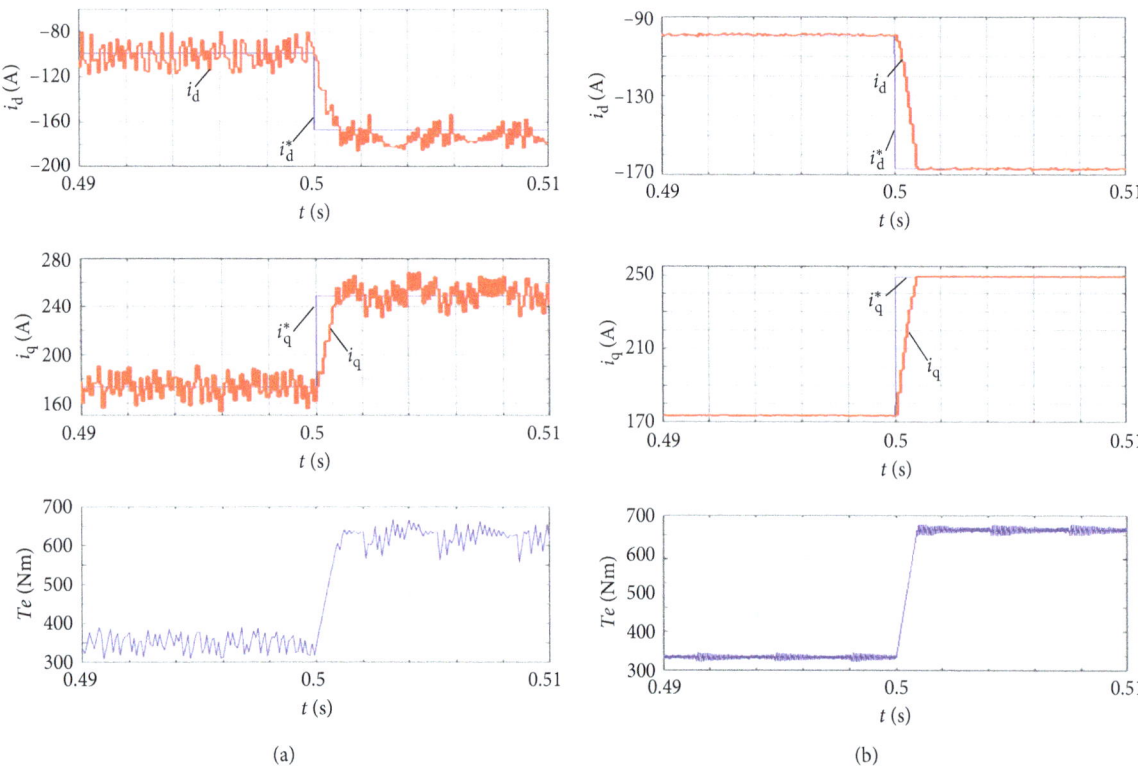

FIGURE 9: d- and q-axis current and the electromagnetic torque waveforms. (a) The conventional MPC-RCMV. (b) The proposed MPC-RCMV with constant switching frequency.

TABLE 4: Comparison results of the peak-to-peak of the fluctuation of i_d, i_q, and T_e.

Peak-to-peak of the fluctuation		The conventional method ($T_s = 100\,\mu s$)	The conventional method ($T_s = 50\,\mu s$)	The proposed method ($T_s = 100\,\mu s$)
i_d	$i_s^* = 200\,A$	36.5 A	18.2 A	1.5 A
	$i_s^* = 300\,A$	34.3 A	18.1 A	1.4 A
i_q	$i_s^* = 200\,A$	40.5 A	19.8 A	0.9 A
	$i_s^* = 300\,A$	33.3 A	17.9 A	0.6 A
T_e	$i_s^* = 200\,A$	91.5 Nm	45.6 Nm	19.9 Nm
	$i_s^* = 300\,A$	97.8 Nm	50.1 Nm	24.6 Nm

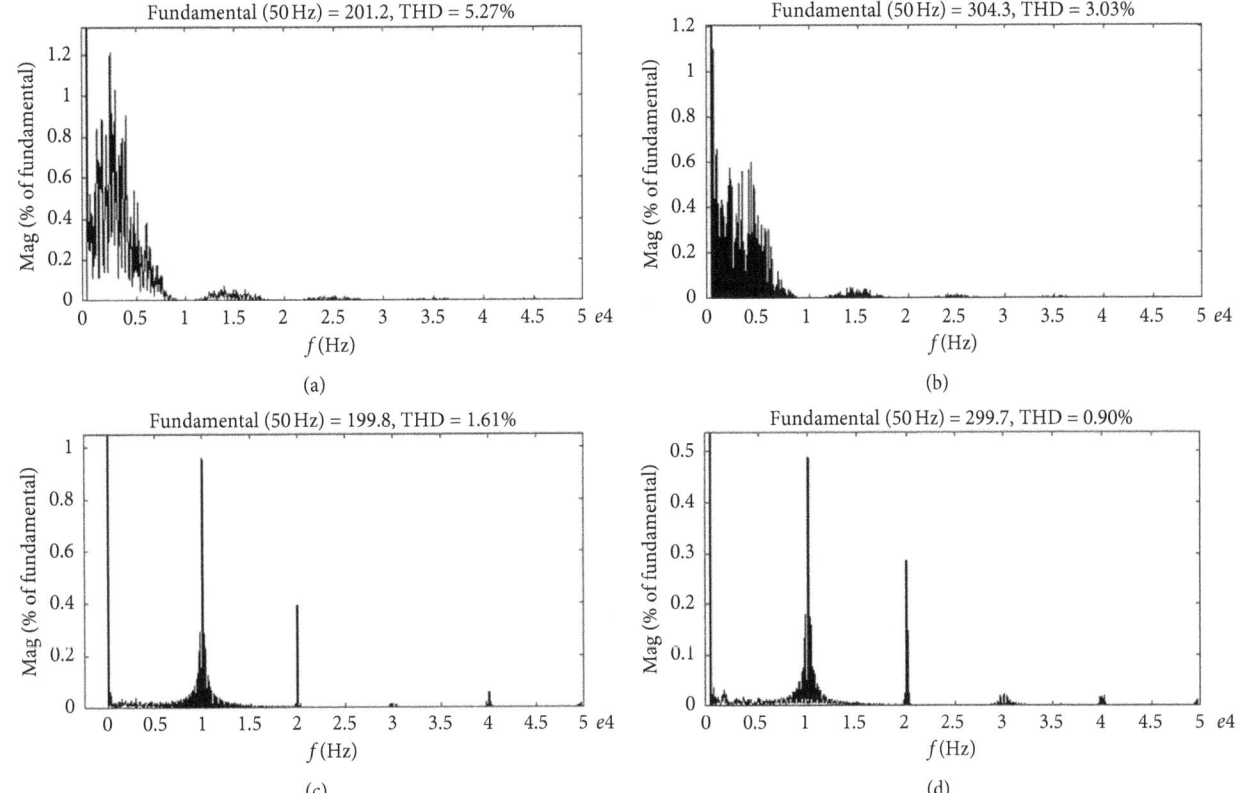

FIGURE 10: The fast Fourier transform analysis of the phase current. (a) The conventional MPC-RCMV (i_s^* = 200 A). (b) The conventional MPC-RCMV (i_s^* = 300 A). (c) The proposed MPC-RCMV (i_s^* = 200 A). (d) The proposed MPC-RCMV (i_s^* = 300 A).

and its multiple, which is helpful for the harmonic filter design. In addition, the total harmonic distortion (THD) of the phase current with the proposed method is smaller than the conventional one.

Accordingly, it can be concluded that the proposed MPC-RCMV with constant switching frequency shows great advantages over the conventional 6 VV strategy, mainly in the current and torque ripple reduction.

In order to illustrate the advantage of the proposed method over the MPC controller without considering the target of CMV reduction, the comparison results between the MPC method with constant switching frequency in [27] and the proposed method are shown in Figure 11. In Figure 11, the MPC method in [27] is adopted before 0.2 s and the proposed method is used after 0.2 s. The line-to-line

voltage, CMV, and three-phase currents are shown in Figure 11. The CMV amplitude of the MPC method in [27] reaches to the half of the dc-link voltage, but it has been limited within sixth of the dc-link voltage with the proposed method. Accordingly, the main advantage of the proposed method is that the CMV can be reduced.

4.2. Experimental Validation. To validate the effectiveness of the proposed MPC-RCMV with constant switching frequency, an experimental prototype has been established as shown in Figure 12. A 540 V dc-link voltage is obtained by the PWM rectifier. The motor 1 is connected with the inverter 1, and the proposed MPC-RCMV with constant switching frequency is implemented by the controller 1. The

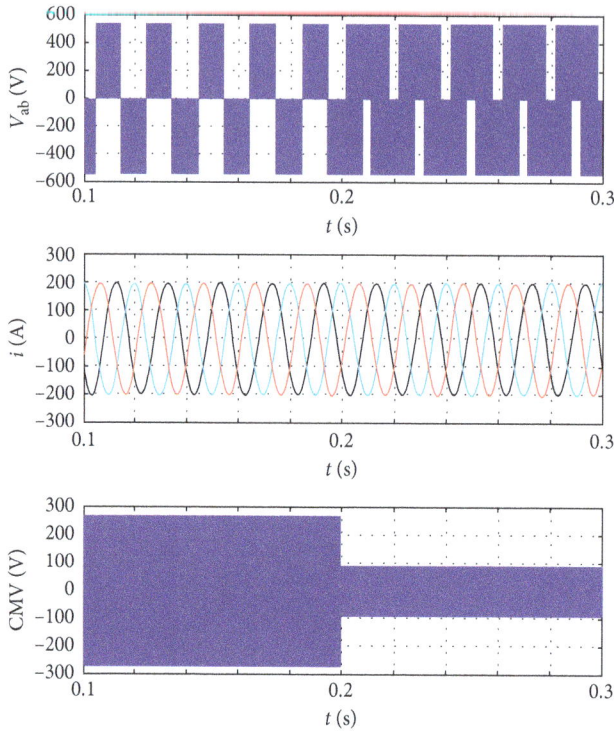

FIGURE 11: Comparison results between the MPC method with constant switching frequency in [27] and the proposed method.

FIGURE 12: The experimental prototype.

motor 2 is a load motor which is controlled by the inverter 2. The parameters of the PMSM are the same as the simulation.

In the experiment, a 150-MIPS fixed-point 32-bit TMS320F2812 DSP board is adopted to carry out the computation. The execution time for the proposed MPC-RCMV methods with necessary protecting and sampling process is 83 μs. To introduce the real-time implementation details, the flowchart of the proposed method in Figure 7(b) is divided into 4 parts, and the execution time of these four parts are 5 μs, 61 μs, 6 μs, and 2 μs, respectively. The other execution time is for the necessary sampling and protecting process.

The d- and q-axis current waveforms for the both conventional and proposed methods, which are measured through a digital-to-analog (DA) chip (TLV5610) on the control board, are shown in Figure 13, where the reference stator current changes from 100 A to 150 A at the middle of the graph. The results in Figure 13(b) indicate that the d- and q-axis current reference can be fast tracked with high precision as the conventional method as shown in Figure 13(a). In addition, it can be seen from Figure 13 that the ripples of both the d- and q-axis current for the proposed method have been greatly reduced compared with the conventional one.

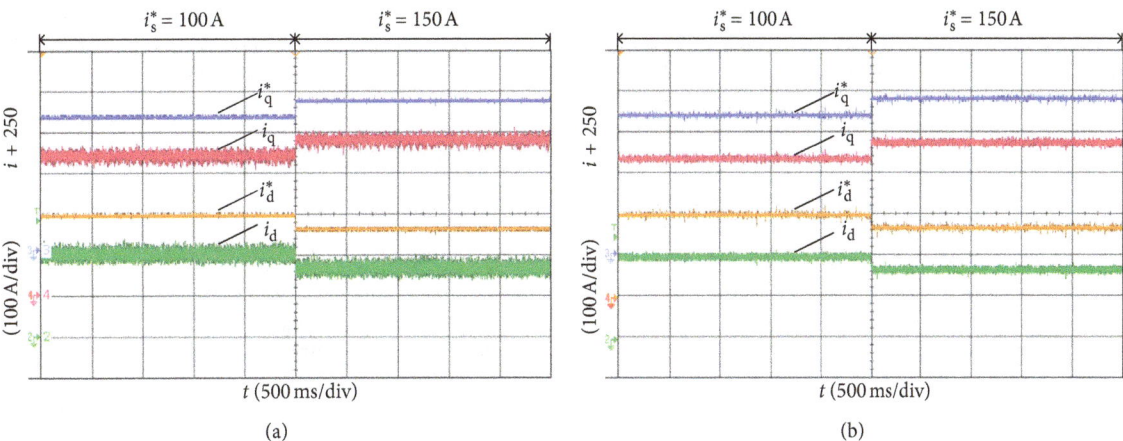

FIGURE 13: d- and q-axis current waveforms. (a) The conventional MPC-RCMV. (b) The proposed MPC-RCMV with constant switching frequency.

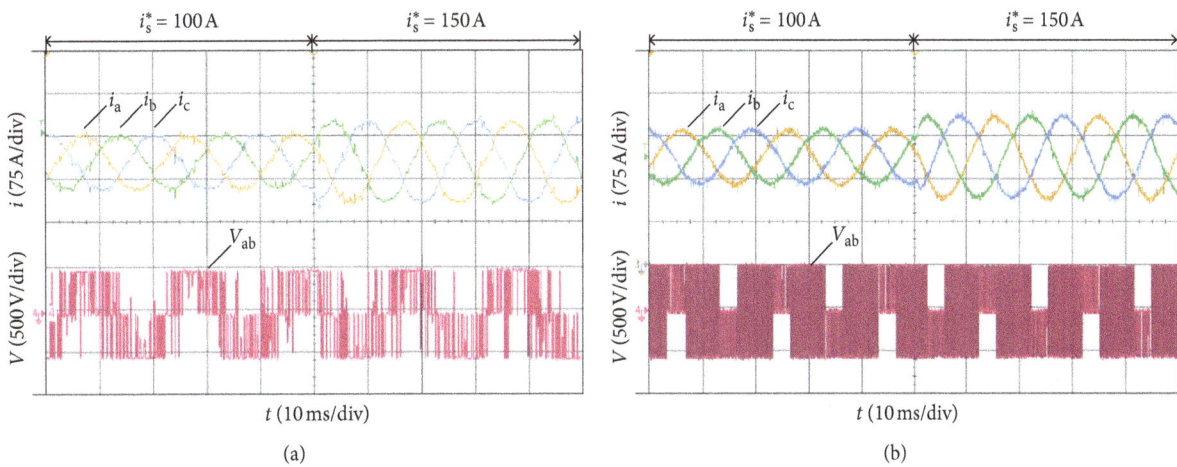

FIGURE 14: Three-phase current and line-to-line voltage curves. (a) The conventional MPC-RCMV. (b) The proposed MPC-RCMV with constant switching frequency.

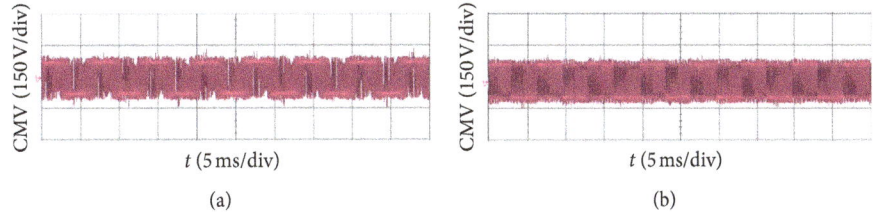

FIGURE 15: CMV curves. (a) The conventional MPC-RCMV. (b) The proposed MPC-RCMV with constant switching frequency.

The three-phase currents and the line-to-line voltage curves for the both methods are shown in Figure 14. The THD values for the conventional method where the reference of the stator current is 100 A and 150 A are 6.3% and 5.1%, respectively. The THD values for the proposed method where the reference of the stator current is 100 A and 150 A are 2.4% and 1.6%, respectively. Thus, the current quality of the proposed method has been greatly improved. The CMV curves shown in Figure 15 indicate that the amplitude of the

CMV has been successfully restricted within sixth of the dc-link voltage. Accordingly, the effectiveness of the proposed method has been verified by the experimental results.

5. Conclusion

A MPC-RCMV method with constant switching frequency for the PMSM drives is studied in detail. In the proposed method, four nonzero VVs are adopted and the switching sequence is designed according to the established current predictive error space vector diagram. Unlike the conventional methods where the switching frequency changes with the work condition of PMSM, the average switching frequency is fixed and equal to $1/T_s$. The simulation and experiment results indicate that the current reference can be fast and accurately tracked, and the current ripples are greatly reduced.

Conflicts of Interest

The authors declare that there are no conflicts of interest regarding the publication of this paper.

Acknowledgments

This paper is partly funded by the National Natural Science Foundation of China under Award U1610113 and National key Research and Development Project of China under award 2016YFC0600804.

References

[1] H. Chen and H. Zhao, "Review on pulse-width modulation strategies for common-mode voltage reduction in three-phase voltage-source inverters," *IET Power Electronics*, vol. 9, no. 14, pp. 2611–2620, 2016.

[2] H. Cao, "A novel power electronic inverter circuit for transformerless photovoltaic systems," *Active & Passive Electronic Components*, vol. 2014, Article ID 329043, 5 pages, 2014.

[3] M. Asefi and J. Nazarzadeh, "Survey on high-frequency models of PWM electric drives for shaft voltage and bearing current analysis," *IET Electrical Systems in Transportation*, vol. 7, no. 3, pp. 179–189, 2017.

[4] N. Mousavi, T. Rahimi, and H. M. Kelk, "Reduction EMI of BLDC motor drive based on software analysis," *Advances in Materials Science and Engineering*, vol. 2016, Article ID 1497360, 9 pages, 2016.

[5] K. S. Narayana, A. K. Rao, and K. Satyanarayana, "Novel AZSPWM algorithms based VCIMD for reduced CMV variations," *International Journal of Power Electronics and Drive Systems (IJPEDS)*, vol. 3, no. 1, pp. 1–8, 2013.

[6] E. Un and A. M. Hava, "A near state PWM method with reduced switching frequency and reduced common mode voltage for three-phase voltage source inverters," *IEEE Transactions on Industry Applications*, vol. 45, no. 2, pp. 782–793, 2009.

[7] A. M. Hava and E. Un, "Performance analysis of reduced common-mode voltage PWM methods and comparison with standard PWM methods for three-phase voltage-source inverters," *IEEE Transactions on Power Electronics*, vol. 24, no. 1, pp. 241–252, 2009.

[8] X. Wu, G. Tan, Z. Ye, Y. Liu, and S. Xu, "Optimized common-mode voltage reduction PWM for three-phase voltage-source inverters," *IEEE Transactions on Power Electronics*, vol. 31, no. 4, pp. 2959–2969, 2016.

[9] P. Chaturvedi, S. Jain, and P. Agarwal, "Carrier-based common mode voltage control techniques in three-level diode-clamped inverter," *Advances in Power Electronics*, vol. 2012, Article ID 327157, 13 pages, 2012.

[10] R. Wang, X. Mu, Z. Wu, L. Zhu, and Z. Chen, "Carrier-based PWM method to reduce common-mode voltage of three-to-five-phase indirect matrix converter," *Mathematical Problems in Engineering*, vol. 2016, Article ID 6086497, 12 pages, 2016.

[11] J. Huang, Q. Liu, X. Wang, and K. Li, "A carrier-based modulation scheme to reduce the third harmonic component of common-mode voltage in a three-phase inverter under high DC voltage utilization," *IEEE Transactions on Industrial Electronics*, vol. 65, no. 3, pp. 1931–1940, 2018.

[12] J. Huang and H. Shi, "Suppressing low-frequency components of common-mode voltage through reverse injection in three-phase inverter," *IET Power Electronics*, vol. 7, no. 6, pp. 1644–1653, 2014.

[13] K. Tian, J. Wang, B. Wu, Z. Cheng, and N. R. Zargari, "A virtual space vector modulation technique for the reduction of common-mode voltages in both magnitude and third-order component," *IEEE Transactions on Power Electronics*, vol. 31, no. 1, pp. 839–848, 2016.

[14] X. Wu, H. Liu, X. Yuan, S. Huang, and D. Luo, "Design and implementation of recursive model predictive control for permanent magnet synchronous motor drives," *Mathematical Problems in Engineering*, vol. 2015, Article ID 431734, 11 pages, 2015.

[15] S. Zhang, H. Cao, Y. Zhang, L. Jia, and Z. Ye, "Data-driven optimization framework for nonlinear model predictive control," *Mathematical Problems in Engineering*, vol. 2017, Article ID 9402684, 15 pages, 2017.

[16] S. K. Hoseini, J. Adabi, and A. Sheikholeslami, "Predictive modulation schemes to reduce common-mode voltage in three-phase inverters-fed AC drive systems," *IET Power Electronics*, vol. 7, no. 4, pp. 840–849, 2014.

[17] M. Preindl, E. Schaltz, and P. Thøgersen, "Switching frequency reduction using model predictive direct current control for high-power voltage source inverters," *IEEE Transactions on Industrial Electronics*, vol. 58, no. 7, pp. 2826–2835, 2011.

[18] L. Guo, X. Zhang, S. Yang, Z. Xie, and R. Cao, "A model predictive control-based common-mode voltage suppression strategy for voltage-source inverter," *IEEE Transactions on Industrial Electronics*, vol. 63, no. 10, pp. 6115–6125, 2016.

[19] S. Kwak and S. Mun, "Common-mode voltage mitigation with a predictive control method considering dead time effects of three-phase voltage source inverters," *IET Power Electronics*, vol. 8, no. 9, pp. 1690–1700, 2014.

[20] S. Mun and S. Kwak, "Reducing common-mode voltage of three-phase VSIs using the predictive current control method based on reference voltage," *Journal of Power Electronics*, vol. 15, no. 3, pp. 712–720, 2015.

[21] S. Kwak and S. Mun, "Model predictive control methods to reduce common-mode voltage for three-phase voltage source inverters," *IEEE Transactions on Power Electronics*, vol. 30, no. 9, pp. 5019–5035, 2015.

[22] X. Cong, H. Zhang, F. Cheng, and H. Zhang, "Model predictive control method to reduce common-mode voltage for permanent-magnet synchronous machine drives," in *Proceedings of 2017 IEEE Transportation Electrification Conference and Expo, Asia-Pacific (ITEC Asia-Pacific)*, pp. 1–6, Harbin, China, August 2017.

[23] L. Zhang and M. Dou, "Multiprotocol communication interface pmsm control on account of industrial configuration

software," *Journal of Electrical and Computer Engineering*, vol. 2014, Article ID 651216, 6 pages, 2014.

[24] F. Khammar and N. Debbache, "Application of artificial intelligence techniques for the control of the asynchronous machine," *Journal of Electrical and Computer Engineering*, vol. 2016, Article ID 8052027, 11 pages, 2016.

[25] M. Tomlinson, H. D. T. Mouton, R. Kennel, and P. Stolze, "A fixed switching frequency scheme for finite-control-set model predictive control—concept and algorithm," *IEEE Transactions on Industrial Electronics*, vol. 63, no. 12, pp. 7662–7670, 2016.

[26] F. Donoso, A. Mora, R. Cardenas, A. Angulo, D. Saez, and M. Rivera, "Finite-set model predictive control strategies for a 3L-NPC inverter operating with fixed switching frequency," *IEEE Transactions on Industrial Electronics*, vol. 65, no. 5, pp. 3954–3965, 2018.

[27] E. Fuentes, C. A. Silva, and R. M. Kennel, "MPC implementation of a quasi-time-optimal speed control for a PMSM drive, with inner modulated-FS-MPC torque control," *IEEE Transactions on Industrial Electronics*, vol. 63, no. 6, pp. 3897–3905, 2016.

[28] C. F. Garcia, C. A. Silva, J. R. Rodriguez, P. Zanchetta, and S. A. Odhano, "Modulated model predictive control with optimized overmodulation," *IEEE Journal of Emerging and Selected Topics in Power Electronics*, p. 1, 2018.

[29] J. Rodriguez and P. Cortes, *Predictive Control of Power Converters and Electrical Drives*, Wiley–IEEE, Hoboken, NJ, USA, 2012.

[30] H. T. Moon, H. S. Kim, and M. J. Youn, "A discrete-time predictive current control for PMSM," *IEEE Transactions on Power Electronics*, vol. 18, no. 1, pp. 464–472, 2003.

[31] Y. Yang, C. Da, X. Zheng, Z. Mi, X. Li, and C. Sun, "Adaptive backstepping based MTPA sensorless control of PM-assisted SynRM with fully uncertain parameters," *Mathematical Problems in Engineering*, vol. 2018, Article ID 8405847, 14 pages, 2018.

[32] A. Ahmed, Y. Sozer, and M. Hamdan, "Maximum torque per ampere control for buried magnet PMSM based on DC-link power measurement," *IEEE Transactions on Power Electronics*, vol. 32, no. 2, pp. 1299–1311, 2016.

16

Predicting Harmonic Distortion of Multiple Converters in a Power System

P. M. Ivry, O. A. Oke, D. W. P. Thomas, and M. Sumner

Department of Electrical and Electronics Engineering, Electrical Systems and Optics Research Group, University of Nottingham, Nottingham, UK

Correspondence should be addressed to P. M. Ivry; preyeivry@mail.ndu.edu.ng

Academic Editor: Raj Senani

Various uncertainties arise in the operation and management of power systems containing Renewable Energy Sources (RES) that affect the systems power quality. These uncertainties may arise due to system parameter changes or design parameter choice. In this work, the impact of uncertainties on the prediction of harmonics in a power system containing multiple Voltage Source Converters (VSCs) is investigated. The study focuses on the prediction of harmonic distortion level in multiple VSCs when some system or design parameters are only known within certain constraints. The Univariate Dimension Reduction (UDR) method was utilized in this study as an efficient predictive tool for the level of harmonic distortion of the VSCs measured at the Point of Common Coupling (PCC) to the grid. Two case studies were considered and the UDR technique was also experimentally validated. The obtained results were compared with that of the Monte Carlo Simulation (MCS) results.

1. Introduction

Controllable small distribution networks containing closed assemblage of distributed generators, Renewable Energy Systems (RES), storage systems, and loads are becoming increasingly common in Electrical Power Systems (EPS). The optimization of RES, such as wind turbines and photovoltaics, has been successful thus far. This can be attributed to the advancement and use of power electronic converters, which help to achieve higher power and voltage level operation in wind turbines and photovoltaics.

In recent years, the Voltage Source Converter (VSC) is popularly used and has enjoyed more attention than other converters due to its better controllability and fast switching responses [1, 2]. Nevertheless, VSCs generate harmonic voltages and currents that are transmitted to the rest of the grid. These harmonics may cause malfunction or damage of the power system and equipment on the system.

On an EPS, harmonics can significantly increase and become difficult to predict where variations are present in certain factors like the operating conditions (output power) or system parameters (grid voltage background distortion) [3]. Furthermore, in a case where many VSCs are connected

to the EPS, the net effect of the harmonics on the EPS's Power Quality (PQ) becomes more challenging to predict as harmonics do not add up arithmetically. This and the inherent stochastic nature of harmonics necessitate the use of statistical techniques in predicting the cumulative harmonic distortion level of power converters in an EPS. The use of statistical techniques was also suggested by the IEEE Probabilistic Aspects Task Force on Harmonics for quantifying harmonic distortions [4].

Statistical techniques, such as the Monte Carlo Simulation (MCS) [5], have been extensively used as a common approach in predicting the harmonic distortion level of power converters [6, 7]. However, it requires tens of thousands of simulations to obtain an accurate prediction and this affects the feasibility of using such approach for systems containing large number of VSCs.

In some previous studies [8–14], an analytical approach was used to predict the level of harmonics generated by power converters. The studies represented harmonic vectors as phasors having random amplitudes and angles. The probability density functions (pdfs) of the phasors were obtained and represented in the rectangular coordinates for the convenience of adding phasors. In cases of a large number of harmonic sources/loads the harmonic phasor's pdfs are then vectorially summed. The studies predicted harmonics in terms of low harmonic orders such as the 3rd, 5th, 7th, and 11th harmonic orders. This may be due to the type of power converter utilized (6/12 pulse converters) and their associated harmonics generated. In a VSC system, the majority of the harmonics appear at the switching frequency and multiplies of the switching frequency [1, 2]. Another way to quantify the harmonics would be to use the Total Harmonic Distortion (THD) [2].

Furthermore, an analytical approach usually entails assumptions to handle the complex interaction of a large number of random harmonic quantities [15, 16] and practical converter systems were usually simplified and represented by mathematical formulas to accommodate the approach as seen in [8–14]. However, other methods which do not require these simplifications or assumptions in designing the systems/converters or in generating random occurrences can be deployed. They include the Monte Carlo Simulation (MCS) [5, 17, 18], Unscented Transform (UT) [19, 20], Point Estimate Method (PEM) [21], and Univariate Dimension Reduction (UDR) [22, 23]. For instance, in Probabilistic Load Flow (PLF) studies, some of these methods have been utilized. The UDR which has proved successful for problems with complex and large number of statistical variation will be utilized in the probabilistic harmonic analysis because of its ability to drastically reduce computational cost, time, and burden. In this study, practical VSC models are simulated with a full switching model.

In VSC design and utilization, the interfacing inductor value depends on the VSC switching frequency to achieve adequate attenuation of harmonics and it is usually chosen with a tradeoff between harmonic attenuation capability and filter cost. Furthermore, various formulas [24–28] can be used in estimating the inductance value with each giving a different filter size, thus contributing to the uncertainty. In most RES, output power is variable. For wind turbine systems, the output power is a function of the wind speed, while, for photovoltaics, output power depends mainly on incident sun rays on the photovoltaic cells. This gives rise to further uncertainty as wind speed/sun light varies with time, day, season, and place. All these uncertainties have to be properly accounted for to ensure harmonic distortion within the EPS is within prescribed limits.

This paper presents a method for predicting current harmonics at the PCC of an EPS in the presence of uncertainties in the filter parameter and operating power of multiple VSCs. The level of harmonic distortion is quantified using statistical evaluators such as the mean and standard deviation. The Univariate Dimension Reduction (UDR) for 3 and 5 points' approximation is utilized. It is utilized as an alternative to the Monte Carlo Simulation (MCS) in predicting the harmonics distortion level of multiple connected VSCs because of its significant reduction of computational cost, time, and burden associated with the MCS. The results obtained and the accuracy of the UDR for each of the cases were compared with the MCS and presented in the sections below. The results for the UDR technique were also validated using a laboratory experiment.

2. Unscented Transform and Univariate Dimension Reduction

2.1. Unscented Transform (UT). The UT works by approximating a nonlinear mapping by a set of selected points called sigma points. The sigma points are developed using the moments of the distribution functions pdf, and the weighted average of the sigma points produces the expectation of the mapping [29]. The UT has been utilized in nonlinear problems in electromagnetic compatibility [19, 20, 30] and medical statistics amongst other fields [22]. This method can be used in approximating a continuous distribution function with pdf $w(x)$ as a discrete distribution using deterministically chosen points called sigma points (S_i) and weights (w_i) such that the moments of both distributions are equal [31]. This is mathematically represented in

$$E\left(x^k\right) = \int x^k w\left(x\right) dx = \sum_i w_i S_i^k. \tag{1}$$

S_i contains the location of the abscissas at which the function $f(x)$ is to be evaluated while w_i are the weighting coefficients which when multiplied by S_i give an approximation to the integral of $f(x)$. k represents the moments of the expectation (E) where $k = 1$ implies the mean value and $k = 2$ implies variance.

The Gaussian quadrature technique is applied to solve (1) such that the integration points for integrating $f(x)$ correspond to the desired sigma points S_i. Hence, for a function $f(x)$ assumed as a polynomial, with pdf (weighting function) $w(x)$, the nonlinear mapping for the expectation is given as

$$E\left[f\left(x\right)\right] = \int_{-\infty}^{\infty} f\left(x\right) w\left(x\right) dx = \sum_i w_i f\left(S_i\right). \tag{2}$$

The sigma points of the distribution function can be obtained as the roots of its associated orthogonal polynomial when (1) or (2) is integrated using Gaussian quadrature. This method can be easily applied, as most common distributions have known classical orthogonal polynomials associated with them. The associated orthogonal polynomials for some distribution functions can be found in [32].

2.2. Univariate Dimension Reduction (UDR). The procedure discussed in the section above is only directly applicable when obtaining individual sigma points and weights. For problems involving more than one variable (N), the simplest technique is based on tensor product of the individual sigma points (n), giving the number of evaluations E_y as

$$E_y = n^N. \tag{3}$$

Unfortunately, the technique is plagued by the *curse of dimensionality* problem as the number of variables increases. For example, using 5 sigma points in (3), 5 variables will require 3,125 evaluations while, for 10 variables, $E_y = 9,765,625$. The dimension reduction technique is thus employed in this work.

The dimension reduction [22, 29, 33, 34] is an approximate technique for estimating the statistical moments of an output function. The technique involves an additive decomposition of an N-dimensional function involving n-dimensional integral into a series sum of D-dimensional functions such that $D < N$. It provides a means of efficiently combining the sigma points and weights for a large number of variables such that the number of evaluation points can be minimized [22, 29]. For $D = 1$ the method is referred to as the Univariate Dimension Reduction (UDR) while it is called bivariate dimension reduction for $D = 2$.

The UDR method has been utilized in stochastic mechanics [23] and probabilistic load flow studies [22, 35, 36]; however, it has been less applied to harmonic analysis. The UDR method is briefly described below while a detailed mathematical derivation of the techniques can be found in [33].

With the UDR, the main function, $f(\mathbf{x})$, is decomposed into a summation of one-dimensional functions such that

$$f(\mathbf{x}) \cong \sum_{i=1}^{N} f\left(\overline{x}_1, \ldots, \overline{x}_{i-1}, x_i, \overline{x}_{i+1}, \ldots, \overline{x}_N\right) - (N-1) f\left(\overline{x}_1, \ldots, \overline{x}_N\right), \tag{4}$$

where μ_i is the mean of the ith random variable.

The resultant function in (4) can easily be integrated as only one randomly distributed variable is present at every instance while the others are held constant at their mean values. The moments of the function are approximately the same as those of the decomposed function as represented in

$$E\left[f(\mathbf{x})\right] \cong E\left[\widehat{f}(\mathbf{x})\right]$$

$$\cong \sum_{i=1}^{N} E\left[f\left(\overline{x}_1, \overline{x}_2, \ldots, x_i, \overline{x}_N\right)\right] \tag{5}$$

$$- (N-1) E\left[f\left(\overline{x}_1, \ldots, \overline{x}_N\right)\right].$$

The same procedure is applied for obtaining the higher order moments.

The number of evaluations (E_v) required for an N-dimensional function using n estimation (sigma points) for the UDR is given in [33]

$$E_v = (n \times N) + 1. \tag{6}$$

For problems where all random variables are symmetrical and n (sigma points) is odd, the estimation points for the UDR can be further reduced to (7) while still maintaining the same level of accuracy. For clarity, this will be referred to as reduced UDR (rUDR).

$$E_v = ((n-1) \times N) + 1. \tag{7}$$

This is clearly a substantial saving in computational time over the alternative of n^N evaluations.

3. Uncertainty Representation

The appropriate design and sizing of the system filter have a great impact on the amount of distortion seen at the grid side of the converter. If not designed properly, it can affect the systems stability. When an RES is integrated to the grid, one of the main concerns is to reduce the harmonic current injected into the grid [24]. In designing harmonic filters for converters, the maximum ripple current should be less than 20% of the rated current [24]. This current is a function of the inductance L_{f1}, the switching frequency, and the DC link voltage. However, to achieve a decrease of the ripple current, the inductance is the most flexible parameter since increasing the switching frequency affects the system efficiency. The value of the inductance can be obtained using

$$L_{f1} = \frac{1}{8} \times \frac{V_{dc}}{f_{sw} \cdot \Delta\widehat{I}_{L\,max}}, \tag{8}$$

where f_{sw} is the switching frequency, V_{dc} is the dc voltage, and $\Delta\widehat{I}_{L\,max}$ is the peak ripple of the maximum rated load current.

In deciding the filter inductance value, another solution is to choose the value relative to the total system inductance. It was suggested in [25, 28] that the total system inductance should be about 10% of the base inductance value. Since no fixed value can be used but that which is determined at the discretion of the designer, the interfacing inductor L_f can be viewed as a stochastic variable such that the optimal size cannot be clearly determined. The value of the inductance will be in a range, howbeit as a percentage of $\Delta\widehat{I}_{L\,max}$. The value of the inductance can thus be represented using the uniform distribution such that an optimal value can be easily predicted. Hence, (9) can be substituted into (8) to account for this variation.

$$\Delta\widehat{I}_{L\,max} = (5 - 20\%) \times I_{rated} \tag{9}$$

and L_{f2} is calculated as

$$L_{f2} = \alpha L_{f1}, \tag{10}$$

where α is a factor limiting L_{f2} to be less than L_{f1}.

The VSC harmonic filter is designed using (8)–(14). The value for the filter capacitor (C_f) can be obtained using

$$C_f = 15\% \times \frac{P_{rated}}{3 \cdot 2\pi f_m \cdot V_{rated}^2}, \tag{11}$$

where C_f is calculated by limiting the total reactive power required by the capacitor to be within 15% of the total rated active power.

It is important to consider resonant frequency of an LC or LCL filter as they usually require a damping element to maintain stability and to ensure optimum operation [24].

The resonant frequency is usually within $10f_n < f_{res} < f_{sw}/2$.

For an LC filter, the resonant frequency is calculated from

$$\omega_{res} = \frac{1}{\sqrt{L_{f1}C_f}}. \tag{12}$$

For an LCL filter, it is

$$\omega_{res} = \sqrt{\frac{L_{f1} + L_{f2}}{L_{f1}L_{f2}C_f}}. \tag{13}$$

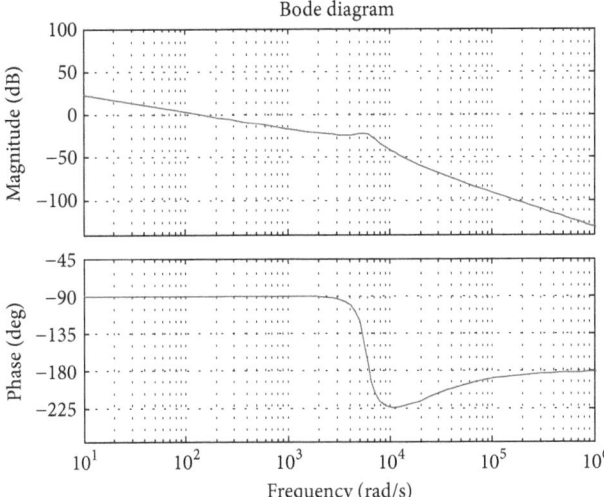

FIGURE 1: Bode plot of designed filter.

The damping resistor is calculated using [24]

$$R_d = \frac{1}{3 \cdot \omega_{res} \cdot C_f}. \tag{14}$$

Figure 1 shows the bode plot of the designed filter and highlights the effectiveness of the filter.

Similarly, other uncertainties that emanate from various RES due to the dependence of their power output on several random factors such as the weather conditions have to be represented. Several models have been proposed in literatures for modeling these uncertainties; wind speed has been modeled using the Weibull distribution, Rayleigh distribution, normal distribution, and so forth [37]. For simplicity, the Power Variation is modeled using the uniform distribution.

4. EPS Structure

The studied RES system comprises the DC Power Source, DC capacitor, RES interfacing VSC, VSC reactors/filter, grid impedance, and 3-phase grid voltage (representing further parts of the grid). Figure 2 shows multiple numbers of VSCs connected in parallel and interfaced to the grid at the PCC via the grid impedance. The system was simulated using PLECS™ [38] and MATLAB Simulink™ [39] simulation tools.

The VSC is modeled as a 2-level VSC and the switching pattern is based on the Sinusoidal Pulse Width Modulation (SPWM) as in [2]. The VSC is controlled based on the commonly used instantaneous power theory [1] using PI controllers. The first PI controller controls the DC capacitor voltage to remain as regulated. The output of the regulated voltage gives the reference I_{dref} that is required to regulate the active current component which in turn controls the active power [1, 3]. The second current PI controller controls the reactive current component I_{qref}. For this study $I_{qref} = I_q = 0$ [1]. The control scheme used in [3] was adopted for each of the VSCs.

5. UDR Harmonic Prediction Process

A summary of the main steps for implementing the UDR technique for predicting current and voltage harmonic distortion at the PCC of the EPS is highlighted below.

Step 1. Identify all randomly varying functions within the system (e.g., power, filter inductance) and obtain their probability distribution functions.

Step 2. Compute the sigma points and weights using Univariate Dimension Reduction technique.

Step 3. Input the EPS data including the sigma points and weights obtained in Step 2.

Step 4. Run the EPS simulation while obtaining necessary statistical data for the output variables (current and voltage THD).

Step 5. Compute the statistical data for the output variables using the measured output data values and weights of the UDR (e.g., mean current THD).

Step 6. Display statistical data of output variables (e.g., mean current and voltage THD).

6. Simulation Results

Two (2) studies are considered in predicting the level of harmonic distortion based on the variation of output power and filter inductor value. In the first study, the effect of filter inductor value L_{f1} on the level of harmonic distortion in the current at the PCC using multiple VSCs is examined. For this system, L_{f1} was calculated using techniques in [24, 40] which gives a value of 5.76 mH which is typical for systems of this level. Based on this and assuming the filter inductor value can be modeled using the uniform distribution, the value was chosen to be between 3.46 mH and 8.06 mH. This range was chosen assuming a ±40% tradeoff between harmonic filter effectiveness and cost of filter inductor.

In the second study, the effect of output Power Variations on the level of harmonic distortion is considered. The output power is assumed to vary randomly within a range following the uniform distribution between 25% and 100% of its rated power. The point here is to show wide variability of the RES output power.

In the considered cases, 1000 simulations were carried out for the MCS to ensure accuracy. This is sufficient as this ensures a 95% confidence interval that the errors in the mean THD values are less than 3% as calculated using (15). The mean and standard deviation values utilized in this calculation were obtained after running the simulations and then used to check against the 95% CI.

$$95\% \ (\text{CI}) = \overline{x} \pm Z \frac{\sigma}{\sqrt{n}}$$

$$\text{SE} = \frac{\sigma}{\sqrt{n}}, \tag{15}$$

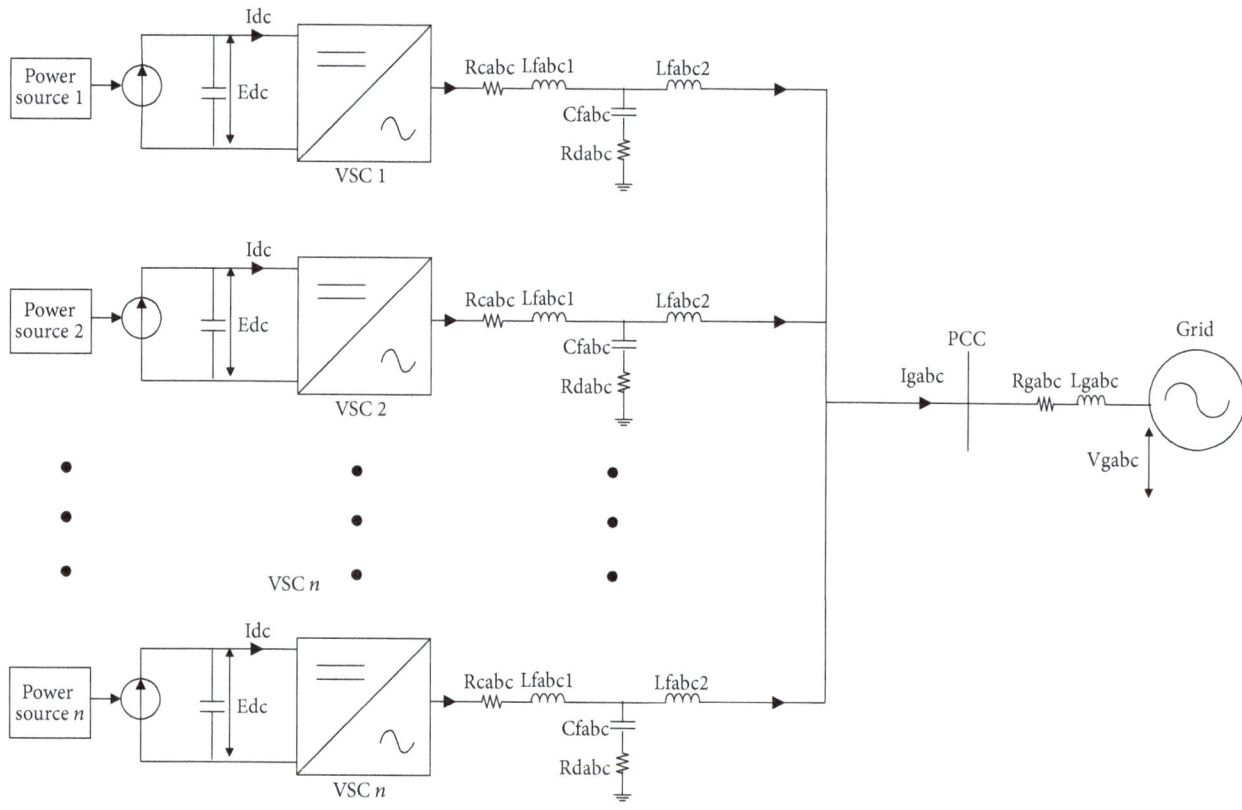

FIGURE 2: Multiple VSCs connected to a common point to the grid.

where CI implies the confidence interval, SE is the standard error of the mean, n is the number of samples, $\overline{x}$ implies the mean value of the samples, σ is standard deviation of the samples, and $Z = 1.96$ the constant representing 95% CI, where the SE for 2 VSCs at 1000 simulations gives 0.04 using the values for 2 VSCs in Figure 3.

To show the adequacy of the chosen number of simulations used, Figure 3 presents the results for the mean current THD for a system with 2 VSCs using 10, 100, 1000, 5000, and 10000. The simulations were carried out using a Windows 7, 32-bit operating system, i3 processor 4 GB RAM PC.

The 3 and 5 points' approximated rUDR were utilized in predicting the THD of the power converters in the EPS. The obtained results were then compared with that of the MCS. The results obtained in terms of the mean and standard deviation of the THD and the accuracy of the methods for each of the cases are discussed in the sections below.

6.1. Effect of Filter Variation on Current THD.
The mean current and voltage THD arising from L_f variations using 1 to 10 VSCs are presented in Figures 5 and 6, respectively. Figure 4 shows the amount of time saved when the rUDR methods were utilized and compared with the MCS. For 10 VSCs the amounts of simulation time for the rUDR 3 points, rUDR 5 points, and MCS were approximately 1100 s, 2100 s, and 52000 s, respectively.

$$\text{Computational Time Saved (\%)} = \frac{t_{\text{MCS}} - t_{\text{rUDR}}}{t_{\text{MCS}}}. \quad (16)$$

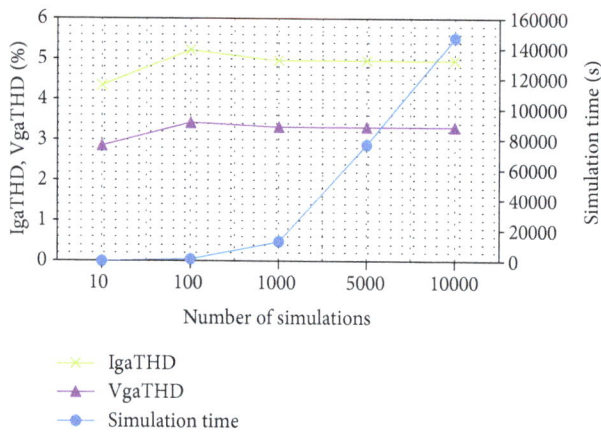

FIGURE 3: THD result for 2 VSCs connected in parallel.

From Figure 5, it can be observed that the impact of the variation of L_f on the current THD progressively reduces as the number of VSC increases. This is as expected, since the total number of filters in the system increases as the number of VSC increases, thus limiting the current distortion. Also, with increased number of VSCs, there is a higher probability of harmonic cancellation due to current diversity and attenuation factor [24, 41, 42].

It is observed that the rUDR methods closely predict the current and voltage THD as the curve follows the same graph trend as that of the MCS. With 10 VSCs, there seems to be

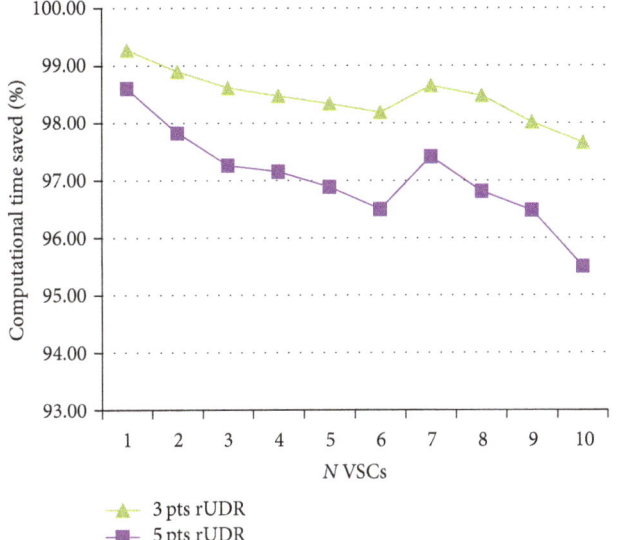

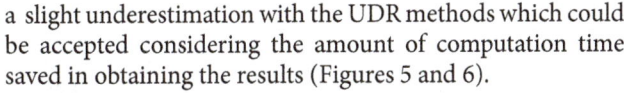

FIGURE 4: Computation time saved by rUDR 3 pts and rUDR 5 pts.

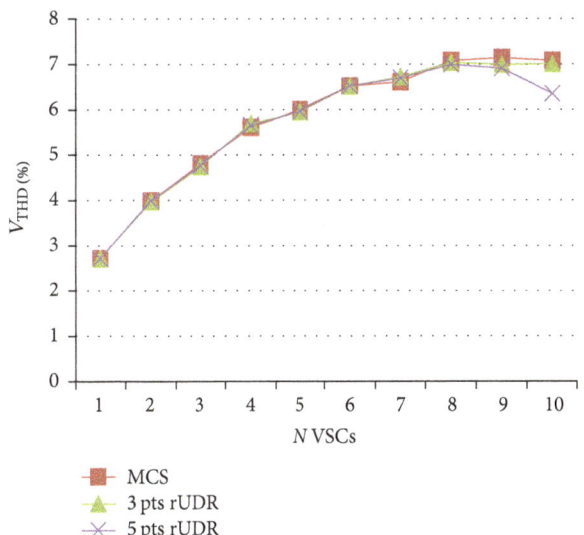

FIGURE 6: Predicted VgaTHD using MCS, rUDR 3 pts, and rUDR 5 pts under Filter Variation.

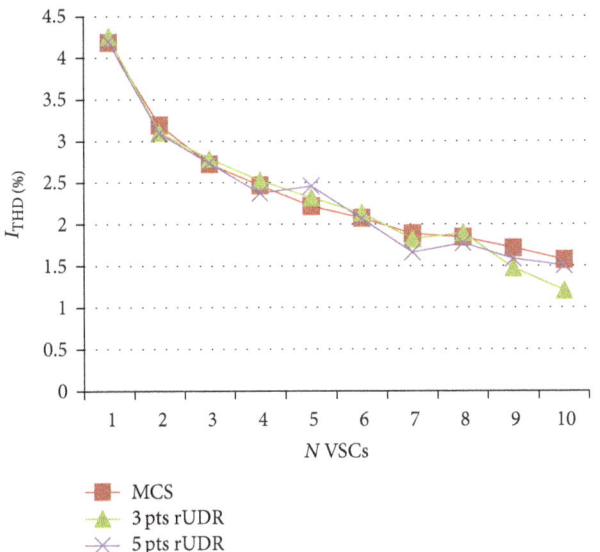

FIGURE 5: Predicted IgaTHD using MCS, rUDR 3 pts, and rUDR 5 pts under Filter Variation.

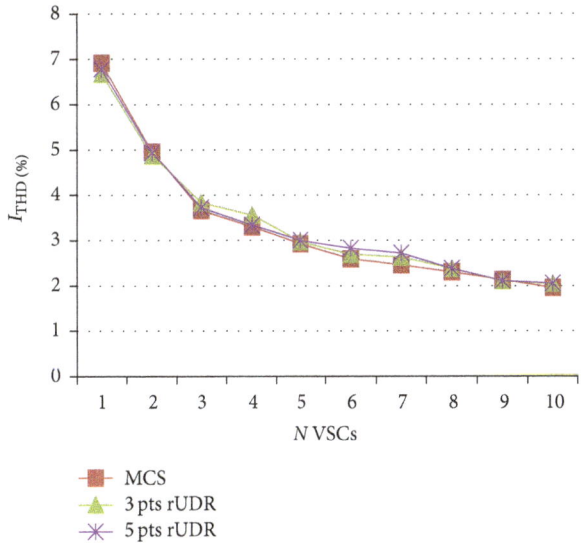

FIGURE 7: Predicted IgaTHD using MCS, rUDR 3 pts, and rUDR 5 pts under Power Variation.

a slight underestimation with the UDR methods which could be accepted considering the amount of computation time saved in obtaining the results (Figures 5 and 6).

6.2. Effect of Power Variation on Current THD. The impact of operating Power Variation on the current and voltage THD seen at the PCC of the EPS was also investigated. The predicted current and voltage THD using the UDR and MCS approach are presented in Figures 7 and 8. The impact of Power Variation on the current THD is more significant than the case of L_{f1} variation as evident in Figure 7. However, as with the previous case, the current THD is observed to reduce with the increase of the number of VSCs. The rUDR methods

produced accurate results and the graph trends are similar to the MCS.

7. Experimental Validation of the UDR Technique

The aim of this experiment is to validate the efficiency of the UDR technique in predicting the harmonic distortion of VSCs in the presence of uncertainty for a small Electrical Power System (EPS). The parameters of which are given in Table 1.

Figure 9 shows the schematic of the laboratory setup. The experiment was conducted using 3 VSCs, a programmable

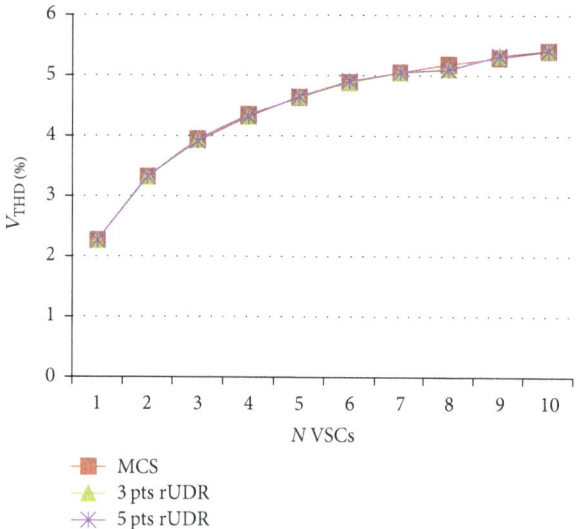

FIGURE 8: Predicted VgaTHD using MCS, rUDR 3 pts, and rUDR 5 pts under Power Variation.

TABLE 1: Microgrid and VSC laboratory parameters.

VSC Parameter	
Rated power, P	12.5 kW
Rated line voltage	415 V
Rated current	32 A
VSC topology	3-phase 2-level VSC
VSC filter configuration	LCL
Electrical power system	
Mains line voltage	415 V
Current	Fuse 20 A
RES source (programmable power/voltage supply)	
Rated power	90 kW
Rated voltage	1000 V
Rated current	300 A
Fuse rating	125 A
Statistical power variation	
Uniform power variation	Range = 2.5 kW–10.0 kW Mean = 6.25 kW
Normal power variation	Mean = 3.44 kW Std = 2.02

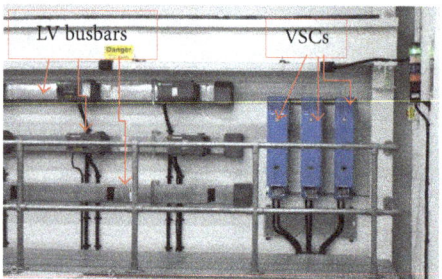

FIGURE 9: Laboratory setup showing 415LV busbars and 3 VSCs.

power/voltage supply (Figure 10) as the RES source, current and voltage measuring device, and 3 isolating transformers.

The 3 VSCs are connected in parallel and supplied from the mains through a 3-phase distribution line and a transformer. The VSCs are isolated using 3 transformers at the grid side of the VSC and connected to the grid side of the programmable power/voltage supply (RES source).

The isolating transformers are necessary to protect the equipment and also to prevent short circuit. The main utility feed represents the grid so the VSCs are operating in the grid tied mode. The voltage and current measurements were done with LabView on a CompactRIO using NI-9227 and NI-9225 units; the current is passed through 100:5 CTs prior to feeding it into the measuring blocks. The time taken to conduct each practical session is presented in Table 2, where E-MCS refers to the Experimental MCS styled approach.

The internal components of the VSCs in Figure 9 were not simulated because of Intellectual Property (IP) protection.

7.1. Case Study 1: Prediction under Uniform Power Variation. Output Power Variations have been earlier shown to affect THD and this study aims to predict the net harmonics of 3 VSCs in a case where the output power uniformly varies randomly and independently (range: 2.5 kW–10.0 kW).

FIGURE 10: Controllable power and voltage source (Triphase).

The VSCs were first varied with thousands of random power input values to imitate the MCS approach. Then the net harmonics are measured and recorded. The UDR sigma points and weights also generated. These were fed into the VSCs as power inputs and the net harmonics measured at the PCC. The results of the E-MCS styled approach and the UDR are then statistically analyzed and compared to measure the efficiency of the UDR technique. The results are given in Table 3 and Figures 11 and 12.

It can be seen in Table 3 and Figures 11 and 12 that the UDR predicted results for I_{THD} and V_{THD} have a good match with the E-MCS approach.

TABLE 2: Approximate run time for each practical session.

	Case 1		Case 2	
	E-MCS	UDR 5 pts	E-MCS	UDR 5 pts
Time (s)	21600	2700	24200	2850
Time saved (%)	—	88	—	88

TABLE 3: THD result of VSCs under uniform variation of power.

	I_{THD} (%)		V_{THD} (%)	
	Mean	Std	Mean	Std
E-MCS	4.46	1.44	1.93	0.52
UDR	5.07	1.67	2.06	0.60
Diff	0.61	0.23	0.13	0.08

TABLE 4: THD result of VSCs under normal variation of power.

	I_{THD} (%)		V_{THD} (%)	
	Mean	Std	Mean	Std
E-MCS	4.61	1.62	1.95	0.51
UDR	4.72	1.19	2.25	0.54
Diff	0.11	−0.43	0.30	0.03

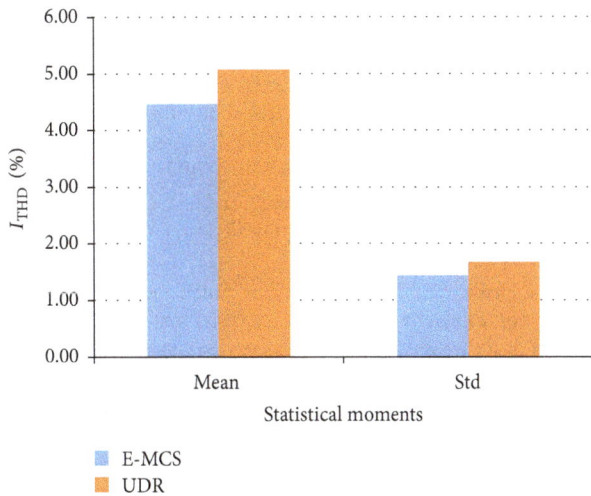

FIGURE 11: Predicted IgaTHD using E-MCS and rUDR 5 pts for Uniform Power Variation.

7.2. Case Study 2: Prediction under Normal Power Variation.

In this case study, the output power was varied randomly following a Gaussian distribution (mean value = 3.44 kW and Std = $2.02e + 3$). Just as in Case 1, the VSCs power was varied with thousands of random power inputs to mimic the MCS approach. The statistical information of the distribution is recorded and the THD measured. Then the 5 pts UDR sigma points and weights were generated and inputted to evaluate the performance of the UDR technique using the E-MCS as a benchmark. The results are given in Table 4 and Figures 13 and 14.

Table 4 and Figures 13 and 14 show a good match between the UDR and the E-MCS mean values for I_{THD} and V_{THD}.

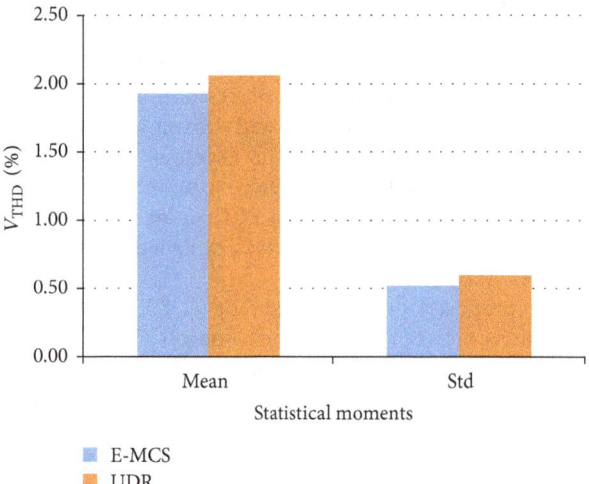

FIGURE 12: Predicted VgaTHD using E-MCS and rUDR 5 pts for Uniform Power Variation.

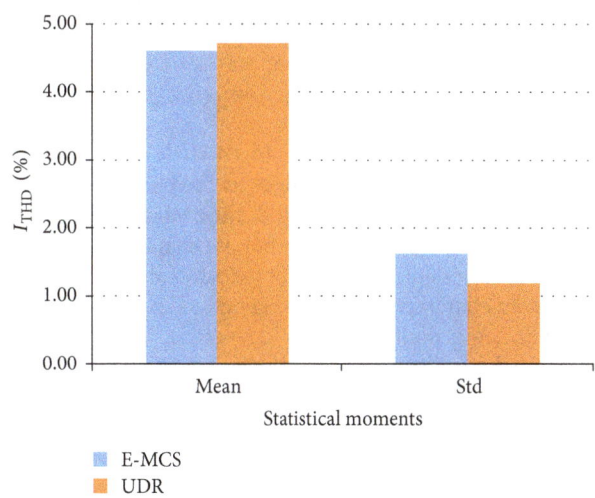

FIGURE 13: Predicted IgaTHD using E-MCS and rUDR 5 pts for Gaussian Power Variation.

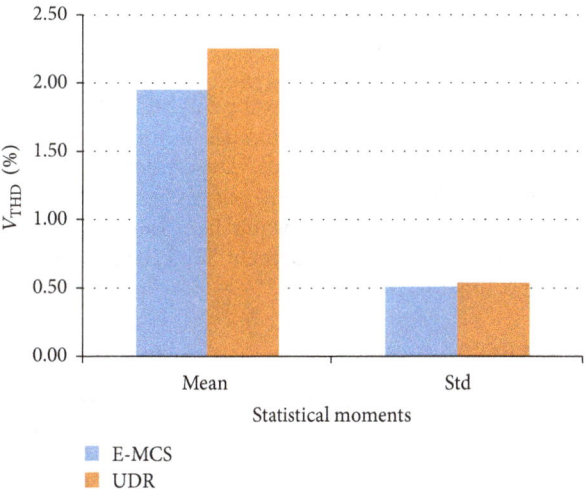

FIGURE 14: Predicted VgaTHD using E-MCS and rUDR 5 pts for Gaussian Power Variation.

8. Conclusion

Current and voltage harmonic distortion level of many connected converters have been successfully predicted using the 3 and 5 points' approximated Univariate Dimension Reduction (UDR) technique. The effects of the variations in filter inductor value and operating power on harmonic distortion levels of VSCs at the PCC of an EPS were also presented. The EPS test system was designed using PLECS and MATLAB simulation tools.

The proposed prediction method (UDR) ensured full interactions between the harmonic sources (VSCs) and the entire EPS in predicting the THD at the PCC unlike most analytical techniques where the harmonic sources are assumed to be independent.

It was observed from the results that the 3 and 5 points' reduced UDR (rUDR) technique effectively predicted the mean and standard deviation of the harmonics at the PCC of the EPS and can be used as an alternative predictive tool for the Monte Carlo Simulation approach. One significant advantage of the rUDR highlighted in this study is its drastic reduction of computational time and burden in predicting harmonics. The UDR technique was also experimentally validated and the results obtained were in good agreement with the E-MCS styled approach.

The study showed that the reduced UDR can be used by power companies/design engineers in the choice of parameters since the effects of real world uncertainties possible in the operation of modern EPS are being taken into account. It also provides possible outcomes of variation of design parameter/system characteristics on the generated harmonics of an EPS containing multiple power converters. This technique can be further utilized on other power converters, nonlinear loads, and harmonic sources and can also be applied to predict other harmonic indices like the individual harmonic distortions of currents and voltages.

Conflicts of Interest

The authors declare that there are no conflicts of interest regarding the publication of this paper.

Acknowledgments

This manuscript is derived from the following thesis. P. M. Ivry, "Predicting Stochastic Harmonics of Multiple Converters in a Power System (Microgrid)," Department of Electrical and Electronic Engineering, University of Nottingham, Nottingham, 2016. The first author would like to acknowledge the Petroleum Technology Development Fund (PTDF), Nigeria, for supporting this research.

References

[1] A. Yazdani and R. Iravani, *Voltage-Sourced Converters in Power Systems: Modeling, Control, and Applications*, John Wiley and Sons, 2010.

[2] N. Mohan, T. M. Undeland, and W. P. Robbins, *Power Electronics, Converters, Applications and Design*, John Wiley and Sons, Third edition, 2003.

[3] P. M. Ivry, M. J. Rawa, D. W. P. Thomas, and M. Sumner, "Power quality of a voltage source converter in a smart grid," in *Proceedings of the IEEE Grenoble Conference PowerTech (POWERTECH '13)*, Grenoble, France, 2013.

[4] Y. Baghzouz, R. F. Burch, A. Capasso et al., "Time-varying harmonics: part I: characterizing measured data," *IEEE Transactions on Power Delivery*, vol. 13, no. 3, pp. 938–944, 1998.

[5] M. H. Kalos and P. A. Whitlock, *Monte Carlo Methods*, Wiley-VCH Verlag GmbH & Co. KGaA, Weinheim, Germany, 2nd edition, 2008.

[6] J. He, J. Jiang, J. Huang, and W. Chen, "Model of EMI coupling paths for an off-line power converter," in *Proceedings of the 19th Annual IEEE Applied Power Electronics Conference and Exposition (APEC '04)*, vol. 2, pp. 708–713, Anaheim, Calif, USA, 2004.

[7] M. Ferber, C. Vollaire, L. Krähenbühl, J.-L. Coulomb, and J. A. Vasconcelos, "Conducted EMI of DC-DC converters with parametric uncertainties," *IEEE Transactions on Electromagnetic Compatibility*, vol. 55, no. 4, pp. 699–706, 2013.

[8] S. R. Kaprielian, A. E. Emanuel, R. V. Dwyer, and H. Mehta, "Predicting voltage distortion in a system with multiple random harmonic sources," *IEEE Transactions on Power Delivery*, vol. 9, no. 3, pp. 1632–1638, 1994.

[9] P. T. Staats, W. M. Grady, A. Arapostathis, and R. S. Thallam, "A statistical method for predicting the net harmonic currents generated by a concentration of electric vehicle battery chargers," *IEEE Transactions on Power Delivery*, vol. 12, no. 3, pp. 1258–1264, 1997.

[10] K.-H. Liu, "Analysis of probabilistic harmonic currents and voltages of electronic power converter contributed in distribution system," *Research Journal of Applied Sciences, Engineering and Technology*, vol. 5, no. 4, pp. 1263–1270, 2013.

[11] J.-H. Teng, R.-C. Leou, C.-Y. Chang, and S.-Y. Chan, "Harmonic current predictors for wind turbines," *Energies*, vol. 6, no. 3, pp. 1314–1328, 2013.

[12] Y. G. Hegazy and M. M. A. Salama, "Calculations of diversified harmonic currents in multiple converter systems," in *Proceedings of the Power Engineering Society Summer Meeting*, vol. 2, pp. 727–731, Seattle, Washington DC, USA, 2000.

[13] Y. J. Wang and L. Pierrat, "Summation of harmonic currents produced by AC/DC static power converters with randomly fluctuating loads," *IEEE Transactions on Power Delivery*, vol. 9, no. 2, pp. 1129–1135, 1994.

[14] E. Ngandui, E. J. Mohammed, and A. Cheriti, "Prediction of harmonics produced by multiple variable speed drives with randomly fluctuating loads," in *Proceedings of the CCECE 2000-Canadian Conference on Electrical and Computer Egineering*, vol. 2, pp. 1157–1161, Seattle, Washington DC, USA, 2000.

[15] H. V. Haghi and M. T. Bina, "A study on probabilistic evaluation of harmonic levels produced by static compensators," in *Proceedings of the Australasian Universities Power Engineering Conference*, pp. 1–6, Sydney, Australia, 2008.

[16] D. G. Infield, P. Onions, A. D. Simmons, and G. A. Smith, "Power quality from multiple grid-connected single-phase inverters," *IEEE Transactions on Power Delivery*, vol. 19, no. 4, pp. 1983–1989, 2004.

[17] MathWorks, "Monte Carlo simulation," 09 March, 2015, http://uk.mathworks.com/discovery/monte-carlo-simulation.html.

[18] P. Jorgensen, J. Christensen, and J. Tande, "Probabilistic load flow calculation using Monte Carlo techniques for distribution network with wind turbines," in *Proceedings of the International*

Conference on Harmonics and Quality of Power, vol. 2, pp. 1146–1151, Athens, Greece, 1998.

[19] D. W. P. Thomas, O. A. Oke, L. R. A. X. De Menezes, and C. Christopoulos, "The use of unscented transforms in modeling the statistical response of nonlinear scatterer in a reverberation chamber," in *Proceedings of the 30th URSI General Assembly and Scientific Symposium (URSIGASS '11)*, pp. 1–4, Turkey, 2011.

[20] L. R. A. X. De Menezes, D. W. P. Thomas, C. Christopoulos, A. Ajayi, and P. Sewell, "The use of unscented transforms for statistical analysis in EMC," in *Proceedings of the IEEE International Symposium on Electromagnetic Compatibility (EMC '08)*, pp. 1–5, Hamburg, Germany, 2008.

[21] J. M. Morales and J. Pérez-Ruiz, "Point estimate schemes to solve the probabilistic power flow," *IEEE Transactions on Power Systems*, vol. 22, no. 4, pp. 1594–1601, 2007.

[22] O. A. Oke, *Enhanced Unscented Transform Method for Probabilistic Load Flow Studies*, Electrical and Electronics Engineering, University of Nottingham, Nottingham, UK, 2013.

[23] S. Rahman and H. Xu, "A univariate dimension-reduction method for multi-dimensional integration in stochastic mechanics," *Probabilistic Engineering Mechanics*, vol. 19, no. 4, pp. 393–408, 2004.

[24] K. H. Ahmed, S. J. Finney, and B. W. Williams, "Passive filter design for three-phase inverter interfacing in distributed generation," in *Proceedings of the 5th International Conference-Workshop Compatibility in Power Electronics (CPE '07)*, pp. 1–9, Poland, 2007.

[25] M. Liserre, F. Blaabjerg, and A. Dell'aquila, "Step-by-step design procedure for a grid-connected three-phase PWM voltage source converter," *International Journal of Electronics*, vol. 91, no. 8, pp. 445–460, 2004.

[26] A. A. Rockhill, M. Liserre, R. Teodorescu, and P. Rodriguez, "Grid-filter design for a multimegawatt medium-voltage voltage-source inverter," *IEEE Transactions on Industrial Electronics*, vol. 58, no. 4, pp. 1205–1217, 2011.

[27] R. Teodorescu, M. Liserre, and P. Rodríguez, *Grid Converters for Photovoltaic and Wind Power Systems*, John Wiley and Sons, 2011.

[28] I. Sefa, N. Altin, and S. Ozdemir, "An implementation of grid interactive inverter with reactive power support capability for renewable energy sources," in *Proceedings of the 3rd IEEE International Conference on Power Engineering, Energy and Electrical Drives (PowerEng '11)*, pp. 1–6, Spain, 2011.

[29] P. M. Ivry, *Predicting stochastic harmonics of multiple converters in a power system (microgrid) [Ph.D. thesis]*, Electrical and Electronic Engineering, University of Nottingham, Nottingham, UK, 2016.

[30] D. W. P. Thomas, O. A. Oke, and C. Smartt, "Statistical analysis in EMC using dimension reduction methods," in *Proceedings of the IEEE International Symposium on Electromagnetic Compatibility (EMC '14)*, pp. 316–321, Raleigh, NC, USA, 2014.

[31] O. A. Oke, D. W. P. Thomas, and G. M. Asher, "A new probabilistic load flow method for systems with wind penetration," in *Proceedings of the IEEE PES Trondheim PowerTech: The Power of Technology for a Sustainable Society (POWERTECH '11)*, pp. 1–6, Norway, 2011.

[32] W. Gautschi, *Numerical Mathematics and Scientific Computation*, Oxford University Press, Oxford, UK, 2004.

[33] H. Xu and S. Rahman, "A generalized dimension-reduction method for multidimensional integration in stochastic mechanics," *International Journal for Numerical Methods in Engineering*, vol. 61, no. 12, pp. 1992–2019, 2004.

[34] I. Lee, K. K. Choi, L. Du, and D. Gorsich, "Dimension reduction method for reliability-based robust design optimization," *Computers and Structures*, vol. 86, no. 13-14, pp. 1550–1562, 2008.

[35] B. Zou and Q. Xiao, "Probabilistic load flow computation using univariate dimension reduction method," *International Transactions on Electrical Energy Systems*, vol. 24, no. 12, pp. 1700–1714, 2014.

[36] F. J. Ruiz-Rodriguez, J. C. Hernández, and F. Jurado, "Probabilistic load flow for photovoltaic distributed generation using the Cornish-Fisher expansion," *Electric Power Systems Research*, vol. 89, pp. 129–138, 2012.

[37] H. Bayem, M. Petit, P. Dessante, F. Dufourd, and R. Belhomme, "Probabilistic characterization of wind farms for grid connection studies," in *Proceedings of the European Wind Energy Conference and Exhibition 2007 (EWEC '07)*, pp. 1482–1489, Tampa, Fla, USA, 2007.

[38] PLECS 3.5.1, "Plexim GmbH," Zurich, 2014.

[39] MATLAB 8.1.0.604 (R2013a), MathWorks, Inc., 2013.

[40] P. M. Ivry, M. J. Rawa, and D. W. P. Thomas, "Factors affecting the harmonics generated by a generic voltage source converter within a microgrid," in *Proceedings of the Saudi Arabia Smart Grid Conference*, Jeddah, 2012, pp. B-65.

[41] E. E. Ahmed and W. Xu, "Assessment of harmonic distortion level considering the interaction between distributed three-phase harmonic sources and power grid," *IET Generation, Transmission and Distribution*, vol. 1, no. 3, pp. 506–515, 2007.

[42] A. Mansoor, W. M. Grady, A. H. Chowdhury, and M. J. Samotyj, "An investigation of harmonics attenuation and diversity among distributed single-phase power electronic loads," *IEEE Transactions on Power Delivery*, vol. 10, no. 1, pp. 467–473, 1995.

17

A Novel Hybrid T-Type Three-Level Inverter based on SVPWM for PV Application

Ayiguzhali Tuluhong⑩,[1,2] Weiqing Wang⑩,[1] Yongdong Li,[2] and Lie Xu[2]

[1]Electrical Engineering, Xinjiang University, Urumqi, Xinjiang 830046, China
[2]State Key Laboratory of Power System, Department of Electrical Engineering, Tsinghua University, Beijing 100084, China

Correspondence should be addressed to Weiqing Wang; wwq59@xju.edu.cn

Academic Editor: Yongheng Yang

We describe several, recently reported, new topologies and compare them with each other, in order to find out the optimal multilevel grid-connected inverter topology. Then, we classify these topologies according to the basic unit which predecessors proposed. Eventually, we propose the hybrid T-type inverter topology structure, which is composed of two best basic units. This structure takes full advantage of the two components, to reduce the harmonic content and the power loss of the converter and improve the conversion efficiency of the system. At the same time, the space vector pulse width modulation (SVPWM) method is used to simulate the proposed topology in the MATLAB/SIMULINK platform, while the loss of each semiconductor switch is calculated using MELCOSIM software. The results show that the proposed structure is superior to the most widely used topology, i.e., the diode clamped and the T-type three-level circuits.

1. Introduction

Nowadays, the contradictions between the consumption pattern of the traditional petrochemical energy and the economic development and environmental protection are becoming more and more prominent. People gradually realize the importance of taking sustainable development road, vigorous developing, and utilizing of the renewable energies. Among the renewable energies, the solar energy represents the largest and the most commonly distributed resource. The photovoltaic power generation technology using the solar cells effectively absorbs the solar energy and changes it into electricity. The grid-connected inverter is the key component and important equipment in a photovoltaic grid-connected system.

In the design of general inverters, synthetically considering the cost-effective factors, insulated-gate bipolar transistor (IGBT) represents the most employed device. However, due to the nonlinearity of the IGBT's conduction voltage drop, it does not significantly increase with the increase of current, thus ensuring that the inverter still presents a relatively low loss and high efficiency at the maximum load condition. However, since the European efficiency is mainly related to the efficiency of the inverter at different light-load, the aforementioned characteristics of the IGBT represent the disadvantage of the photovoltaic grid-connected inverter. For light loads, the turn-on voltage drop of the IGBT does not significantly reduce, which in turn reduces the European efficiency of the inverter. In contrast, due to linear conduction voltage drop of the MOSFET, it provides lower turn-on voltage drop for light loads. Considering the excellent dynamic characteristics and high frequency work ability, MOSFET becomes the first choice for photovoltaic inverting [1].

Three-level inverter has been widely used in the middle and high voltage large capacity AC speed regulating fields, since its output has higher power quality, lower harmonic contents, better electromagnetic compatibility, lower switching losses, and other advantages. However, it still suffers from some key problems, including the simplification of the three-level algorithm, neutral point voltage control in the overmodulation region, and the stability of the system at high voltage. In view of the above problems, this paper studies

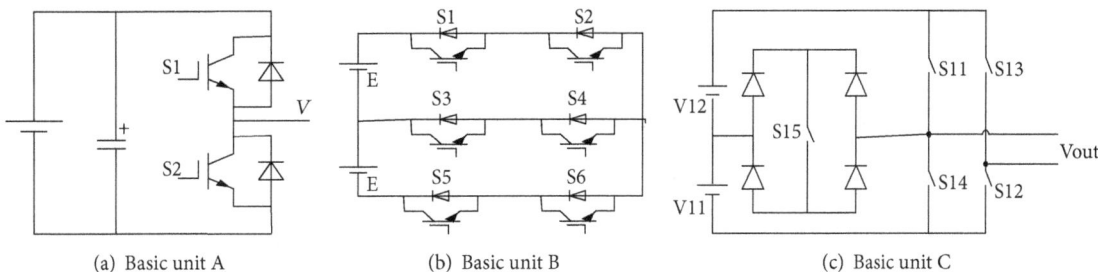

(a) Basic unit A (b) Basic unit B (c) Basic unit C

FIGURE 1: Basic unit of the inverter.

the structure and principle of the three-level inverter, the control of the neutral point voltage of the capacitor, and the realization of the SVPWM algorithm.

In this paper, we study novel T-type inverter topology in PV system using SVPWM control algorithm. The structure is organized as follows: Section 2.1 introduces basic cells of the new multilevel PV inverters and classifies them. Section 2.2 presents and compares new types of multilevel inverters. Section 2.3 analyzes and compares switch losses and conversion efficiency of diode clamped T-type and proposed hybrid T-type. Section 2.4 details SVPWM control algorithm. Section 3 gives simulation results. Finally, Section 4 concludes the paper.

2. Materials and Methods

2.1. Basic Unit. Inverter basic unit refers to the minimum component that meets the demands of the topological multilevel grid-connected inverter. Odeh [2] proposed one basic concept of the basic unit of multilevel grid-connected inverter.

As shown in Figure 1(a), after achieving the basic unit, it may be extended in form of series-parallel and parallel-series connections to present higher voltage levels. Odeh [3] analyzes the connection between the topologies of the multilevel grid-connected inverters and proposes a regular pattern that is followed when general multilevel grid-connected inverter topology simplifies to other multilevel grid-connected inverter topology. For example, the main switches must be all maintained, all switching devices must be deleted symmetrically, diodes and capacitors must be distributed symmetrically, and so on. With the development of power electronic devices, in order to unify the pressure drop of the power electronic switch transistors, a more practical basic unit is put forward.

Figure 1(b) shows the stacked commutation cells or three-pole cells. As the employed conduction strategy, one of the two transistors in the outer leg is in a frequent switching state only during half-cycle, while the other transistor is switched only once during the fundamental wave period, which greatly reduces the switching loss. This conduction strategy also avoids the voltage-equalization problem caused by the simultaneous turn-on and turn-off of the series devices. Figure 1(c) shows the structure of a multilevel inverter basic unit reported by Draxe et al. [4]. This inverter is formed using

five switches with antiparallel diodes, where S11 to S14 are arranged as the traditional CHB inverter, while, however, S15 is added to increase the output voltage levels by selecting the appropriate voltage source. This structure produces higher voltage levels with minimum number of switches by optimizing the circuit layout and reducing the gate drive circuitry.

Recently reported inverter topologies aim to reduce the number of the power electronic devices to improve the conversion efficiency. Figure 2 shows the classification of the inverters based on their topologies. As we see from this figure, the largest portion of the proposed inverter topologies formed by the basic unit A, due to its relatively simple construction. The topologies composed of basic units B and C have not yet appeared because of their complex basic constitution unit.

The basic unit of the inverter consists of a DC power supply and a pair of switches. By using the same basic unit, taking Figure 1(a), for example, in the form of series-parallel combinations a new circuit topology, as shown in Figure 3, called single phase H-bridge topology, is obtained. Or by series-parallel combinations between different basic units also a new one is got. For instance, three parallel basic units A and one basic unit B constitute the three-level T-type inverter topology, as shown in Figure 4. Different multilevel inverters can be obtained by multiseries and multiparallel connections of multiple basic units. And the construction of the other topologies in Figure 2 can be deduced by the same analogy.

2.2. New Type Inverter Topology. The work in [5] proposed the split capacitor H-bridge (SC-HB) inverter topology. Adding a simple DC-DC converter, this topology overcomes the capacitance-voltage balance problem, while reducing the leakage current, with improved efficiency. The work in [2] presented the improved cascaded H-bridge topology which consists of a half-bridge level latching circuit and a main inverter H-bridge. Thus, it reduces the number of the devices, switching loss, and harmonic content. An asymmetric cascaded H-bridge topology, with different, and proportional, DC voltage source values is reported in [4]. This reduces the capacity and number of the devices, as well as the costs. A sine-wave pulse width modulation (SPWM) three-phase multilevel inverter topology may be achieved by inserting two auxiliary switches in each phase bridge to change the basic two-level to three-level inverter to synthesize higher levels [3]. In comparison with traditional diode clamps, flying

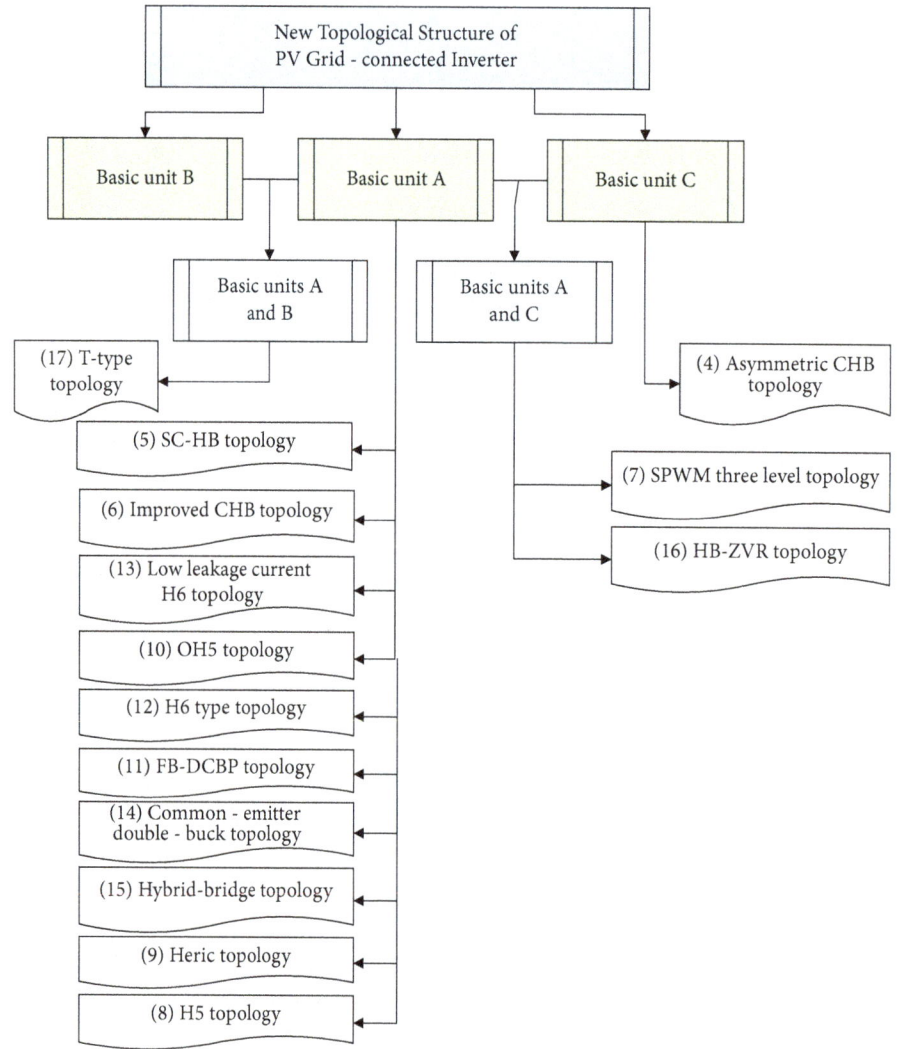

FIGURE 2: Per-phase circuit configuration of the conventional multilevel inverter.

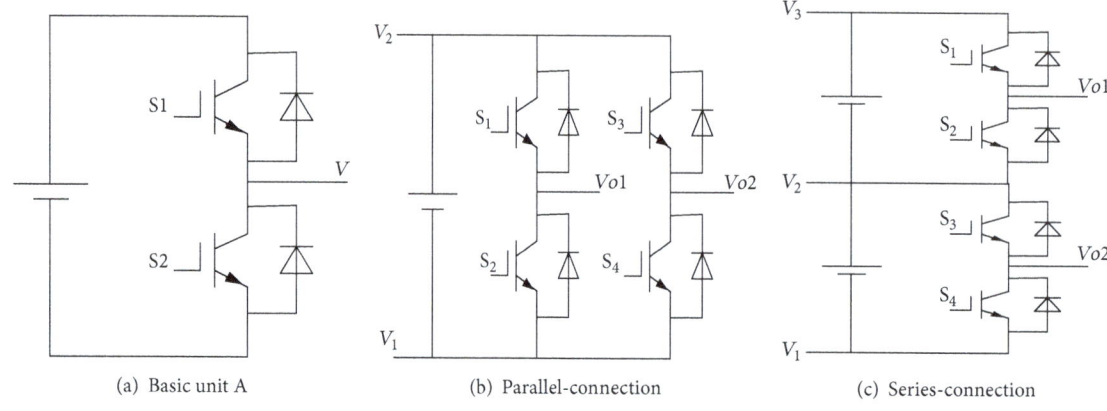

(a) Basic unit A   (b) Parallel-connection   (c) Series-connection

FIGURE 3: Parallel-series-connection topology of the same basic units.

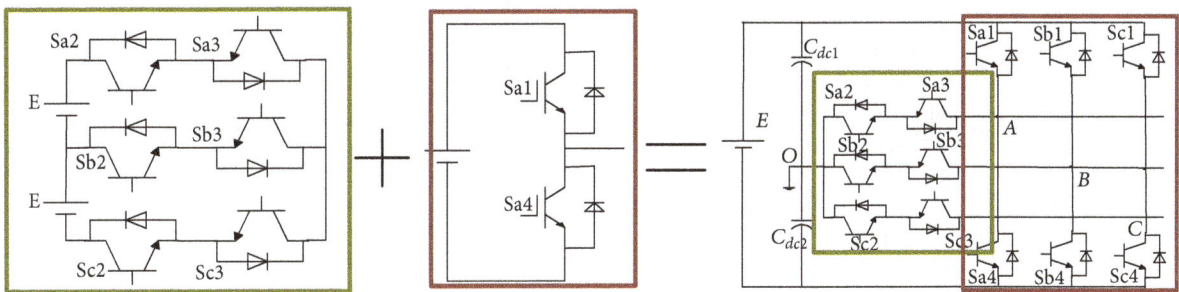

FIGURE 4: Parallel-series-connection topology of the different basic units.

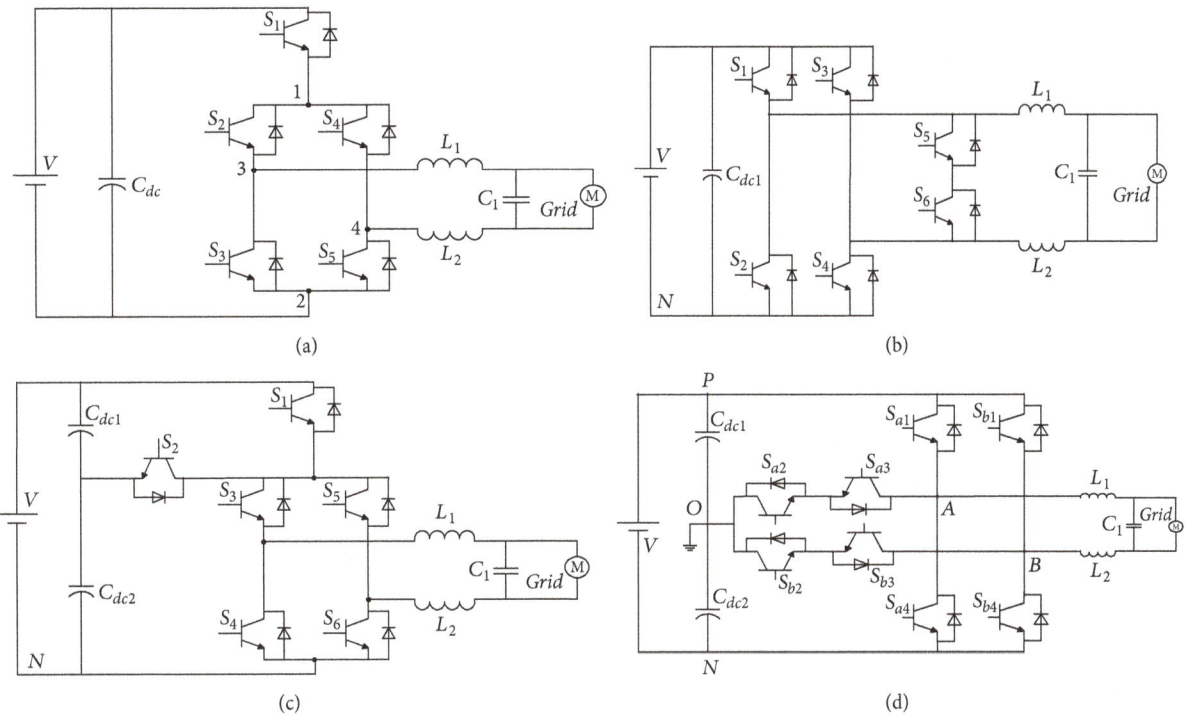

FIGURE 5: Some new topology structure. (a) H5 topology, (b) Heric topology, (c) OH5 topology, and (d) T-type topology.

capacitors, and cascaded H-bridge inverters, this topology uses only a DC power supply and lower power electronics providing the same number of the levels.

Figure 5(a) shows the H5 topology [6], composed of a freewheeling circuit by adding a high frequency switch in the positive end of the DC input side, where the efficiency may reach up to 98.1%. However, it suffers from big loss and heat and unbalanced thermal stress. Figure 5(b) shows the structure of the Heric topology [7], where a new freewheeling circuit is added, which itself is composed of a set of bidirectional switch branches on the AC side, based on the traditional full-bridge inverter topology. The efficiency of this topology may reach more than 98%. The OH5 topology is shown in Figure 5(c) [8], which uses a switch and capacitor to form the bidirectional clamping circuit on the basis of the H5 topology, which greatly inhibits the leakage current, and improves the conversion efficiency. However, in such a structure, the clamping circuit is constrained by the dead time of the switch, and a large on-state loss is presented. The work

in [9] proposes the FB-DCBP topology, which achieves the complete elimination of the leakage current by adding two controllable switch and clamp diodes in the DC side to format the clamping circuit. The leakage current suppression capability of this topology is stronger than that of the H5 and Heric topologies but suffers from large on-state loss and high number of devices, providing the maximum efficiency of 97.4%.

H6-type topology is proposed in [10], and it is a variant of the Heric topology. It does not require setting dead time between the switches of the same bridge arm and does not pass through the body diode of the switch. Therefore, all switches may be formed by MOSFET, but the topology requires extra two freewheeling diodes, so the cost is increased. Moreover, the efficiency is lower than the Heric topology, and the efficiency is up to 97.6%. The low-leakage current H6 topology is reported in [11], which is a compromise between the H5 and Heric topologies. A new freewheeling circuit is formed based on the H5 topology by adding a switch to make the number of switches through two half-frequency cycles of the network current at power transfer mode not the same, yielding reduced on-state loss. The efficiency of the H6 topology is lower than that of the Heric topology and higher than that of H5 topology. The common mode is inferior to H5 topology and superior to Heric topology. A common emitter double BUCK topology may be considered [12], composed of four controllable switches and two diodes. The leakage current of this structure is basically zero, and there is no high frequency dead zone between the switches. In addition, this structure presents no break-through phenomenon, high reliability, and small switching loss, where the efficiency may reach more than 98.5%. However, the topological magnetic component utilization is low and cannot output reactive current.

The hybrid bridge topology proposed in [13] consists of six controllable switches and two freewheeling diodes. This topology moved the switch on the AC side of the Heric topology to the middle of the bridge arm A, similar to H6-type topology. HB-ZVR (H-bridge zero-voltage state rectifier) topology is proposed in [14], introducing an AC bypass circuit, which itself is composed of a IGBT and a group of diode rectifiers, which is clamped to the midpoint of the two capacitors on the DC bus, to achieve low common mode current and high efficiency inverter topology. The HB-ZVR topology solves the problem that bidirectional switches S5 and S6 of the Heric topology cannot turn on, and this topology will always find a way in the bidirectional switch. Figure 5(d) shows the T-type inverter topology [15], formed by a set of switches Sa3/Sa4 as a bidirectional switch to achieve the main switch Sa1/Sa2 clamping function. It uses the clamp diodes or clamp bit capacitance to improve the midpoint clamping circuit, reducing the number of devices and uneven distribution of the loss. In the topological selection, T-type three-level circuit leverages many advantages of the nonisolation technology as well as the multilevel technology. Therefore, it is very suitable for the photovoltaic grid-connected power generation occasions; however, it is required to effectively suppress the circuit leakage current and system efficiency.

By comparing the above different new topologies, we see that the proposed multilevel inverter topology optimizes the performances of the inverter by adding auxiliary/clamping circuit, using hybrid switch or asymmetric structure.

Figure 6 shows the topology of the hybrid T-type inverter, which is on the basis of T-type structure and composed of nine MOSFET switches, i.e., Sa2, Sb2, Sc2,..., Sa4, Sb4, Sc4. We choose IGBT for Sa1, Sb1, and Sc1, since the reverse recovery ability of the body diode in the field effect transistor is poor; therefore they act in low frequencies. The high frequency MOSFET semiconductor switches, i.e., Sa2, Sb2, Sc2,..., Sa4, Sb4, Sc4, provide good switching characteristics and low on-resistance. Moreover, due to the low speed characteristics of the built-in diode of MOSFETs, MOSFET cannot be used in the upper bridge arm. We take advantage of the two devices, to reduce the harmonic content and the power loss of the converter and improve the conversion efficiency of the system.

2.3. Loss Analysis of the Proposed Topology. The high power inverter works in high voltage and large current situations,

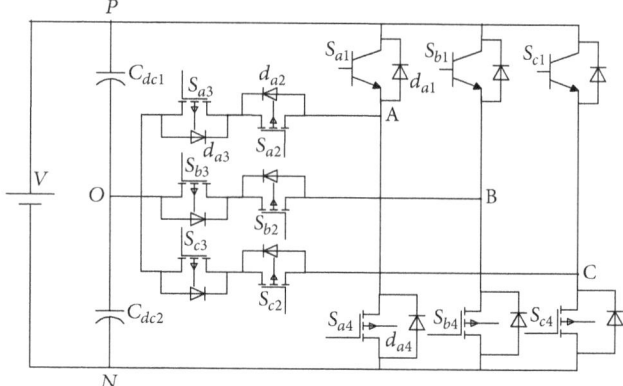

FIGURE 6: The proposed hybrid T-type inverter.

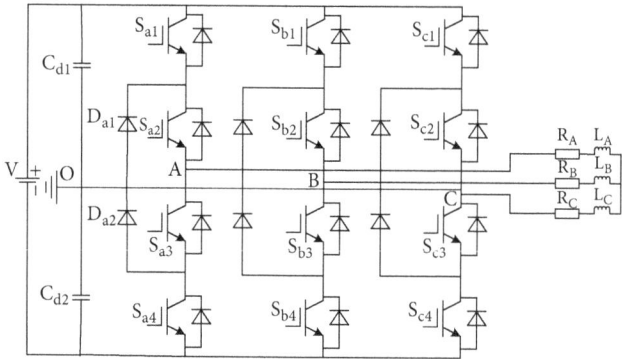

FIGURE 7: Diode clamped three-level inverter.

where various losses caused by the opening device are relatively large. Besides, optimization of the switching characteristics of the power electronic devices yields bigger conduction losses of the system, especially for soft switching technology, where the switching loss of the power electronic devices is reduced, and the source of the power loss is converted to the conduction loss [16]. Therefore, it is a key step to accurately calculate the state loss for the design of grid-connected inverter systems. Take A phase of Figures 6 and 7 as an example, we compare the various losses of diode clamped (Figure 7) and hybrid T-type (Figure 6) inverter losses and their conversion efficiency. We assume that the output current of the grid-connected inverter is an ideal sine wave; the output voltage of the inverter integrates the conduction period in the period T. Then, the conduction losses per device may be expressed as

$$P_{s_{a1}} = \frac{1}{2\pi} \left[\int_0^{\pi-\theta} d(\omega t) U_{com}(i) \left(I_{com} \sin(\omega t) d(\omega t) \right) \right] \quad (1)$$

where $P_{s_{a1}}$ is the conduction loss of the per device, $d(\omega t)$ is the duty cycle, $U_{com}(i)$ represents the conduction voltage drop of the IGBT, I_{com} denotes the conduction current peak of the IGBT, and ω is the angle speed. The integral interval from 0 to $\pi-\theta$ is a chopper phase of the semiconductor switch Sa1 in a fundamental period. The conduction loss of the antiparallel diode in the IGBT reads

$$P_{da1} = \frac{1}{2\pi} \int_{\pi-\theta}^{\pi} (-m * \sin(\omega t + \phi)) \left[U_{od} I_{com} \sin(\omega t) \right.$$
$$\left. + R_{od} (I_{com} \sin(\omega t))^2 \right] d(\omega t) \quad (2)$$

where P_{da1} is the conduction loss of the antiparallel diode in the IGBT, m is the modulation index, Φ is the output power factor angles, and R_{od} and U_{od} are the conduction pressure drop constants of the antiparallel diode.

When IGBT/MOSFET is turned off, the current flows from the clamped diodes and then reads the following clamped diode conduction loss:

$$P_{Da1} = \frac{1}{2\pi} \int_0^{\pi-\theta} (1 - m * \sin(\omega t + \phi))$$
$$\cdot \left[U_{dt} I_{com} \sin(\omega t) \right.$$
$$\left. + R_{dt} (I_{com} \sin(\omega t))^2 \right] d(\omega t) + \frac{1}{2\pi} \quad (3)$$
$$\cdot \int_0^{\pi-\theta} (1 + m * \sin(\omega t + \phi)) \left[U_{dt} I_{com} \sin(\omega t) \right.$$
$$\left. + R_{dt} (I_{com} \sin(\omega t))^2 \right] d(\omega t)$$

where U_{dt} and R_{dt} are the conduction pressure drop constants of the clamped diode.

We considered the DC side voltage of 700 V, carrier frequency fc of 5 kHz, gate resistance of 1.65 ohm, output frequency equal to 60 Hz, modulation rate of unity, and power factor of 0.8. We used MELCOSIM software to simulate the loss. Table 1 presents the calculated loss of different components.

In Table 1, the power losses of Sa1 and Sa4 as well as Sa2 and Sa3 IGBT switches are identical. The four antiparallel diode power dissipations, da1, da2, da3, and da4, are also identical. Moreover, clamping diodes, Da1 and Da2, have the same loss. Table 2 lists the overall loss of the two topologies, which may be calculated based on the number of components contained in two topologies. This table also presents the total power of the inverter, which is 10 kW, and the conversion efficiency of each topology.

The conversion efficiency is lower than 90% because of the higher switching frequency. We chose the frequency of 10 Khz MOSFET, and the switching frequency is relatively high; at the same time the switching loss itself is higher than at power frequency, so the conversion efficiency is lower than 90%.

Table 3 shows the comparison between the NP and T-type three-level inverters.

2.4. SVPWM Control Method.

The theoretical basis of the SVPWM is the mean equivalent principle, that is, combining the fundamental voltage vectors in a switching cycle to make the average value equal to the given voltage vector. At a certain moment, the voltage vector rotates into a certain region, which may be achieved by a different combination over time of two adjacent nonzero vectors and zero vectors that make up this region. The action time of the two vectors

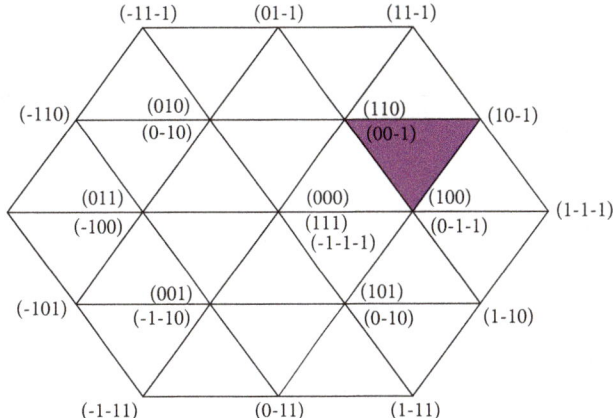

FIGURE 8: Three-level space vector diagram.

is repeatedly applied in one sampling period, so that it controls the action time of each voltage vector. This rotates the voltage space vector in accordance with the circular trajectory and approaching ideal flux circle through the actual magnetic flux, generated by different switching states of the inverter. Then, it determines the inverter switch state by the comparison, at the end form of the PWM waveform.

SVPWM presents lower total harmonic distortion (THD) compared to other control strategies, and it may refine the steady and dynamic state performances of the PV grid-connected system, simultaneously. Meanwhile, output results of the inverters with SVPWM control strategy provide better power quality than that of the inverters with other control strategies. The SVPWM control strategy with three-phase three-level voltage source inverter is effective and feasible [17]. The detailed equations of the SVPWM strategies based on the proposed topology are as follows.

In this paper we used decomposition hexagon method.

Idea: decompose the multilevel space vector into a combination of multiple two-level space vectors to achieve greatly simplified PWM calculation method.

The three-level space vector diagram is shown in Figure 8. Any reference vector must fall within a small triangle. The vertex of this triangle is the basic voltage vector that composes this reference vector.

The reference vector falling into the shadow in Figure 8 can be decomposed into an offset vector and a two-level vector, as shown in Figure 9.

Steps

2.4.1. Sector Judgments. Based on Clark's transformation, the normalized output vectors transformed from abc to $\alpha\beta$ reference frame may be expressed as

$$\begin{pmatrix} V_\alpha \\ V_\beta \\ V_o \end{pmatrix} = \begin{pmatrix} \frac{2}{3} & -\frac{1}{3} & -\frac{1}{3} \\ 0 & \frac{1}{\sqrt{3}} & -\frac{1}{\sqrt{3}} \\ 1 & 1 & 1 \end{pmatrix} \begin{pmatrix} V_A \\ V_B \\ V_C \end{pmatrix} \quad (4)$$

TABLE 1: The loss of each device in hybrid T-type inverter.

Switch type	device	Loss (W)		
		Conduction loss	Switching loss	Total loss
IGBT	S_{a1}	90.45	26.46	116.92
	S_{a2}	31.78	1.03	32.82
Freewheeling diode	d_{a1}	40.59	0.47	41.07
Clamped diode	D_{a1}	2.23	9.93	12.17

TABLE 2: The power loss and conversion efficiency of two topologies.

component	The number of DC	Loss (W)			conversion efficiency
		IGBT	diode	Total loss	
NPC	1	998.88	310.72	1309.6	0.8690
Hybrid T-type	1	898.44	319.44	1217.88	0.8782

TABLE 3: Similarities and differences of diode clamped and T-type three-level circuit topologies.

Compared items	Diode clamped three-level inverter	T-type three-level inverter
Switch pressure	Sa1~Sa4: 0.5 UPV	Sa1/Sa2: Upv; Sa3/Sa4: 0.5 Upv
Commutation path	Long commutation path and short commutation path	All paths are consistent
Efficiency	The higher switching frequency (>16 kHZ) increases efficiency	The lower switching frequency (<16 kHZ) increases efficiency
Modulation strategy	Traditional control strategy of current PI loop	Traditional control strategy of current PI loop
Number of components	4 switches plus 2 diodes	4 switches
Drive power	4 groups	3 groups

TABLE 4: The relations between N and the sector where V_{ref} is located.

N	1	2	3	4	5	6
sector	II	VI	I	IV	III	V

where V_α, V_β, V_o are voltages on the two-phase stationary coordinate system and V_A, V_B, V_C are voltages on the three-phase stationary coordinate system.

We define $N = A+2B+4C$, and the value of N determines in which sector the reference vector V_{ref} is located. So the relations between N and the corresponding sector are shown in Table 4.

2.4.2. Basic Vector Dwell Time Calculation. We assume V_{ref} is located at sector I, which yields the following equation:

$$V_\alpha T_s = T_1 \left| V_1 \right| + T_2 \left| V_2 \right| \cos \frac{\pi}{3}$$

$$V_\beta T_s = T_2 \left| V_2 \right| \sin \frac{\pi}{3} \tag{5}$$

$$T_s = T_1 + T_2 + T_{0,7}$$

where T_s is the sampling period; T_1, T_2, and T_0 represent the dwell times of the basic vectors V_1, V_2, and V_0, respectively.

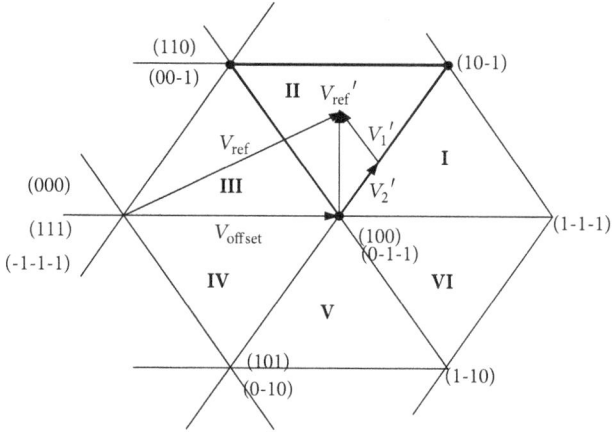

FIGURE 9: Decomposed two-level space vector diagram.

According to the above expressions, we can obtain the following equation:

$$T_1 = \frac{\sqrt{3} T_S}{2 V_{dc}} \left(\sqrt{3} V_\alpha - V_\beta \right)$$

$$T_2 = \frac{\sqrt{3} T_s V_\beta}{V_{dc}} \tag{6}$$

$$T_{0,7} = T_s - T_1 - T_2$$

TABLE 5: The relations between vector dwelling time and its sector.

sector	I	II	III	IV	V	VI
T_1	Z	Y	-Z	-X	X	-Y
T_2	Y	-X	X	Z	-Y	-Z

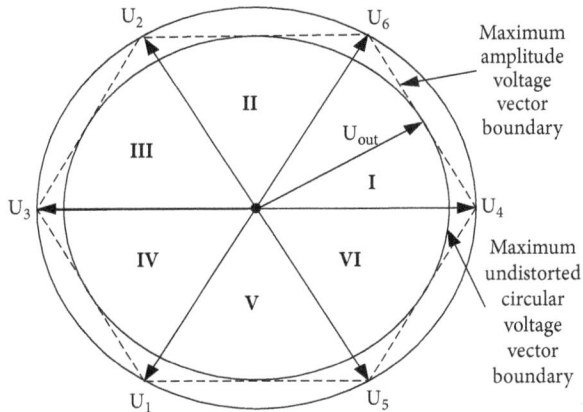

FIGURE 10: Voltage vector amplitude boundary in SVPWM mode.

Following the same principle, the dwell time of each vector, we achieve V_{ref} of different sectors. To facilitate the solution, we may define it as (7). The value of T_1 and T_2 at different sectors could be set according to Table 5.

$$X = \sqrt{3}\frac{T_S}{V_{dc}}V_\beta$$

$$Y = \frac{T_s}{V_{dc}}\left(\frac{\sqrt{3}}{2}V_\beta + \frac{\sqrt{3}}{2}V_\alpha\right) \qquad (7)$$

$$Z = \frac{T_s}{V_{dc}}\left(\frac{\sqrt{3}}{2}V_\beta - \frac{\sqrt{3}}{2}V_\alpha\right)$$

2.4.3. Vector Switching Point Calculation. When the synthesized voltage vector endpoint falls between the regular hexagon and the circumscribed circle, as shown in Figure 10, the overmodulation has occurred and the output voltage will be distorted. So we use a proportional scaling algorithm to control the overmodulation. The vector dwell time that first occurs in each sector is defined as T_{Nx}, and the vector dwell time that occurs after is defined as T_{Ny}. When $T_x + T_y \leqslant T_{NPWM}$, the vector endpoint is within the regular hexagon and no overmodulation occurs. When $T_{Nx} + T_{Ny} > T_{NPWM}$, the vector endpoint is beyond the regular hexagon and overmodulation occurs. The output waveform will be seriously distorted and must take the following measures.

Suppose that the nonzero vector dwell time is T'_{Nx}, T'_{Ny}, when the endpoint of the voltage vector trace is pulled back to the inscribed circle of the regular hexagon, and then there is a proportional relationship:

$$\frac{T'_{Nx}}{T_{Nx}} = \frac{T'_{Ny}}{T_{Ny}} \qquad (8)$$

Therefore, T'_{Nx}, T'_{Ny}, T_{N0}, T_{N7} can be obtained by the following formula:

$$T'_{Nx} = \frac{T_{Nx}}{T_{Nx} + T_{Ny}}T_{NPWM}$$

$$T'_{Ny} = \frac{T_{Ny}}{T_{Nx} + T_{Ny}}T_{NPWM} \qquad (9)$$

$$T_0 = T_7 = 0$$

According to the above process, the action time of two adjacent voltage space vectors and zero-voltage vectors in each sector can be obtained. The operation relationship is shown in Figure 11 when U_{ref} is in sector I. After the U_{ref} sector and the corresponding effective voltage vector are determined, according to the PWM modulation principle, the value of each corresponding comparator is calculated, and the operation relationship is as follows:

$$t_{aon} = \frac{\left(T_s - T_x - T_y\right)}{2}$$

$$t_{bon} = t_{aon} + T_x \qquad (10)$$

$$t_{con} = t_{bon} + T_y$$

Other sectors follow the same above principle.

Here, T_{cm1}, T_{cm2}, T_{cm3} denote the transistor's switching time. And the relation between sector switching point and its appropriate sector is tabulated in Table 6.

3. Results and Discussion

This study put forward a novel hybrid T-type inverter topology which is composed of basic units A and B on the basis of previous research studies. We established a three-phase three-level hybrid T-type photovoltaic grid-connected inverter topology model, which is shown in Figure 12, using MATLAB platform. Considering the A-phase bridge leg, for example, it consists of one half-bridge IGBT, one half-bridge MOSFET, and two neutral point MOSFETs. Switches Sa1 and Sa2 work in the mutual intermittent state, and switches Sa3 and Sa4 only work near the current zero-crossing point with high frequency. This topology is rather suited for the photovoltaic nonisolated AC system applications.

The topological structure is on the basis of T-type structure, changing the nine switches into MOSFET, i.e., Sa2, Sb2, Sc2⋯Sa4, Sb4, and Sc4. We choose IGBT for Sa1, Sb1, and Sc1; since the reverse recovery ability of the body diode in the field effect transistor is poor, therefore they act in low frequencies. The high frequency MOSFET semiconductor switches, i.e., Sa2, Sb2, Sc2,..., Sa4, Sb4, Sc4, provide good switching characteristics and low on-resistance. Moreover, due to the low speed characteristics of the built-in diode of MOSFETs, MOSFET cannot be used in the upper bridge arm. We take advantage of the two devices, to reduce the harmonic content and the power loss of the converter and improve the conversion efficiency of the system.

Regarding the topology selection, the three-level circuit combines the advantages of the nonisolation and multilevel

TABLE 6: The relationship between the vector switching point and its corresponding sector.

sector	I	II	III	IV	V	VI
T_a	T_{bon}	T_{aon}	T_{aon}	T_{con}	T_{con}	T_{bon}
T_b	T_{aon}	T_{con}	T_{bon}	T_{bon}	T_{aon}	T_{con}
T_c	T_{con}	T_{bon}	T_{con}	T_{aon}	T_{bon}	T_{aon}

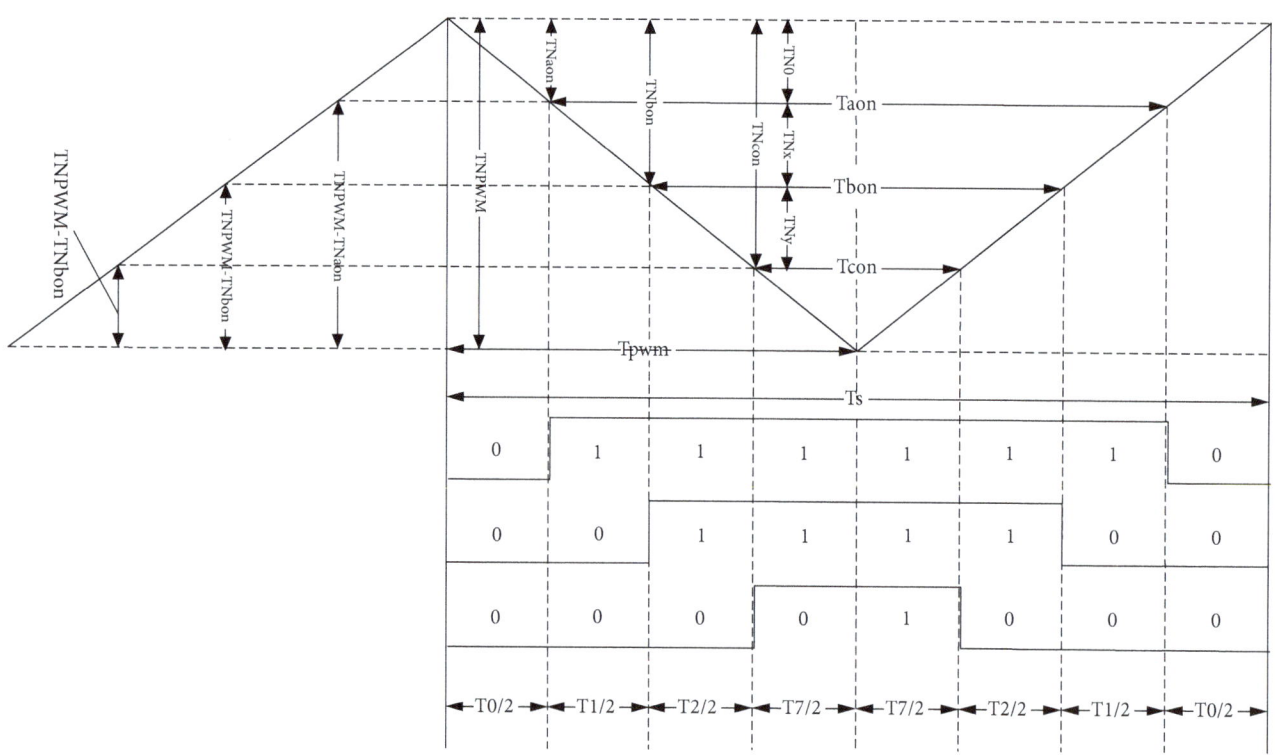

FIGURE 11: Operation relationship when U_{ref} is in sector I.

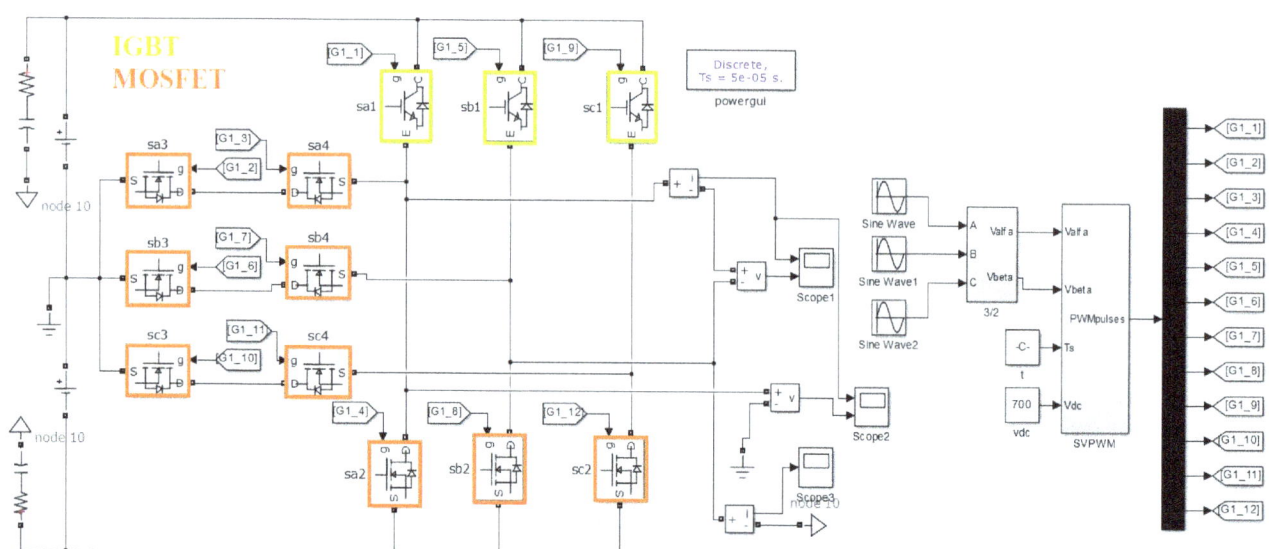

FIGURE 12: The simulation model of the SVPWM controlled novel T-type three-phase three-level inverter.

TABLE 7: The classification of the influence of the switching state of the short and medium vectors on the direction of the neutral point current.

Positive short vector switching status	io	Negative short vector switching status	io	Medium vector switching status	io
100	ia	211	- ia	210	ib
221	ic	110	- ic	120	ia
010	ib	121	- ib	021	ic
122	ia	011	- ia	012	ib
001	ic	112	- ic	102	ia
212	ib	101	- ib	201	ic

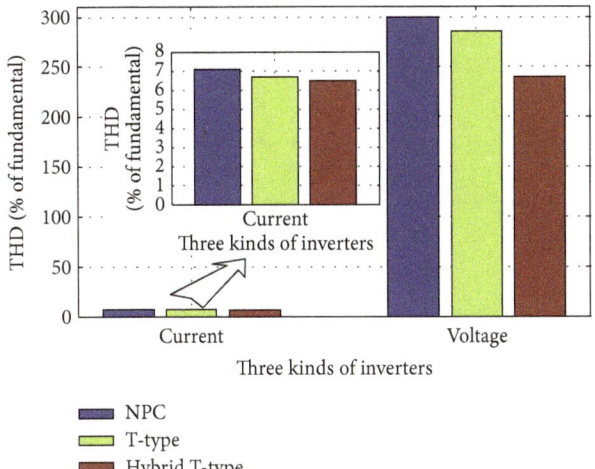

FIGURE 13: The current and voltage THD comparison of three types of inverters.

technologies, which is very suitable for the photovoltaic grid-connected power generation.

System simulation parameters are as follows: The DC-link voltage is 700 V, with the frequency of 50 Hz, the value of the support capacitor reads 3300 uF, with the AC-side inductance of 3 mH, the AC-side resistance is 0.1 Ω, and the switching frequency is 10 kHz. The value of the stray capacitance is $10e - 6$ F, and the small resistor is connected in parallel with the stray capacitance of $R = 10e - 6$ ohm.

Table 7 shows the neutral point current when the positive and negative short vector and medium vector act. It can be seen that positive and negative short vector and medium vectors will cause neutral voltage fluctuations.

The neutral point voltage control method is based on the SVPWM. According to the influence of the medium and short vectors on the neutral point voltage offset and by selecting the appropriate switching state and the most suitable switching sequence for the neutral point voltage, the midpoint voltage offset during each control cycle has been minimized.

So selecting the excellent transistor sequence is really important. To solve this problem we run through all the transistors in every possible combination and permutation on the condition of the following switch rules.

(1) The switching states of the devices in every bridge leg are independent. (2) The switching states of any two adjacent switches of each bridge leg are complementary (e.g., if Sa1 is turned on, then Sa2 must be switched off). (3) According to the principle of complementarity, if the switching state of any of the devices in the same bridge leg is determined, the state of the other switching devices of that bridge leg can also be confirmed [18].

The influence of harmonic current on the power grid is greater than the harmonic voltage and it is the fundamental cause of most of the problems, and the neutral voltage fluctuations are proportional to the amount of the harmonic current. So, as can be seen from Table 8, giving comprehensive consideration, choose the following optimal transistor sequence: Sa1-Sa3-Sa4-Sa2-Sb1-Sb3-Sb4-Sb2-Sc1-Sc3-Sc4-Sc2.

The control method is achieved by the SVPWM with 12 trigger pulses. Switching sequence reads Sa1-Sa3-Sa4-Sa2-Sb1-Sb3-Sb4-Sb2-Sc1-Sc3-Sc4-Sc2.

Figure 13 shows the output voltage and current harmonic content of the three types of inverter topology structure, using SVPWM control method. From Figure 13 we could see that current and phase voltage THD (total harmonic distortion) of the NPC inverter is the biggest, the T-type inverter is the middle one, and the proposed hybrid T-type is the smallest one. Meanwhile, the amplitude of the three kinds of inverter current harmonics is not obvious; however, the voltage THD results of the three kinds of inverter are sharply comparable.

The topology and control strategy of the two circuits are the same, except the devices used. The T-type topology consists of 12 IGBTs, while the hybrid T-type topology consists of 9 MOSFETs and 3 IGBTs. And the off-delay time and dead time of the IGBT are longer than those of the MOSFET.

Although the proportion of dead time is often very small relative to one switching cycle, the dead zone will make the three-phase control system deviate from the ideal mathematical model. When the switching frequency becomes higher, the dead-zone effect will gradually accumulate, and when it accumulates to a certain degree, it will distort the AC-side voltage waveform, which will affect the waveform quality of the input current. It will also cause fluctuations in the DC voltage, which degrades the accuracy of the entire system. In addition, the dead time can cause common mode voltage waveform distortion, resulting in higher harmonics [19]. So the THD of the novel T-type inverter is the lowest.

From Figure 14 we see that the common mode current of the proposed novel T-type inverter is smaller than the previous T-type inverter topology. In conclusion the proposed

TABLE 8: Comparison of output parameters under various switching sequence (simulation time: 0.1 s).

Sequence of the transistors	Power factor	Current harmonics%	voltage harmonics%
Sa1- Sa2- Sb1- Sb2- Sc1- Sc2- Sa3- Sa4- Sb3- Sb4- Sc3- Sc4	0.7209	27.27	158.99
Sa1- Sa2- Sb1- Sb2- Sc1- Sc2- Sa4- Sa3- Sb4- Sb3- Sc4- Sc3	0.7263	21.73	236.75
Sa1- Sc2- Sb1- Sa2- Sc1- Sb2- Sa3- Sa4- Sb3- Sb4- Sc3- Sc4	0.647	11.61	149.13
Sa1- Sc2- Sb1- Sa2- Sc1- Sb2- Sa4- Sa3- Sb4- Sb3- Sc4- Sc3	0.6158	7.81	224.95
Sa1- Sa2- Sa3- Sa4- Sb1- Sb2- Sb3- Sb4- Sc1- Sc2- Sc3- Sc4	0.5799	0.63	156.03
Sa1- Sa2- Sa4- Sa3- Sb1- Sb2- Sb4- Sb3- Sc1- Sc2- Sc4- Sc3	0.447	0.92	230.51
Sa1- Sa3- Sa4- Sa2- Sb1- Sb3- Sb4- Sb2- Sc1- Sc3- Sc4- Sc2	0.7132	6.61	248.44
Sa1- Sa4- Sa3- Sa2- Sb1- Sb4- Sb3- Sb2- Sc1- Sc4- Sc3- Sc2	0.6225	6.91	158.98

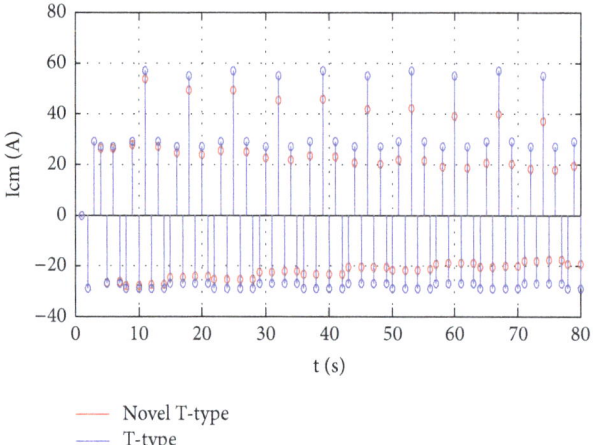

— Novel T-type
— T-type

FIGURE 14: The obtained CM-current of the novel and previous T-type inverters.

hybrid T-type inverter has priority compared to the NPC and T-type inverter.

The instantaneous common mode voltage and ground leakage current could be given by the following equation:

$$u_{cm} = \frac{u_{AN} + u_{BN} + u_{CN}}{3}$$
$$i_{cm} = C\frac{d_{u_{cm}}}{dt} \tag{11}$$

where u_{AN}, u_{BN}, and u_{CN} are the pulse voltages between the branch midpoint and the dc bus minus terminal, respectively

It can be seen from the equation that the common mode current i_{cm} is proportional to the change rate of the common mode voltage u_{cm}.

According to the above interpretation it can be got that the dead time of the proposed topology is the shortest because of the smallest number of IGBTs compared with the other two topologies, so that the common mode voltage and current are also the smallest.

4. Conclusions

We introduced a variety of new multilevel photovoltaic grid-connected inverter topologies, compared them with each other, and classified them according to the basic unit which predecessors proposed. We proposed the hybrid T-type inverter topology, which is composed of two best basic units. This structure makes full advantages of the two devices, which leads to reducing the harmonic content and power loss of the converter and improving the conversion efficiency of the system. By comparing the new topology, the new multilevel inverter topology is formed by the basic constituent elements to increase the auxiliary/embedded circuit or with a hybrid switch or with asymmetric structure to optimize the inverter performance. We use the space vector pulse width modulation (SVPWM) method to build the SIMULINK models of the proposed hybrid T-type and NPC three-level inverter in the MATLAB software and compare them with each other. The simulation results show that the proposed topology presents lower output harmonics than the NPC topology. We then provided further loss analysis for these two topologies, using the MELCOSIM software to calculate the loss of different components in each topology as well as the total power loss. The outcome is that the loss of the T-type topology is significantly lower than that of the NPC topology, and the conversion efficiency is higher than that of the NPC topology. Moreover the simulation results show that the common mode current of the proposed novel T-type inverter is smaller than the previous T-type inverter topology.

The proposed inverter topology has a certain practicality and economy to meet the actual needs.

Conflicts of Interest

All the authors declare that there are no conflicts of interest regarding the publication of this paper.

Acknowledgments

The work was supported by the National Natural Science Foundation of China (Grants no. 51667020 and no. 61364010), the Outstanding Doctor Graduate Student Innovation Project of Xinjiang University (no. XJUBSCX-2016018), and Graduate Student Innovation Project of Xinjiang Uygur Autonomous Region (no. XJGRI2017025).

References

[1] F. Michael and E. Temesi, "High Efficient Topologies for Next Generation Solar Inverter," *Inverter World*, vol. 03, pp. 31–36, 2009.

[2] C. I. Odeh, "A cascaded multi-level inverter topology with improved modulation scheme," *Electric Power Components and Systems*, vol. 42, no. 7, pp. 768–777, 2014.

[3] C. I. Odeh, "Sinusoidal pulse-width modulated three-phase multi-level inverter topology," *Electric Power Components and Systems*, vol. 43, no. 1, pp. 1–9, 2015.

[4] K. P. Draxe, M. S. B. Ranjana, and K. M. Pandav, "A cascaded asymmetric multilevel inverter with minimum number of switches for solar applications," in *Proceedings of the 2014 Power and Energy Systems Conference: Towards Sustainable Energy, PESTSE 2014*, 15, 13 pages, ind, March 2014.

[5] M. Azri, N. A. Rahim, and W. A. Halim, "A highly efficient single-phase transformerless H-bridge inverter for reducing leakage ground current in photovoltaic grid-connected system," *Electric Power Components and Systems*, vol. 43, no. 8-10, pp. 928–938, 2015.

[6] H. Schmidt, C. Siedle, and J. Ketterer, "Inverter for transforming a DC voltage into an AC current or an AC voltage," Tech. Rep. EP1369985(A2), Germany, 2003-12-10.

[7] D. Jovcic and K. Ahmed, "Method of converting a direct current voltage form a source of direct current voltage, more specifically from a photovoltaic source of direct current voltage, into a alternating current voltage," Tech. Rep. US7411802(B2), United States, 2008.

[8] H. Xiao, S. Xie, Y. Chen, and R. Huang, "An optimized transformerless photovoltaic grid-connected inverter," *IEEE Transactions on Industrial Electronics*, vol. 58, no. 5, pp. 1887–1895, 2011.

[9] R. González, J. López, P. Sanchis, and L. Marroyo, "Transformerless inverter for single-phase photovoltaic systems," *IEEE Transactions on Power Electronics*, vol. 22, no. 2, pp. 693–697, 2007.

[10] W. Yu, J.-S. Lai, H. Qian, and C. Hutchens, "High-efficiency MOSFET inverter with H6-type configuration for photovoltaic nonisolated AC-module applications," *IEEE Transactions on Power Electronics*, vol. 26, no. 4, pp. 1253–1260, 2011.

[11] L. Zhang, K. Sun, and L. L. Feng, "H6 non-isolated full bridge grid-connected PV inverters with low leakage currents," in *Pro-

ceedings of the CSEE 32*, vol. 32.15, pp. 1–7, 2012.

[12] S. V. Araújo, P. Zacharias, and R. Mallwitz, "Highly efficient single-phase transformerless inverters for grid-connected photovoltaic systems," *IEEE Transactions on Industrial Electronics*, vol. 57, no. 9, pp. 3118–3128, 2010.

[13] X. Zhang, L. L. Sun, and P. Xu, "Research on the restraining of common mode current in the single-phase transformerless grid-connected PV system," *Acta Energiae Solaris Sinica 30.9*, vol. 30, no. 9, pp. 1202–1208, 2009.

[14] T. Kerekes, R. Teodorescu, P. Rodríguez, G. Vázquez, and E. Aldabas, "A New high-efficiency single-phase transformerless PV inverter topology," *IEEE Transactions on Industrial Electronics*, vol. 58, no. 1, pp. 184–191, 2011.

[15] U.-M. Choi, F. Blaabjerg, and K.-B. Lee, "Reliability improvement of a T-type three-level inverter with fault-tolerant control strategy," *IEEE Transactions on Power Electronics*, vol. 30, no. 5, pp. 2660–2673, 2015.

[16] C. Yang and Y. Liu, "The contrast of conduction dissipation of T-type 3-level inverter and diode-clamped 3-level inverter," *Autocontrol 02*, vol. 02, pp. 61–66, 2015.

[17] W. Kun, Y. Xiaojie, W. Chenchen, and Z. Minglei, "An equivalent dual three-phase SVPWM realization of the modified 24-sector SVPWM strategy for asymmetrical dual stator induction machine," in *Proceedings of the 2016 IEEE Energy Conversion Congress and Exposition, ECCE 2016*, Milwaukee, WI, USA, September 2016.

[18] T. Mingting, *Research on Three-Phase T-Type Three-Level Non-Isolated Grid-Connected Inverter*, Hefei University of Technology, 2013.

[19] J.-W. Choi and S.-K. Sul, "Inverter output voltage synthesis using novel dead time compensation," *IEEE Transactions on Power Electronics*, vol. 11, no. 2, pp. 221–227, 1996.

18

Accurate Load Modeling based on Analytic Hierarchy Process

Zhenshu Wang,[1] **Xiaohui Jiang,**[1] **Shaorun Bian,**[2] **Yangyang Ma,**[1] **and Bowen Fan**[1]

[1]*Key Laboratory of Power System Intelligent Dispatch and Control of Ministry of Education, Shandong University, Jinan 250061, China*
[2]*Jinan Power Supply Company of National Grid, Jinan 250100, China*

Correspondence should be addressed to Zhenshu Wang; zhenshuwang@sdu.edu.cn

Academic Editor: George S. Tombras

Establishing an accurate load model is a critical problem in power system modeling. That has significant meaning in power system digital simulation and dynamic security analysis. The synthesis load model (SLM) considers the impact of power distribution network and compensation capacitor, while randomness of power load is more precisely described by traction power system load model (TPSLM). On the basis of these two load models, a load modeling method that combines synthesis load with traction power load is proposed in this paper. This method uses analytic hierarchy process (AHP) to interact with two load models. Weight coefficients of two models can be calculated after formulating criteria and judgment matrixes and then establishing a synthesis model by weight coefficients. The effectiveness of the proposed method was examined through simulation. The results show that accurate load modeling based on AHP can effectively improve the accuracy of load model and prove the validity of this method.

1. Introduction

The main models of power system include generator, power line, and load models. Compared with generator models and power line models, load models have lower accuracy. With the scale of power grid expanding and center of power load increasing, load model will have more and more significant meaning in power system digital simulation and dynamic security analysis [1–5].

In the development of power system, lots of load modeling methods have been proposed, mainly including component-based modeling approach [6, 7], measurement-based modeling approach [8, 9], and fault-based modeling approach [10]. Measurement-based modeling approach considers the power load as a whole. Parameters of load model can be identified by voltage, frequency, current, and phase angle data of acquiring bus of field devices and experiment. It is a simple and practical way to solve load modeling problems. Load characteristic data are important factor affecting load model results [11].

Power distribution network impedance and reactive power compensation are not considered in traditional composite load models. To overcome this problem, the synthesis load model (SLM) that considers the impact of power distribution network and compensation capacitor is built in [12–14]. The SLM overcomes the deficiencies of traditional composite load model. It can easily simulate motor load, static load, distribution network, and compensation capacitor in power system. However, the randomness characteristics of load power cannot be described precisely.

With the rapid development of electrified railway, proportion of electrified railway traction load in power load is increasing every year. Electric locomotive, the main composition of traction power system, is a high-power rectifier load. It has significant impact on power system security and stability [15]. Electric traction load has an uncertain interference degree to power system at different times [16, 17]. Dynamic equations and simulation models of traction motor circuit are deduced in [18, 19]. The SLM considers the impact of power distribution network impedance and compensation capacitor, while randomness of power load is well described by traction power system load model (TPSLM).

Single load model cannot describe power load accurately and effectively. The accurate load modeling based on analytic hierarchy process (AHP) is proposed in this paper. To analyze qualitative problems quantitatively, internal connections and differences of these two models are used as criteria of AHP. Weight coefficients of these two models can be calculated,

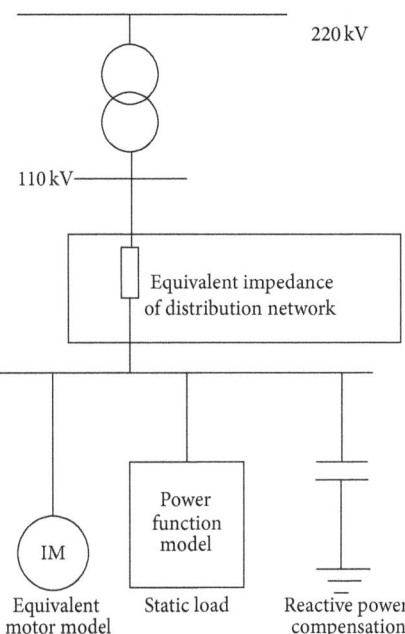

$110\,kV$

$220\,kV$

Equivalent impedance of distribution network

Power function model

IM

Equivalent motor model

Static load

Reactive power compensation

FIGURE 1: The structure diagram of SLM.

after different criteria hierarchies are formed by analyzing internal connections and differences. Distribution network and traction power system synthesis load model will establish after synthesizing these two load models by weight coefficients.

2. Distribution Network and Traction Power System Synthesis Load Model

2.1. The SLM. The SLM is proposed in [12, 13], which include equivalent static load, equivalent motor load, distribution network, and compensation capacitor. The SLM structure is shown in Figure 1.

Ignoring the influence of frequency variation, three-order electromechanical transient model is used to describe equivalent motor model:

$$
\begin{aligned}
T'_{d0}\frac{dE'_d}{dt} &= -E'_d - \left(X - X'\right)I_q + s\omega_B E'_q T'_{d0}, \\
T'_{d0}\frac{dE'_q}{dt} &= -E'_q + \left(X - X'\right)I_d - s\omega_B E'_d T'_{d0}, \\
T_j\frac{ds}{dt} &= T_m - T_e, \\
I_d &= \frac{1}{R_s^2 + X'^2}\left[R_s\left(U_{ld} - E'_d\right) + X'\left(U_{lq} - E'_q\right)\right], \\
I_q &= \frac{1}{R_s^2 + X'^2}\left[R_s\left(U_{lq} - E'_q\right) - X'\left(U_{ld} - E'_d\right)\right], \\
P_{im} &= U_d I_d + U_q I_q, \\
Q_{im} &= U_q I_d - U_d I_q,
\end{aligned}
\tag{1}
$$

where E'_d, E'_q are the d-axis and q-axis motor transient electromotive force in the synchronous rotating coordinate system; U_{ld}, U_{lq} are the d-axis and q-axis bus voltage, while I_d, I_q represent the d-axis and the q-axis stator currents; s is the rotor slip of motor; ω_B is the electrical angular frequency base value; mechanical torque $T_m = K_L[\alpha + (1 - \alpha)(1 - s)^p]$; electromagnetic torque $T_e = E'_d I_d + E'_q I_q$; the motor load rate coefficient $K_L = P^*_{M0}/(\alpha + (1 - \alpha)(1 - s_0)^P)$; α is the coefficient of drag torque that has no relationship with rotate speed; T_j is the rotor inertia time constant; $X' = X_s + X_m X_r/(X_m + X_r)$ is the rotor transient reactance; $X = X_s + X_m$ is the steady rotor reactance; R_s is the stator resistance; R_r is the rotor resistance; X_s is the stator reactance; X_r is the rotor reactance; X_m is the mutual inductance reactance; $T'_{d0} = (X_r + X_m)/R_r$ is the rotor circuit time constant; the parameter P_{M0} is used to define the initial active power proportion of the equivalent motor in the composite load model; P_{im}, Q_{im} are the active power and reactive power of equivalent induction motor model.

The power function model is used to describe static load in Figure 1:

$$
\begin{aligned}
P_s &= P_{s0}\left(\frac{U}{U_0}\right)^{n_p}, \\
Q_s &= Q_{s0}\left(\frac{U}{U_0}\right)^{n_q},
\end{aligned}
\tag{2}
$$

where P_s, Q_s, and U are the active power, reactive power, and load bus voltage magnitude of actual equivalent static load; P_{s0}, Q_{s0}, and U_0 are the active power, reactive power, and load bus voltage magnitude of reference point operated stably; n_p, n_q are the active voltage and reactive voltage feature coefficients, respectively.

The capacitance compensation model is used to describe the reactive power compensation in Figure 1:

$$
Q_c = -\frac{U^2}{X_c} = -\frac{U^2}{X_{c0}}f
\tag{3}
$$

in which Q_c is reactive power from the grid; f represents frequency of power system; X_c is compensation reactance; X_{c0} is compensation capacitor.

To sum up, power balances of SLM in bus bars are

$$
\begin{aligned}
P_1 &= P_d + P_s + P_{im}, \\
Q_1 &= Q_d + Q_s + Q_{im} - Q_c
\end{aligned}
\tag{4}
$$

in which P_1, Q_1 are the active power and reactive power of SLM; P_d, Q_d are the active power and reactive power of distribution network; P_s, Q_s are the active power and reactive power of static load; P_{im}, Q_{im} are the active power and reactive power of equivalent induction motor model, respectively; Q_c is the reactive power of compensation capacitor.

2.2. The TPSLM. The TPSLM mainly includes traction power supply networks and traction trains. Traction train loads include traction motor, locomotive auxiliary load, and carriage load. The locomotive traction power supply is single

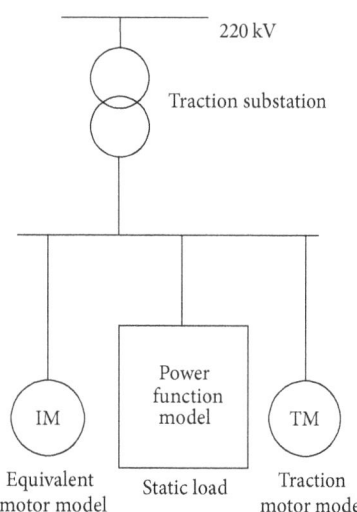

FIGURE 2: The structure diagram of TPSLM.

phase alternating current, in which the traction motor is supplied by the locomotive rectifier, which belongs to the single phase load. The induction motor in the auxiliary unit of the locomotive is powered by the splitter, which is a three-phase load. Carriage load includes air conditioning, electric water heater, and a little proportion of light load [20, 21].

According to the research on the structure and load characteristics of traction power supply system, the load model of traction power supply system with "induction motor parallel traction motor and constant impedance" is proposed in [20]. The TPSLM structure is shown in Figure 2.

In Figure 2, static load model uses the power function model; induction motor load model uses the three-order electromechanical transient model.

Expressions of traction motor load model are as follows:

$$L\frac{dI_d}{dt} = U_{tm} - RI_d - C_e K_s \omega I_d,$$
$$J\frac{d\omega}{dt} = C_T K_s \omega I_d^2 - T_L, \tag{5}$$
$$P_{tm} = U_{tm} I_d,$$

where $I_d \omega$ are the current and rotating speed of equivalent traction motor; L, R are the inductance and resistance of traction motor loop; J, K_s are the rotary inertia and excitation coefficient of traction motor; U_{tm} is traction substation voltage; considering $C_T = C_e K_s$, C_e and C_T are the voltage coefficient and torque coefficient of traction motor; the mechanical load torque T_L has influence on the mechanical properties of traction motor. The operating mode switch of electric locomotive is frequent. According to [20], $T_L = T_{L0}(A\omega^2 + B\omega + C)$; considering $A + B + C = 1$; T_{L0} is the traction motor load rate.

To sum up, power balances of TPSLM in bus bars are

$$P_2 = P_s + P_{tm} + P_{im}$$
$$Q_2 = Q_s + Q_{im}, \tag{6}$$

where P_2, Q_2 are the active power and reactive power of the traction power system load model; P_s, Q_s are the active power and reactive power of static load; P_{im}, Q_{im} are the active power and reactive power of equivalent induction motor model; P_{tm} is the active power of traction motor model.

2.3. Distribution Network and Traction Power System Synthesis Load Model Structure. The SLM considers the impact of power distribution network and compensation capacitor, while randomness of power load is more precisely considered by TPSLM. The distribution network and traction power system synthesis load model (DTSLM) interacts with SLM and TPSLM; its structure is shown in Figure 3.

The SLM uses system capacity as a reference value which is three-phase bus-bar voltage, while TPSLM is more complex.

Traction substation is supplied by power grid and the voltage is 110 kV typically. Contact net, which is supply by traction power system, connects traction substation. The voltage of contact net is mainly 25 kV. Voltage bases of TPSLM and SLM are not consistent. Voltage reference of SLM is bus-bar side, while the equivalent traction motor model and induction motor model are based on traction side. Equivalent "system transformation" is essential:

$$U_{tm} = K_{vt}U$$
$$U_{im} = K_{vi}U, \tag{7}$$

where K_{vt} is the transform coefficient between traction motor voltage and bus voltage, K_{vi} is the transform coefficient between induction motor voltage and bus voltage; U_{tm}, U_{im} are traction motor voltage and induction motor voltage under voltage base of the traction power system load model; U is root mean square of system voltage.

To sum up, power balances of DTSLM in bus-bar are

$$P = P_1 + P_2,$$
$$Q = Q_1 + Q_2 \tag{8}$$

in which P, Q are the active power and reactive power of DTSLM; P_1, Q_1 are the active power and reactive power of SLM; P_2, Q_2 are the active power and reactive power of TPSLM, respectively.

3. The AHP

The AHP is a scientific decision-making method in system engineering, which has been proposed by Satty in the 1970s [22]. It is an effective method to solve multiobjective and multilevel decision-making problems; weight coefficients can be calculated by digitizing some factors. However, the AHP is difficult to quantitative analysis completely [23].

Power plant site selection and integrated resource planning evaluation are considered by a lot of influence factors, such as social, technical, and operation in [24–27]. In [17, 28], a method based on AHP is proposed to improve power quality evaluation system and analyze high-speed rail major impacts on power supply grid. Compared with traditional

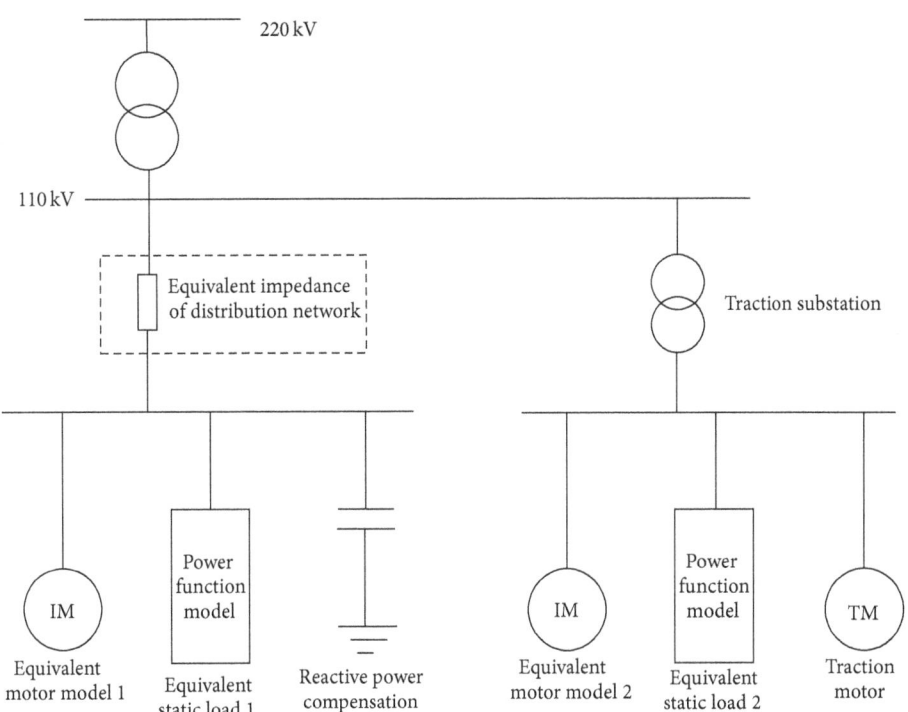

FIGURE 3: The structure diagram of DTSLM.

TABLE 1: Comparison scale.

Numerical scale	Verbal scale
1	Element i has equal importance as element j
3	Element i is moderately important compared to element j
5	Element i is strongly important compared to element j
7	Element i is very strongly important compared to element j
9	Element i is extremely important compared to element j
2, 4, 6, and 8	Intermediate value between adjacent scale values

decision-making methods, influences of qualitative factors are fully considered in AHP. The AHP decomposes factors of objective hierarchy into several levels. To analyze the problem synthetically, human experiences and judgments are integrated in criteria hierarchy. It has solved problems of quantitative information and lack of flexibility. The "1–9 scale method," as shown in Table 1, is combined with expert judgment to form a judgment matrix. Quantify results are obtained by simple mathematical calculation. It is scientific way to optimize complex mathematical models and circumvent heavy computation demands of traditional decision-making methods.

The main procedure of AHP can be divided into four steps:

(1) Structure the decision hierarchy from top with the objective hierarchy to bottom with the scheme hierarchy. Objective hierarchy divides decision-making factors into several categories to be the criteria hierarchy. Issues of affiliation form a top-down relationship.

(2) Form judgment matrix using the "1–9 scale method." The core of the AHP is forming judgment matrix using integers and reciprocal between 1 and 9. Weight coefficients of each hierarchy can be calculated by judgment matrix.

(3) Test consistency of judgment matrix.

(4) Calculate weight coefficient of scheme hierarchy under the standard of criteria hierarchy layer by layer.

4. Accurate Load Modeling Based on AHP

4.1. Established AHP Structure. The main idea of the AHP is to establish a hierarchical model based on the decision-making problem, which is divided into three levels, namely, the objective hierarchy, the criterion hierarchy, and the scheme hierarchy. The establishment of the hierarchical structure of the graph is shown in Figure 4.

4.2. Judgment Matrix of Criteria Hierarchy

(1) Objective Hierarchy. Including one goal, the expected results for the problem or the ideal goal, the goal of modeling load is to accurately characterize the load characteristics.

(2) Criteria Hierarchy. Including four evaluation criteria, active power fitting degree, reactive power fitting degree,

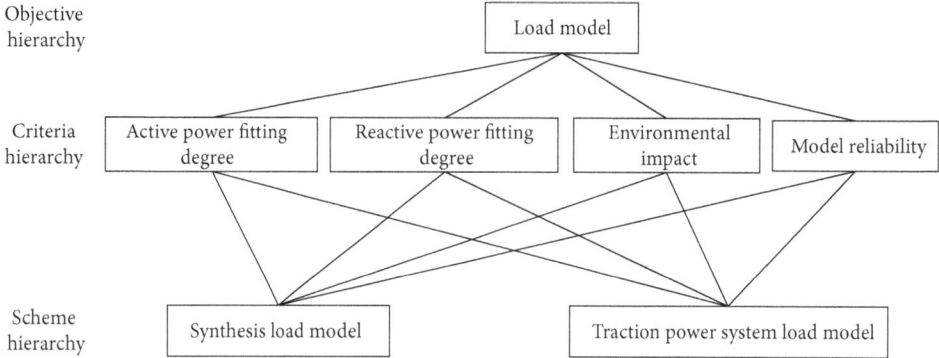

FIGURE 4: Hierarchy model structure of AHP.

TABLE 2: Criteria hierarchy judgment matrix.

a_{ij}	A	R	E	M
A	a_{11}	a_{12}	a_{13}	a_{14}
R	a_{21}	a_{22}	a_{23}	a_{24}
E	a_{31}	a_{32}	a_{33}	a_{34}
M	a_{41}	a_{42}	a_{43}	a_{44}

A is active power fitting degree; R is reactive power fitting degree; E is environmental impact; M is model reliability.

TABLE 3: Judgment matrix of each criterion.

b_{ij}	SLM	TPSLM
SLM	b_{11}	b_{12}
TPSLM	b_{21}	b_{22}

environmental impact, and model reliability. Considering the importance of four criteria, active power fitting degree and reactive power fitting degree, which can describe the accuracy and rationality of the load model precisely, are more important than the other two criteria. With the change of season, climate, and living habit, the composition of power load will also change, which can reflect the randomness of power load. Model reliability is credibility of these two models. According to accuracy, scaling factor in the judgment matrix can be set. Criteria hierarchy judgment matrix is shown in Table 2. a_{ij} is a comparison between scheme i and scheme j; symmetry element a_{ji} is a comparison between scheme j and scheme i; diagonal element a_{ii} is a comparison between scheme i and itself, and its value is equal to 1 apparently.

(3) Scheme Hierarchy. It includes two kinds of load models, respectively, considering the SLM and TPSLM. The SLM can accurately simulate distribution network and reactive power compensation system, while randomness of power system can be described by TPSLM.

4.3. Judgment Matrix of Each Criterion

(1) Active Power Fitting Degree. On the basis of measured data or test data, model parameters can be identified by measurement-based modeling approach. Calculate active power fitting degree based on obtained difference between measured values and calculated values of load model. Judgment matrix can be formed by standards of the "1–9 scale method." As the real-time data of the field acquisition

or experiment is different, the calculated values and the measured values of the model will be changed, so that the composition and size of the actual load of the power system can be reflected in real time and accurately.

(2) Reactive Power Fitting Degree. Implementation steps are similar to that of active power fitting degree. Calculate the reactive power fitting degree based on difference between models measured values and calculated values of load model and form judgment matrix of reactive power fitting degree.

(3) Environmental Impact. The power load has characteristics that are uncertain and time-varying. With the change of season, climate and living habit, the composition of power load will also change. Electrified railway is widely used in modern railway transportation. That has increased the proportion of traction load in power system. Considering environment factors, the SLM cannot describe randomness of power load precisely, so the TPSLM accounts for a larger proportion.

(4) Model Reliability. The model reliability is not only a subjective judgment of identification results, but also an improvement of existing models. Judgment matrix of SLM and TPSLM in criteria hierarchy is formed by literature and experience. In [29], sensitivity of traction load model under voltage disturbances is higher than that of constant impedance model, while it is lower than that of the induction motor model. So SLM accounts for a larger proportion considering model reliability factors.

Judgment matrix of each criterion is shown in Table 3.

Judgment matrix of active power fitting degree and reactive power fitting degree will be calculated after identifying parameters of SLM and TPSLM.

4.4. Parameter Identification of Single Load Model. From (1) to (4), a total of 15 independent parameters of the SLM need to be identified:

$$\theta_1 = \left[R_D, X_D, s_0, T'_{d0}, R_s, X_s, R_r, X_r, T_j, a, P, n_p, n_q, X_{C0}, \right.$$
$$\left. P_{MP} \right]. \tag{9}$$

According to [30, 31], the sensitivity of parameters $[T'_{d0}, R_s, R_r, X_r, T_j, a, P]$ is low, and these parameters use the typical values:

$$T'_{d0} = 0.576,$$

$$R_s = 0,$$

$$R_r = 0.02,$$

$$X_r = 0.12, \tag{10}$$

$$T_j = 2,$$

$$\alpha = 0.15,$$

$$P = 2.$$

To sum up, 8 parameters in SLM need to be identified:

$$\theta'_1 = \left[R_D, X_D, s_0, X_s, n_p, n_q, X_{C0}, P_{MP} \right]. \tag{11}$$

From (5) to (6), it can be seen that 21 parameters in TPSLM need to be identified:

$$\theta_2 = \left[s_0, T'_{d0}, R_s, X_s, R_r, X_r, T_j, a, P, n_p, n_q, K_{vt}, K_{vi}, K_{tm}, \right.$$
$$\left. K_{im}, R, C_T, L, J, A, B \right], \tag{12}$$

where K_{tm}, K_{im} are the proportion of the traction and induction motor initial active power, respectively. Equivalent load random fluctuation caused by natural factors uniform for mechanical traction motor load torque changes in [13, 14]. T_L is traction motor load torque, considering $A = 0.5$, $B = 0.3$, and $C = 0.2$; $T_L = T_{L0}(A\omega^2 + B\omega + C) = T_{L0}(0.5\omega^2 + 0.3\omega + 0.2)$. Induction motor and reactive power compensation parameters reference the typical values in [30, 31].

The objective function of measurement-based modeling approach is

$$\min \varepsilon(\theta) = \min \frac{1}{n} \sum_{k=1}^{n} \left[(P_c - P_m)^2 + (Q_c - Q_m)^2 \right] \tag{13}$$

in which P_c, Q_c are the active power and reactive power of measurement; P_m, Q_m are the calculated power of load model. n is number of samples. When the difference between the calculated values of the load model and the measured values of the simulation is in a certain error range, parameters of the load model are identified.

In [32], load model is evaluated by relative error in the following formula:

$$\varepsilon = 100 \times \frac{\left((1/N) \sum_{k=1}^{N} (y(k) - \hat{y}(k))^2 \right)^{1/2}}{\left((1/N) \sum_{k=1}^{N} y(k)^2 \right)^{1/2}}, \tag{14}$$

where N is the number of samples; $y(k)$, $\hat{y}(k)$ are the measured power of simulation and calculated power of load models. If relative error is less than 5%, this load model is considered to be acceptable.

4.5. Calculated Weight Coefficients of Judgment Matrix. There are four ways to calculate weight coefficients of judgment matrix which are geometric average method, arithmetic average method, feature vector method, and least square method. Feature vector method is the most common way to calculate weight coefficients, which has advantage of simple calculation process and high calculating speed. That uses judgment matrix normalized eigenvectors of maximum characteristic root as weight coefficients. Feature vector method is used to calculated weight coefficients of judgment matrix in this paper.

Feature vector method uses weight coefficients ω to revise multiply judgment matrix A:

$$A\omega = \lambda_{\max}\omega \tag{15}$$

in which $\lambda_{\max}$ is the maximum characteristic root; the positive ω as eigenvectors of maximum characteristic root is the only one; weight coefficients of analytic hierarchy process will be calculation after normalizing eigenvectors.

The criteria hierarchy weight coefficient is ω_0, ω_1, ω_2 are weight coefficient of active power fitting and reactive power fitting degree, and ω_3, ω_4 are weight coefficient of environmental impact and model reliability, respectively.

The final weight coefficients of load models are as follows:

$$\omega = \left[(\omega_1, \omega_2, \omega_3, \omega_4) \cdot \omega_0 \right]^T. \tag{16}$$

To sum up, flow diagram of accurate load modeling based on AHP is shown in Figure 5.

5. Simulation Example

To illustrate load modeling based on AHP that can effectively improve the accuracy of load model, a simulation based on Matlab/Simulink is carried out in this paper. Locomotive auxiliary load and carriage load, which is equal to three-phase induction motor in parallel power function model, are connected to high voltage side of traction transformer. The faults are chosen as three-phase short circuit with the duration of 0.04 s. The simulation system is depicted in Figure 6; as for the system parameters, one can refer to [33].

Load modeling data must be the responses of voltage, current, active power, and reactive power. Data of SLM, TPSLM, and DTSLM under the disturbance occurring outside the region can be measured from simulation system in Figure 6, which are measured values of load characteristic. Model parameters and values of fitting curves of load models are calculated by parameter identification. According to parameter identification results, judgment matrixes and weight coefficients are calculated. The validity and effectiveness of the method can be illustrated by analyzing the test results.

Parameters identification of SLM results is shown in Table 4. Power fitting curves of SLM are shown in Figures 7(a) and 7(b).

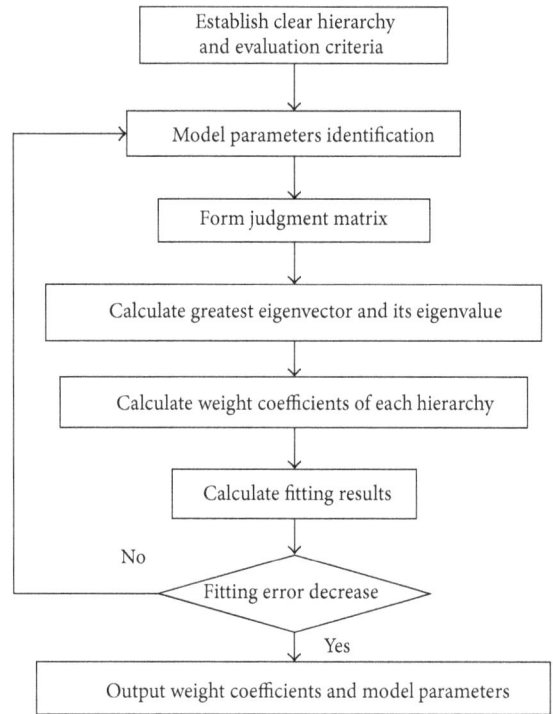

FIGURE 5: Flow diagram of load modeling based on AHP.

TABLE 4: Parameter of SLM.

P_{MP}	0.3179	k_D	1.8799
X_s	0.3357	X_D	0.0817
s_0	0.0177	n_p	2.1976
X_{C0}	1.7748	n_q	1.8346
ε_p (%)	0.39	ε_q (%)	0.68

TABLE 5: Parameter of TPSLM.

K_{vt}	0.3089	K_{vi}	0.9985
K_{tm}	0.7872	K_{im}	0.0998
X_s	0.1736	s_0	0.0162
n_p	1.2499	n_q	1.5658
R_s	0.1319	X_r	0.1852
J	6.7924	R	0.0423
L	5.9969	C_T	3.0074
ε_p (%)	1.08	ε_q (%)	0.42

Under the circumstance that model parameters identification result is reasonable, the average error of active power, reactive power, and total error is 6.9308×10^{-5}, 9.9626×10^{-5}, and 1.5893×10^{-4}, respectively. Relative error is less than 5%. Within the error rang allowed, model fitting is relatively accurate.

Parameters identification results of TPSLM are shown in Table 5. Power fitting curves of TPSLM are shown in Figures 8(a) and 8(b).

Under the circumstance that model parameters identification result is reasonable, the average error of active power,

TABLE 6: The criteria hierarchy judgment matrix.

a_{ij}	A	R	E	M
A	1	1	4	5
R	1	1	4	5
E	1/4	1/4	1	2
M	1/5	1/5	1/2	1

A is active power fitting degree; R is reactive power fitting degree; E is environmental impact; M is model reliability.

TABLE 7: Judgment matrix of active power fitting degree.

a_{ij}	SLM	TPSLM
SLM	1	5
TPSLM	1/5	1

TABLE 8: Judgment matrix of reactive power fitting degree.

a_{ij}	SLM	TPSLM
SLM	1	1/2
TPSLM	2	1

TABLE 9: Judgment matrix of environmental impact.

a_{ij}	SLM	TPSLM
SLM	1	1/5
TPSLM	5	1

TABLE 10: Judgment matrix of model reliability.

a_{ij}	SLM	TPSLM
SLM	1	3
TPSLM	1/3	1

reactive power, and total error is 8.7753×10^{-5}, 1.3880×10^{-4}, and 2.2655×10^{-4}, respectively. Relative error is less than 5%. Within the error rang allowed, TPSLM considers more randomness of power load.

According to the relationship between models, judgment matrix of criteria hierarchy can be calculated, which is shown in Table 6. After load model parameters identification, judgment matrixes of each criterion can be calculated, which are shown in Tables 7, 8, 9, and 10.

Final weight coefficients are calculated by feature vector method. The criteria hierarchy weight coefficient $\omega_0 = [0.4061, 0.4061, 0.1151, 0.0727]^T$; weight coefficient of active power fitting degree $\omega_1 = [0.8333, 0.1667]^T$; weight coefficient of reactive power fitting degree $\omega_2 = [0.3333, 0.6667]^T$; weight coefficient of environmental impact $\omega_3 = [0.1667, 0.8333]^T$; weight coefficient of model reliability $\omega_4 = [0.7500, 0.2500]^T$.

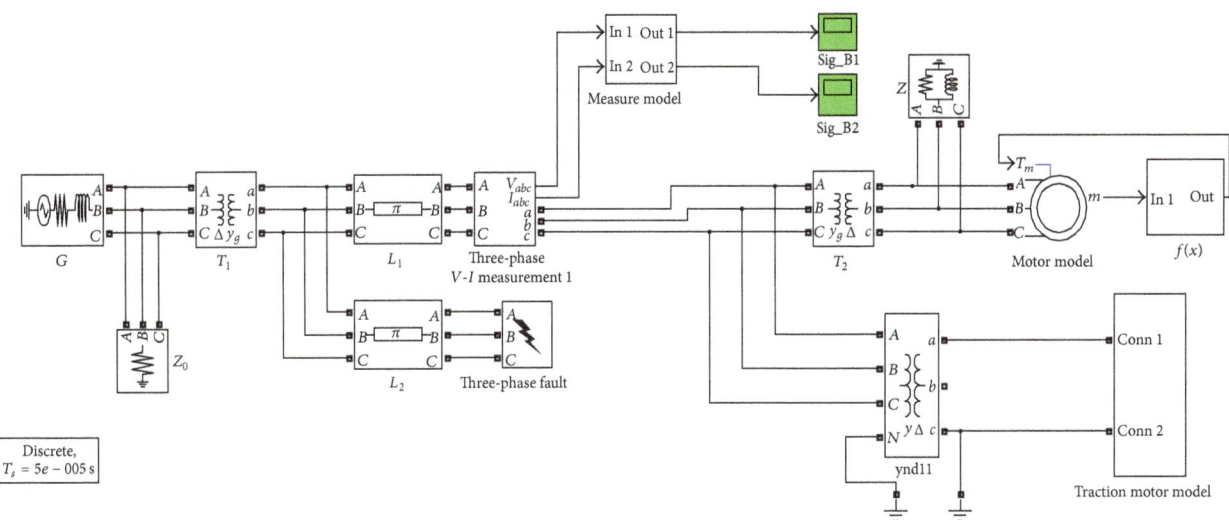

FIGURE 6: Load modeling based on AHP simulation system in Matlab/Simulink.

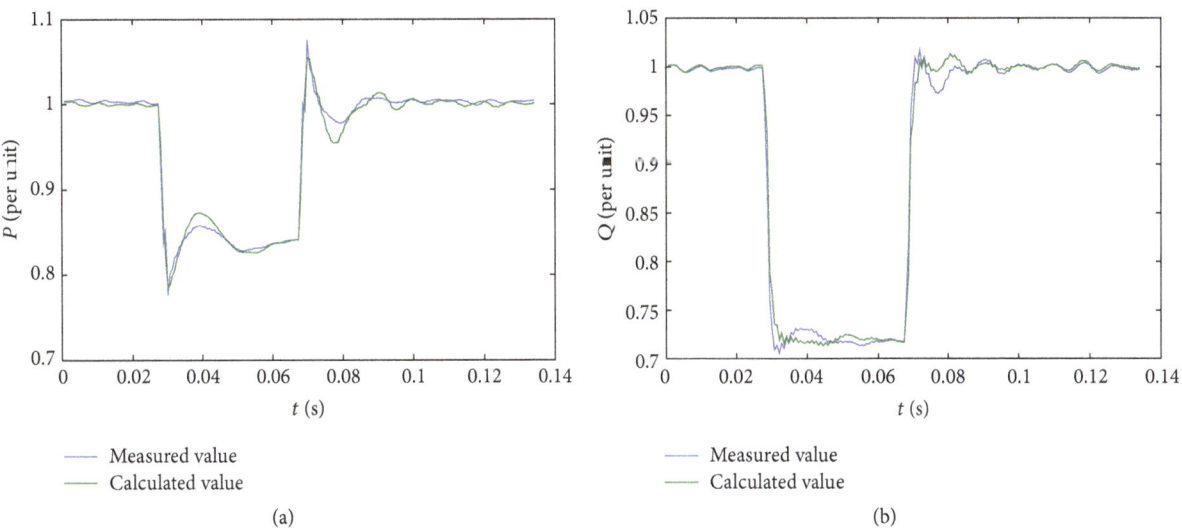

FIGURE 7: Power fitting curves of SLM.

The final weight coefficients of load models are as follows:

$$\omega = \left[(\omega_1, \omega_2, \omega_3, \omega_4) \cdot \omega_0 \right]^T$$

$$= \begin{bmatrix} 0.8333 & 0.3333 & 0.1667 & 0.7500 \\ 0.1667 & 0.6667 & 0.8333 & 0.2500 \end{bmatrix} \begin{bmatrix} 0.4061 \\ 0.4061 \\ 0.1151 \\ 0.0727 \end{bmatrix} \quad (17)$$

$$= [0.5475, 0.4525]^T ,$$

where ω are weight coefficients of each model in accurate load modeling based on AHP. The final load model will be calculated by weighting SLM and TPSLM.

Power fitting curves of accurate load model based on AHP are shown in Figures 9(a) and 9(b).

Comparison of average error is shown in Table 11. From model identification parameters of Tables 4 and 5, it indicates that the accurate load modeling based on AHP can decrease error obviously under reasonable model parameters and weight coefficients. From power fitting curves of load models in Figures 7, 8, and 9, the accurate load modeling based on AHP can fit measured values and calculated values more precisely. Simulation results prove that 54.75% of SLM and 45.25% of TPSLM could accurately describe the characterization of load model at this moment.

6. Conclusion

The method of accurate load modeling based on AHP is proposed in this paper. On the basis of SLM, TPSLM, which

TABLE 11: Comparison of average error.

Model error	Active power error	Reactive power error	Total error
SLM	6.9308×10^{-5}	9.9626×10^{-5}	1.5893×10^{-4}
TPSLM	1.3880×10^{-4}	8.7753×10^{-5}	2.2655×10^{-4}
DTSLM	3.9884×10^{-5}	5.9118×10^{-5}	9.9002×10^{-5}

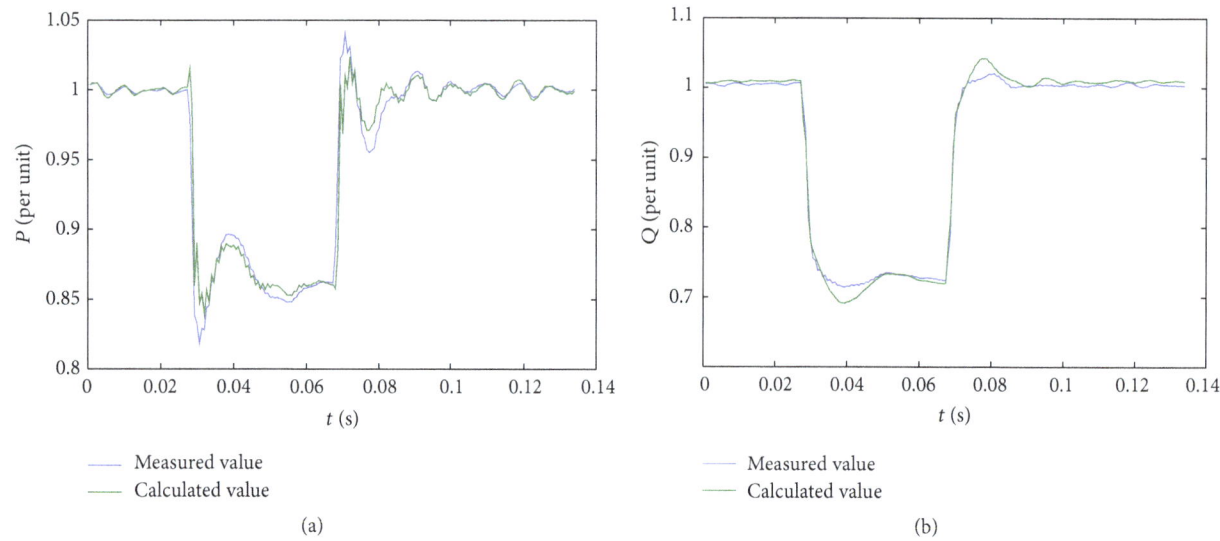

FIGURE 8: Power fitting curves of TPSLM.

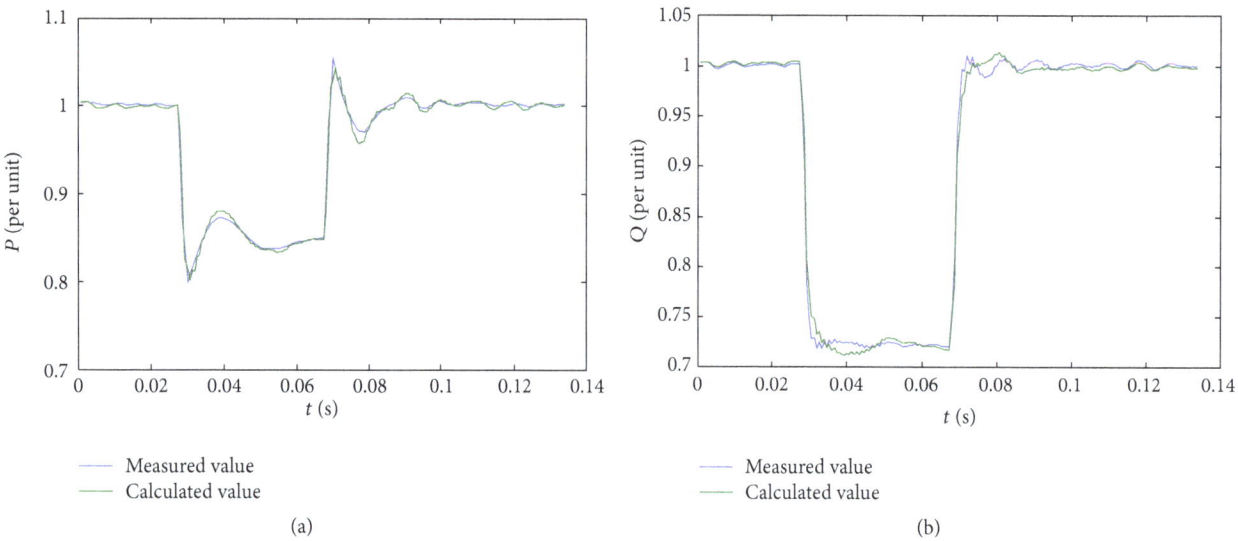

FIGURE 9: Power fitting curves of accurate load model based on AHP.

considers the randomness of power load, is fused by AHP. This method can adjust the weight coefficients with the change of power load, so the randomness problem of power load can be solved. The effectiveness of the proposed method was examined through simulation. Parameter identification results by measurement-based modeling approach show that load modeling based on AHP can effectively improve the accuracy of load model.

Competing Interests

The authors declare that there are no competing interests regarding the republication of this paper.

Acknowledgments

This work is supported by the National Natural Science Foundation of China (Grant no. 51377099) and is also supported by

Shandong Provincial Science and Technology Development Plan (no. 2014GGX101001).

References

[1] Z. Wang, S. Bian, M. Lei, C. Zhao, Y. Liu, and Z. Zhao, "Feature extraction and classification of load dynamic characteristics based on lifting wavelet packet transform in power system load modeling," *International Journal of Electrical Power and Energy Systems*, vol. 62, pp. 353–363, 2014.

[2] T. Yong, H. Junxian, and L. Wenzhou, "The modeling of distribution network and var compensator and induction motor in the load model for power system digital simulation," *Proceedings of the CSEE*, vol. 25, no. 3, pp. 8–12, 2005.

[3] W. Hongbin, D. Ming, L. Shenghu et al., "Probabilistic analysis on influences of generator model and load model on transient stability," *Power System Technology*, vol. 28, no. 1, pp. 19–21, 2004.

[4] P. Ju, F. Wu, Z.-Y. Shao et al., "Composite load models based on field measurements and their applications in dynamic analysis," *IET Generation, Transmission and Distribution*, vol. 1, no. 5, pp. 724–730, 2007.

[5] K. E. Wong and M. E. Haque, "Dynamic load modelling of a paper mill for small signal stability studies," *IET Generation, Transmission and Distribution*, vol. 8, no. 1, pp. 131–141, 2014.

[6] W. W. Price, K. A. Wirgau, A. Murdoch, J. V. Mitsche, E. Vaahedi, and M. El-Kady, "Load modeling for power flow and transient stability computer studies," *IEEE Transactions on Power Systems*, vol. 3, no. 1, pp. 180–187, 1998.

[7] Z. Wang, S. Bian, Y. Liu, and Z. Liu, "The load characteristics classification and synthesis of substations in large area power grid," *International Journal of Electrical Power & Energy Systems*, vol. 48, no. 1, pp. 71–82, 2013.

[8] C.-J. Lin, Y.-T. Chen, C.-Y. Chiou et al., "Dynamic load models in power systems using the measurement approach," *IEEE Transactions on Power Systems*, vol. 8, no. 1, pp. 309–315, 1993.

[9] Z. Yan, Z. Wen, C. Xiaodong, and L. Yutian, "Real-time optimal voltage control using measurement-based aggregate load model," *Electric Power Systems Research*, vol. 116, pp. 293–300, 2014.

[10] Y. Tang, H. Zhang, J. Hou, and D. Zhang, "Study on essential principle and methods for load modeling," *Power System Technology*, vol. 31, no. 4, pp. 1–5, 2007.

[11] P. Ju, C. Qin, F. Wu, H. Xie, and Y. Ning, "Load modeling for wide area power system," *International Journal of Electrical Power & Energy Systems*, vol. 33, no. 4, pp. 909–917, 2011.

[12] Y. Tang, H. Zhang, J. Hou, and D. Zhang, "A synthesis load model with distribution network," *Power System Technology*, vol. 31, no. 5, pp. 34–38, 2007.

[13] Y. Tang, H.-B. Zhang, D.-X. Zhang, and J.-X. Hou, "A synthesis load model with distribution network for power transmission system simulation and its validation," in *Proceedings of the IEEE Power & Energy Society General Meeting (PES '09)*, pp. 1–7, IEEE, Calgary, Canada, July 2009.

[14] Y. Zhang, R. Chen, Y. Xiao, W. Zhao, X. Li, and G. Zhang, "Measurement-based modeling of traction power supply system," in *Proceedings of the International Conference on Advanced Power System Automation and Protection (APAP '11)*, pp. 946–950, Beijing, China, October 2011.

[15] Z.-H. Lu, J.-W. Li, and J. Zhou, "The impact of electrified railway on electric power system," *Relay*, vol. 32, no. 11, pp. 33–36, 2004.

[16] G.-D. Zhang, X.-R. Li, and P.-Q. Li, "Simulation research on load characteristics of electric locomotive traction motor," *Electric Drive for Locomotives*, vol. 1, pp. 21–24, 2009.

[17] J. Bai, W. Gu, X. Yuan, Q. Li, B. Chen, and X. Wang, "Power quality warning of high-speed rail based on multi-features similarity," *Journal of Electrical Engineering and Technology*, vol. 10, no. 1, pp. 92–101, 2015.

[18] J. Chen, J. Zhu, and Y. Guo, "Modeling and performance analysis of energy regeneration system in electric vehicle with permanent magnet DC motor driving system," in *Proceedings of the International Conference on Electrical Machines and Systems (ICEMS '07)*, pp. 2051–2056, Seoul, South Korea, October 2007.

[19] Y.-C. Liang and V. J. Gosbell, "DC machine models for SPICE2 simulation," *IEEE Transactions on Power Electronics*, vol. 5, no. 1, pp. 16–20, 1990.

[20] X. Li, G. Zhang, X. Zhu, Z. Hu, and H. Hu, "A load model of traction power supply system," *Automation of Electric Power Systems*, vol. 33, no. 16, pp. 71–95, 2009.

[21] L. Monjo, L. Sainz, and J. Rull, "Statistical study of resonance in AC traction systems equipped with Steinmetz circuit," *Electric Power Systems Research*, vol. 103, pp. 223–232, 2013.

[22] T. L. Saaty, "Decision making with the analytic hierarchy process," *International Journal of Services Sciences*, vol. 1, no. 1, pp. 83–98, 2008.

[23] T. L. Saaty, *The Analytic Hierarchy Process*, McGraw-Hill, New York, NY, USA, 1980.

[24] Y.-F. Zhao and J.-F. Chen, "Analytic hierarchy process and its application in power system," *Electric Power Automation Equipment*, vol. 24, no. 9, pp. 85–87, 2004.

[25] W. Guangsheng, H. Zhaoguang, and W. Pingyang, "Decision support system for thermal power plant siting (SDSS)," in *Proceedings of the IEEE Region 10 Conference on Computer, Communication, Control and Power Engineering (TENCON '93)*, vol. 5, pp. 385–387, Beijing, China, October 1993.

[26] R. R. Clarke, "Validation and legitimation of an analytic hierarchy approach to integrated resource planning for electric utilities," in *Proceedings of the 32nd Intersociety Energy Conversion Engineering Conference*, vol. 3, pp. 2197–2201, August 1997.

[27] R. R. Clarke, "Choosing an integrated resource plan for electric utilities: an analytic hierarchy approach," in *Proceedings of the Intersociety Energy Conversion Engineering Conference*, vol. 3, pp. 1592–1597, August 1996.

[28] S. A. Farghal, M. S. Kandil, and A. Elmitwally, "Quantifying electric power quality via fuzzy modelling and analytic hierarchy processing," *IEE Proceedings—Generation, Transmission and Distribution*, vol. 149, no. 1, pp. 44–49, 2002.

[29] J. Li and X. Li, "Transient stability calculation considering traction load model," *Electric Power Automation Equipment*, vol. 33, no. 4, pp. 109–113, 2013.

[30] P. Ju and D.-Q. Ma, *Load Modeling of Power System*, China Electric Power Press, Beijing, China, 2008.

[31] Z. Wang, S. Bian, X. Liu, K. Yu, and Y. Shi, "Research on load model parameter identification based on the CQDPSO algorithm," *Transactions of China Electrotechnical Society*, vol. 29, no. 12, pp. 211–217, 2014.

Permissions

All chapters in this book were first published in JECE, by Hindawi Publishing Corporation; hereby published with permission under the Creative Commons Attribution License or equivalent. Every chapter published in this book has been scrutinized by our experts. Their significance has been extensively debated. The topics covered herein carry significant findings which will fuel the growth of the discipline. They may even be implemented as practical applications or may be referred to as a beginning point for another development.

The contributors of this book come from diverse backgrounds, making this book a truly international effort. This book will bring forth new frontiers with its revolutionizing research information and detailed analysis of the nascent developments around the world.

We would like to thank all the contributing authors for lending their expertise to make the book truly unique. They have played a crucial role in the development of this book. Without their invaluable contributions this book wouldn't have been possible. They have made vital efforts to compile up to date information on the varied aspects of this subject to make this book a valuable addition to the collection of many professionals and students.

This book was conceptualized with the vision of imparting up-to-date information and advanced data in this field. To ensure the same, a matchless editorial board was set up. Every individual on the board went through rigorous rounds of assessment to prove their worth. After which they invested a large part of their time researching and compiling the most relevant data for our readers.

The editorial board has been involved in producing this book since its inception. They have spent rigorous hours researching and exploring the diverse topics which have resulted in the successful publishing of this book. They have passed on their knowledge of decades through this book. To expedite this challenging task, the publisher supported the team at every step. A small team of assistant editors was also appointed to further simplify the editing procedure and attain best results for the readers.

Apart from the editorial board, the designing team has also invested a significant amount of their time in understanding the subject and creating the most relevant covers. They scrutinized every image to scout for the most suitable representation of the subject and create an appropriate cover for the book.

The publishing team has been an ardent support to the editorial, designing and production team. Their endless efforts to recruit the best for this project, has resulted in the accomplishment of this book. They are a veteran in the field of academics and their pool of knowledge is as vast as their experience in printing. Their expertise and guidance has proved useful at every step. Their uncompromising quality standards have made this book an exceptional effort. Their encouragement from time to time has been an inspiration for everyone.

The publisher and the editorial board hope that this book will prove to be a valuable piece of knowledge for researchers, students, practitioners and scholars across the globe.

List of Contributors

Mikias Hailu Kebede
Electrical and Computer Engineering Department, Debre Berhan University, Debre Berhan, Ethiopia

Getachew Bekele Beyene
Addis Ababa Institute of Technology (AAiT), Addis Ababa University, Addis Ababa, Ethiopia

Sen Ye, Youbing Zhang and Luyao Xie
College of Information Engineering, Zhejiang University of Technology, Hangzhou, China

Haiqiang Lu
Jiaxing Heng Chuang Electric Equipment Co., Ltd., Jiaxing, China

Hairong Wu, Jian Wu, Jiwei Feng and Shifu Liu
School of Mechanical and Automotive Engineering, Liaocheng University, Liaocheng City, Shandong Province 252000, China

Guangfei Xu
School of Mechanical and Automotive Engineering, Liaocheng University, Liaocheng City, Shandong Province 252000, China
Modern Agricultural Demonstration Garden, Liaocheng Academy of Agricultural Sciences, Liaocheng City, Shandong Province 252000, China

Linglong Bu
Modern Agricultural Demonstration Garden, Liaocheng Academy of Agricultural Sciences, Liaocheng City, Shandong Province 252000, China

Xue Han
Economic & Technology Research Institute, State Grid Shandong Electric Power Company, Jinan City, Shandong Province 250000, China

Zhaohong Zheng, Tianxia Zhang and Jiaxiang Xue
School of Mechanical and Automotive Engineering, South China University of Technology, Guangzhou 510641, China

Shi-Zhou Xu and Feng-You He
Department of Information and Electrical Engineering, China University of Mining and Technology, No. 1 Daxue Road, Xuzhou, Jiangsu 221116, China

Jiayu Wang, Zhifei Shan and Jie Min
College of Electrical Engineering & New Energy, China ree Gorges University, Yichang, China

Yewen Wei and Shuailong Dai
College of Electrical Engineering & New Energy, China ree Gorges University, Yichang, China
Hubei Province Collaborative Innovation Center for New Energy Microgrid, CTGU, Yichang, China

Ling Yang, Yandong Chen, An Luo, Leming Zhou, Xiaoping Zhou, Wenhua Wu, Wenjuan Tan and Zhiwei Xie
National Electric Power Conversion and Control Engineering Technology Research Center, Hunan University, Changsha 410082, China

Kunshan Huai
Guangzhou Power Supply Bureau, Guangzhou 510620, China

Lei Shao, Xu Zhou, Ji Li, Hongli Liu and Xiaoqi Chen
Tianjin Key Laboratory for Control Theory & Application in Complicated Systems, Tianjin University of Technology, Tianjin 300384, China

Carlos Enrique Imbaquingo, Eduard Sarrá, Nicola Isernia and Alberto Tonellotto
Yu-Hsing Chen, Catalin GabrielDincan, Philip Kjær, Claus Leth Bak, andXiongfeiWang Department of Energy Technology, Aalborg University, Aalborg, Denmark

Wen-ning Yan, Ke-jun Li, Zhuo-diWang and Xin-hanMeng
School of Electrical Engineering, Shandong University, Jinan 250061, China

Jianguo Zhao
State Grid of China Technology College, Jinan 250002, China

Yeison Alberto Garcés Gomez
Universidad Católica de Manizales and Universidad Nacional de Colombia, Colombia

Nicolás Toro García
Universidad Nacional de Colombia Sede Manizales, Colombia

Fredy E. Hoyos
Universidad Nacional de Colombia Sede Manizales, Colombia
Universidad Nacional de Colombia Sede Medellin, Colombia

R. Sarjila, K. Ravi, J. Belwin Edward, K. Sathish Kumar and Avagaddi Prasad
School of Electrical Engineering, VIT University, Vellore, Tamil Nadu, India

Wei Sun and Minquan Ye
Department of Business Administration, North China Electric Power University, Baoding 071000, China

Youssef Errami, Abdellatif Obbad and Smail Sahnoun
Laboratory of Electronics, Instrumentation and Energy, Team of Exploitation and Processing of Renewable Energy, Department of Physics, Faculty of Science, Chouaib Doukkali University, El Jadida, Morocco

Mohammed Ouassaid and Mohamed Maaroufi
Department of Electrical Engineering, Mohammadia School of Engineers, University Mohammed V, Rabat, Morocco

Hao Li, Shuo Chen and Xiang Wu
School of Electrical and Power Engineering, China University of Mining and Technology, Xuzhou 86221008, China

Guojun Tan
School of Electrical and Power Engineering, China University of Mining and Technology, Xuzhou 86221008, China
China Mining Drives & Automation Co., Ltd., Xuzhou 221008, China

P. M. Ivry, O. A. Oke, D. W. P. Thomas and M. Sumner
Department of Electrical and Electronics Engineering, Electrical Systems and Optics Research Group, University of Nottingham, Nottingham, UK

Zhenshu Wang, Xiaohui Jiang, Yangyang Ma and Bowen Fan
Key Laboratory of Power SystemIntelligent Dispatch and Control of Ministry of Education, Shandong University, Jinan 250061, China

Shaorun Bian
Jinan Power Supply Company of National Grid, Jinan 250100, China

Index